T0234669

DIRAC

Dirac

A SCIENTIFIC BIOGRAPHY

HELGE KRAGH

The right of the
University of Cambridge
to print and sell
all manner of books
was granted by
Henry VIII in 1534.
The University has printed
and published continuously
since 1584.

CAMBRIDGE UNIVERSITY PRESS

Cambridge

New York   Port Chester   Melbourne   Sydney

Published by the Press Syndicate of the University of Cambridge
The Pitt Building, Trumpington Street, Cambridge CB2 1RP
40 West 20th Street, New York, NY 10011, USA
10 Stamford Road, Oakleigh, Melbourne 3166, Australia

First published 1990

Library of Congress Cataloging-in-Publication Data

Kragh, Helge, 1944-
Dirak : a scientific biography / Helge Kragh.
P.   cm.
Bibliography: p.
Includes indexes.
ISBN 0-521-38089-8
1. Dirac, P. A. M. (Paul Adrien Maurice), 1902-
2. Physicists—Great Britain—Biography.   I. Title.
DC16.D57K73   1990
530'.092—dc20
[B]   89-17257
                                        CIP

ISBN 0-521-38089-8 hard covers

Transferred to digital printing 2003

TO BODIL

CONTENTS

PREFACE

ONE of the greatest physicists who ever lived, P. A. M. Dirac (1902–84) made contributions that may well be compared with those of other, better known giants of science such as Newton, Maxwell, Einstein, and Bohr. But unlike these famous men, Dirac was virtually unknown outside the physics community. A few years after his death, there have already appeared two memorial books [Kursunoglu and Wigner (1987) and Taylor (1987)], a historically sensitive biographical memoir [Dalitz and Peierls (1986)], and a detailed account of his early career in physics [Mehra and Rechenberg (1982+), vol. 4]. These works, written by scientists who knew Dirac personally, express physicists' homage to a great colleague. In some respects it *may* be an advantage for a biographer to have known his subject personally, but it is not always or *necessarily* an advantage. I have never met Dirac.

The present work, though far from claiming completeness, aims to supplement the volumes mentioned above by providing a more comprehensive and coherent account of Dirac's life and contributions to science. Because Dirac was a private person, who identified himself very much with his physics, it is natural to place emphasis on his scientific work, which, after all, has secured his name's immortality. Most of the chapters (2, 3, 5–6 and 8–11) are essentially accounts of these contributions in their historical context, but a few chapters are of a more personal nature. Taking the view that a scientific biography should deal not only with the portrayed scientist's successes but also with his failures, I present relatively detailed accounts (Chapters 8, 9, and 11) of parts of Dirac's work that are today considered either failures or less important but that nevertheless commanded his commitment and occupied his scientific life. Other chapters (1, 4, 7, and 12) are almost purely biographical. Chapter 12 attempts a portrait of the person of whom Bohr once remarked, "of all physicists, Dirac has the purest soul." In addition to describing Dirac's life and science, I have also, in Chapters 13 and 14, attempted to consider

his views of physics in its more general, philosophical aspect. Two appen-
dixes, including a bibliography, deal with Dirac's publications from a
quantitative point of view.

During work on this book, I have consulted a number of libraries and
archives in search of relevant material and have used sources from the
following places: Bohr Scientific Correspondence, Niels Bohr Institute,
Copenhagen; Archive for History of Quantum Physics, Niels Bohr Insti-
tute; Schrödinger Nachlass, Zentralbibliothek für Physik, Vienna; Bethe
Papers, Cornell University Archive, Ithaca; Manuscript Division,
Library of Congress, Washington; Dirac Papers, Churchill College, Cambridge, now moved to Florida State University, Tallahassee; Centre of
History of Physics, American Institute of Physics, New York; Ehrenfest
Archive, Museum Boerhaave, Leiden; Nobel Archive, Royal Swedish
Academy of Science, Stockholm; Sussex University Library; and Ständiger Arbeitsausschuss für die Tagungen der Nobelpreisträger, Lindau. I am
grateful for permission to use and quote material from these sources. The
many letters excerpted in the text are, if written in English, quoted literally; this accounts for the strange English usage found in letters by Pauli,
Gamow, Ehrenfest, Heisenberg, and others. I would like to thank the following people for providing information and other assistance: Karl von
Meyenn, Sir Rudolf Peierls, Abraham Pais, Luis Alvarez, Sir Nevill Mott,
Silvan Schweber, Helmuth Rechenberg, Olivier Darrigol, Kurt Gottfried,
Ulrich Röseberg, Aleksey Kozhevnikov, Richard Eden, Finn Aaserud,
and Carsten Jensen. Special thanks to Robert Corby Hovis for his careful
editing of the manuscript and many helpful suggestions.

November 1988   Helge Kragh
                                                      Ithaca, New York

CHAPTER 1

EARLY YEARS

PAUL DIRAC signed his scientific papers and most of his letters P. A. M. Dirac, and for a long time, it was somewhat of a mystery what the initials stood for. Dirac sometimes seemed reluctant to take away that mystery. At a dinner party given for him when he visited America in 1929 – when he was already a prominent physicist – the host decided to find out the first names of his honored guest. At each place around the table, he placed cards with different guesses as to what P.A.M. stood for, such as Peter Albert Martin or Paul Alfred Matthew. Having studied the cards, Dirac said that the correct name could be obtained by a proper combination of the names on cards. After some questioning, the other guests were able to deduce that the full name of their guest of honor was Paul Adrien Maurice Dirac.[1]

Dirac got his French-sounding name from his father, Charles Adrien Ladislas Dirac, who was Swiss by birth. Charles Dirac was born in 1866 in Monthey in the French-speaking canton Valais, and did not become a British citizen until 1919. At age twenty he revolted against his parents and ran away from home. After studies at the University of Geneva, he left around 1890 for England, where he settled in Bristol. In England Charles made a living by teaching French, his native language, and in 1896 he was appointed a teacher at the Merchant Venturer's Technical College in Bristol. There he met Florence Hannah Holten, whom he married in 1899. Florence was the daughter of a ship's captain and was twelve years younger than Charles. The following year they had their first child, Reginald Charles Felix, and two years later, on August 8, 1902, Paul Adrien Maurice was born. At that time, the family lived in a house on Monk Road.[2] The third child of the Dirac family was Beatrice Isabelle Marguerite, who was four years younger than Paul.

For many years, Charles Dirac seems to have retained his willful isolation from his family in Switzerland; they were not even informed of his marriage or first children. However, in 1905 Charles visited his mother

1

in Geneva, bringing his wife and two children with him. At that time, Charles's father had been dead for ten years. Like his brother and sister, Paul was registered as Swiss by birth, and only in 1919, when he was seventeen years old, did he acquire British nationality.

Paul's childhood and youth had a profound influence on his character throughout his entire life, an influence that resulted primarily from his father's peculiar lack of appreciation of social contacts. Charles Dirac was a strong-willed man, a domestic tyrant. He seems to have dominated his family and to have impressed on them a sense of silence and isolation. He had a distaste for social contacts and kept his children in a virtual prison as far as social life was concerned. One senses from Paul Dirac's reminiscences a certain bitterness, if not hatred, toward his father, who brought him up in an atmosphere of cold, silence, and isolation. "Things contrived early in such a way that I should become an introvert," he once pathetically remarked to Jagdish Mehra.[3] And in another interview in 1962, he said, "In those days I didn't speak to anybody unless I was spoken to. I was very much an introvert, and I spent my time thinking about problems in nature."[4] When his father died in 1936, Paul felt no grief. "I feel much freer now," he wrote to his wife.[5] In 1962 he said:[6]

In fact I had no social life at all as a child. . . . My father made the rule that I should only talk to him in French. He thought it would be good for me to learn French in that way. Since I found I couldn't express myself in French, it was better for me to stay silent than to talking English. So I became very silent at that time – that started very early . . .

Paul also recalled the protocol for meals in the Dirac house to have been such that he and his father ate in the dining room while his mother, who did not speak French well, ate with his brother and sister in the kitchen. This peculiar arrangement, which contributed to the destruction of the social relationship within the family, seems to have resulted from Charles's strict insistence that only French should be spoken at the dinner table.

Unlike other great physicists – Bohr, Heisenberg, and Schrödinger, for example – Paul Dirac did not grow up under conditions that were culturally or socially stimulating. Art, poetry, and music were unknown elements during his early years, and discussions were not welcomed in the house on Monk Road. Whatever ideas he had, he had to keep them to himself. Perhaps, as Paul once intimated, Charles Dirac's dislike of social contacts and the expression of human feelings was rooted in his own childhood in Switzerland. "I think my father *also* had an unhappy childhood," Paul said.[7]

Paul lived with his parents in their home in Bristol until he entered

Cambridge University in 1923. The young Dirac was shy, retiring, and uncertain about what he wanted from the future. He had little to do with other boys, and nothing at all to do with girls. Although he played a little soccer and cricket, he was neither interested in sports nor had any success in them. "He haunted the library and did not take part in games," recalled one of his own schoolmates. "On the one isolated occasion I saw him handle a cricket bat, he was curiously inept."[8] One incident illustrates the almost pathological antisocial attitude he carried with him from his childhood: In the summer of 1920, he worked as a student apprentice in Rugby at the same factory where his elder brother Reginald was employed. The solitary Paul, who had never been away from home, often met his brother in the town, but when they met, they did not even talk to each other! "If we passed each other in the street," remarked Paul, "we didn't exchange a word."[9]

Charles Dirac's bringing-up of his children must have been emotionally crippling for them. Paul's father resented any kind of social contact, and his mother wished to protect him from girls. As a young man, Paul never had a girlfriend and seems to have had a rather Platonic conception of the opposite sex for a long time. According to Esther Salaman, an author and good friend of Dirac, he once confided to her: "I never saw a woman naked, either in childhood or youth. . . . The first time I saw a woman naked was in 1927, when I went to Russia with Peter Kapitza. She was a child, an adolescent. I was taken to a girls' swimming-pool, and they bathed without swimming suits. I thought they looked nice."[10] He was not able to revolt against his father's influence and compensated for the lack of emotional and social life by concentrating on mathematics and physics with a religious fervor. Paul's relationship with his father was cold and strained; unable to revolt openly, his subconscious father-hatred manifested itself in isolation and a wish to have as little personal contact with his father as possible. Charles Dirac may have cared for his children and especially for Paul, whose intelligence Charles seems to have been proud of; but the way in which he exercised his care only brought alienation and tragedy. He was highly regarded as a teacher and was notorious for his strict discipline and meticulous system of punishment. Ambitious on behalf of his children, he wanted to give them as good an education as possible. But his pathological lack of human understanding and his requirement of discipline and submission made him a tyrant, unloving and unloved. Charles Dirac died in 1936; his wife, five years later.

Even more than Paul, his brother Reginald suffered from the way the Dirac children were brought up. Both the lack of social contact and a feeling of intellectual inferiority to his younger brother made Reginald depressed. He wanted to become a doctor, but his father forced him to study engineering, in which he graduated with only a third-class degree in 1919. His life ended tragically in 1924 when, on an engineering job in Wolverhampton, he committed suicide.

Young Paul was first sent to the Bishop Road primary school and then, at the age of twelve, to the school where his father was a teacher, the Merchant Venturer's College. Unlike most schools in England at the time, this school did not emphasize classics or the arts but concentrated instead on science, practical subjects, and modern languages. Paul did well in school without being particularly brilliant. Only in mathematics did he show exceptional interest and ability. This subject fascinated him, and he read many mathematics books that were advanced for his age. The education he received was a good and modern one, but it lacked the classical and humanistic elements that were taught at schools on the Continent and at other British schools. Heisenberg, Pauli, Bohr, Weyl, and Schrödinger received a broader, more traditional education than did Dirac, who was never confronted with Greek mythology, Latin, or classical poetry. Partly as a result of his early education and his father's influence, his cultural and human perspectives became much narrower than those of his later colleagues in physics. Not that Dirac ever felt attracted to these wider perspectives or would have wanted a more traditional education; on the contrary, he considered himself lucky to have attended the Merchant Venturer's College. "[It] was an excellent school for science and modern languages," he recalled in 1980. "There was no Latin or Greek, something of which I was rather glad, because I did not appreciate the value of old cultures."[11]

In the compulsory school system, Paul was pushed into a higher class and thus finished when he was only sixteen years old. But this early promotion was not because he was regarded as extraordinarily brilliant for his age, as he recalled in 1979:[12]

All the young men had been taken away from the universities to serve in the army. There were some professors left, those who were too old to serve in the army and those who were not physically fit; but they had empty classrooms. So the younger boys were pushed on, as far as they were able to absorb the knowledge, to fill up these empty classrooms.

Paul had no particular idea about what profession to go into and seems to have been a rather silent and dependent boy who just did as he was told. "I did not have much initiative of my own," he told Mehra. "[My] path was rather set out for me, and I did not know very well what I wanted."[13] In 1918, Paul entered the Engineering College of Bristol University as a student of electrical engineering – not because he really wanted to become an engineer, but because this seemed the most natural and smooth career. His elder brother Reginald had also studied at the

Engineering College, which was located in the same buildings that housed the Merchant Venturer's College. Paul was thus in familiar surroundings.

The lack of initiative and independence that characterized Paul's personality at the time may partly explain why he did not choose to study mathematics, the only subject he really liked. He also believed that as a mathematician he would have to become a teacher at the secondary school, a job he did not want and in which he would almost certainly have been a failure. A research career was not in his mind.

During his training as a student of electrical engineering, Paul came into close contact with mathematics and the physical sciences. He studied all the standard subjects (materials testing, electrical circuits, and electromagnetic waves) and the mathematics necessary to master these and other technical subjects. He enjoyed the theoretical aspects of his studies but felt a vague dissatisfaction with the kind of engineering mathematics he was taught. Although his knowledge of physics and mathematics was much improved at Bristol University, it was, of course, the engineering aspects of and approaches to these subjects that he encountered there. Many topics were not considered relevant to the engineer and were not included in the curriculum. For example, neither atomic physics nor Maxwell's electrodynamic theory was taught systematically. And, of course, such a modern and "irrelevant" subject as the theory of relativity was also absent from the formal curriculum.

During his otherwise rather dull education as an engineer, one event became of decisive importance to Paul's later career: the emergence into public prominence of Einstein's theory of relativity, which was mainly caused by the spectacular confirmation of the general theory made by British astronomers in 1919. In that year, Frank Dyson and Arthur Eddington announced that solar eclipse observations confirmed the bending of starlight predicted by Einstein.[14] The announcement created a great stir, and suddenly relativity (at the time fourteen years old) was on everybody's lips. Dirac, who knew nothing about relativity, was fascinated and naturally wanted to understand the theory in a deeper way than the newspaper articles allowed. He recalled:[15]

It is easy to see the reason for this tremendous impact. We had just been living through a terrible and very serious war. ... Everyone wanted to forget it. And then relativity came along as a wonderful idea leading to a new domain of thought. It was an escape from the war. ... At this time I was a student at Bristol University, and of course I was caught up in this excitement produced by relativity. We discussed it very much. The students discussed it among themselves, but had very little accurate information to go on. Relativity was a subject that everybody felt himself competent to write about in a general philosophical way. The philosophers just put forward the view that everything had to be considered relatively to something else, and they rather claimed that they had known about relativity all along.

In 1920–1, together with some of his fellow engineering students, Dirac attended a course of lectures on relativity given by the philosopher Charlie D. Broad, at the time a professor at Bristol. These lectures dealt with the philosophical aspects of relativity, not with the physical and mathematical aspects, which Dirac would have preferred. Although he did not appreciate Broad's philosophical outlook, the lectures inspired him to think more deeply about the relationship between space and time. Ever since that time, Dirac was firmly committed to the theory of relativity, with which he soon became better acquainted. His first immersion in the subject was Eddington's best-selling *Space, Time and Gravitation*, published in 1920, and before he completed his subsequent studies in mathematics at Bristol University, he had mastered both the special and general theories of relativity, including most of the mathematical apparatus.

While Paul did very well in the theoretical engineering subjects, he was neither interested in nor particularly good at the experimental and technological ones. Probably he would never have become a good engineer, but his skills were never tested. After graduating with first-class honors in 1921, he looked for employment but was unable to find a job. Not only were his qualifications not the best, but at the time the unemployment rate was very high in England because of the economic depression. After some time with nothing to do, Paul was lucky enough to be offered free tuition to study mathematics at Bristol University. He happily accepted.

From 1921 to 1923, Dirac studied mathematics, specializing in applied mathematics. Although he did no research of his own, he studied diligently and was introduced to the world of pure mathematical reasoning, which was very different in spirit from the engineering approach encountered in his earlier studies. The mathematicians at Bristol were not much oriented toward research, but Dirac had excellent teachers in Peter Fraser and H. R. Hassé, who soon recognized his outstanding abilities. Fraser particularly impressed Dirac, who described him many years later as "a wonderful teacher, able to inspire his students with real excitement about basic ideas in mathematics."[16] Both Fraser and Hassé were Cambridge men and thought that Dirac ought to continue for graduate studies at that distinguished university. Dirac completed his examinations at Bristol University with excellent results in the summer of 1923. Thanks to a grant from the Department of Scientific and Industrial Research (DSIR), he was able to enroll at Cambridge in the fall of 1923.

This was not the first time that Dirac visited Cambridge. After graduating in engineering in the summer of 1921, at his father's request he went to the famous university city to be examined for a St. John's College

Exhibition Studentship. He passed the examination and was offered the studentship, which was worth seventy pounds per year. But since he was unable to raise additional funds and his father was unable or unwilling to support him at Cambridge, he had to return to his parents in Bristol. It was only when he was awarded a DSIR studentship in addition, in 1923, that Dirac was finally able to attend Cambridge.

At Cambridge a new chapter in his life began, leading to his distinguished career as a physicist. He was away from his parents and the scarcely stimulating intellectual environment of Bristol, and at first he was not sure that he was really capable of succeeding in a research career. Cambridge, with its great scientific traditions, was a very different place from Bristol. The twenty-one–year–old Dirac arrived at a university that housed not only established scientists such as Larmor, Thomson, Rutherford, Eddington, and Jeans, but also rising stars including Chadwick, Blackett, Fowler, Milne, Aston, Hartree, Kapitza, and Lennard-Jones.[17] Dirac was admitted to St. John's College but during some periods lived in private lodgings because there were not enough rooms at the college. During most of 1925, he lived at 55 Alpha Road, only a few hundred meters from St. John's.

As a research student he had to have a supervisor who would advise on, or determine, the research topic on which he would work. With his limited scientific experience and lack of acquaintance with most of the Cambridge physicists, Dirac wanted to have Ebenezer Cunningham as his supervisor and to pursue research in the theory of relativity. He knew Cunningham from his earlier examination in 1921 and knew that he was a specialist in electromagnetic theory and the author of books and articles on electron theory and relativity.[18] Cunningham, who taught at St. John's College from 1911 to 1946, was only forty-two years old in 1923. He had been a pioneer of relativity in England, but found it difficult to follow the new physics of the younger generation and did not want to take on any more research students. "I just felt they'd run away from me. I was lost," he said.[19] Consequently, Dirac was assigned to Ralph Fowler. This was undoubtedly a happy choice, since Cunningham belonged to the old school of physics whereas Fowler was one of the few British physicists who had an interest and competence in advanced atomic theory. However, Fowler's field was not relativity, and at first Dirac felt disappointed not to have Cunningham as his supervisor.

Fowler was the main exponent of modern theoretical physics at Cambridge and the only one with a firm grip on the most recent developments in quantum theory as it was evolving in Germany and Denmark. He had good contacts with the German quantum theorists and also particularly with Niels Bohr in Copenhagen. In addition, he was about the only contact between the theorists and the experimentalists at the Cavendish.

However, as a supervisor for research students he was somewhat undisciplined. He was often abroad, and when at Cambridge he was difficult to find. Alan Wilson, who was a research student under Fowler in 1926–7, recalled that "Fowler, like the rest of us, worked in his college rooms – in Trinity – and if you wanted to consult him you had to drop in half a dozen times before you could find him in. He lived in Trumpington and did most of his work there."[20] The retiring Dirac probably did not consult Fowler often. Fowler's main interests were in the quantum theory of atoms and in statistical mechanics, including the application of these fields to astrophysics. In the summer of 1923, Dirac was largely ignorant of atomic theory and statistical physics, fields he found much less interesting than those he knew most about, electrodynamics and relativity. But as Fowler's research student, he was forced to learn the new subjects and soon discovered that they were far from uninteresting:[21]

Fowler introduced me to quite a new field of interest, namely the atom of Rutherford, Bohr and Somerfeld. Previously I had heard nothing about the Bohr theory, it was quite an eyeopener to me. I was very much surprised to see that one could make use of the equations of classical electrodynamics in the atom. The atoms were always considered as very hypothetical things by me, and here were people actually dealing with equations concerned with the structure of the atom.

Dirac worked hard to master the new students and to improve his knowledge of subjects he had learned at Bristol on a level not commensurate with the higher standards of Cambridge. He did well. Most of atomic theory he learned either from Fowler or by studying research papers in British and foreign journals available in the Cambridge libraries. He knew sufficient German to read articles in the *Zeitschrift für Physik,* the leading vehicle for quantum theory, and to read Arnold Sommerfeld's authoritative *Atombau und Spektrallinien.* Within a year, Dirac became fully acquainted with the quantum theory of atoms. As to mathematical methods, he scrutinized Whittaker's *Analytical Dynamics,* which became the standard reference work for him. From this book, written by a mathematician and former Cambridge man, he learned the methods of Hamiltonian dynamics and general transformation theory, both of which became guiding principles in his later work in quantum theory. At the same time he improved his knowledge of the theory of relativity by studying Eddington's recently published *The Mathematical Theory of Relativity,* and also by attending Eddington's lectures, which in one term covered special and general relativity and tensor analysis. Occasionally, Dirac had the opportunity to discuss questions with Eddington himself, an experience of which he remarked, "It was really a wonderful thing to meet the man who was the fountainhead of relativity so far as England was concerned."[22]

In addition to studying Eddington's book and attending his lectures, Dirac also followed Cunningham's course of lectures on electromagnetic theory and special relativity.[23] Forty years later, Cunningham recalled a time when he had worked out a long calculation on the electrical and magnetic components of the radiation field:[24]

I said to the class one day, I remember, "This is an extraordinarily simple result in the end, but why? Why should it work out like this?" A week later, a young man who had only been in Cambridge a year or two, a year I think, came up to me and said, "Here you are." That was Dirac.

Another student who followed Cunningham's course in 1923 was John Slater, a postdoctoral research student from Harvard who was in Europe on a traveling fellowship. But, characteristic of the remoteness of students from each other in Cambridge in those days, it was years later before Slater and Dirac realized they had attended the same course.[25] The Belgian George Lemaître, who a few years later would revolutionize cosmology, was a research student under Eddington in 1923–4 and also attended some of the same courses as Dirac, but it was a decade later before he and Dirac became acquainted (see also Chapter 11). Still another student who followed some of the same courses as Dirac – including those of Fowler, Cunningham, and Eddington – was Llewellyn H. Thomas, who received his B.A. in 1924 and stayed in Cambridge until 1929. He recalled the young Dirac as a quiet man who made no major impression at Cambridge until he published his papers on quantum mechanics. "He is a man of few words," Thomas said in 1962. "If you ask him a question, he'd say oh, that's very difficult. Then a week later he'd come back with the complete answer completely worked out."[26]

Thanks to the stimulating Cambridge environment, Dirac's scientific perspective became much wider. For the first time, he came in contact with the international research fronts of theoretical physics. As he met more people and established contacts with loose social groups, Dirac gradually became a little less shy and introverted. He recalled attending the combined tea parties and geometry colloquia that took place weekly at the home of Henry Baker, the professor of geometry:[27]

These tea parties did very much to stimulate my interest in the beauty of mathematics. The all-important thing there was to strive to express the relationships in a beautiful form, and they were successful. I did some work on projective geometry myself and gave one of the talks at one of the tea parties. This was the first lecture I ever gave, and so of course I remember it very well.

Although there were several attractive academic clubs at Cambridge – such as the Observatory Club, the Trinity Mathematical Society, and the Cavendish Society – Dirac restricted his interest to two: the $\nabla^2 V$ (del-squared) Club, which he joined in May 1924, and the Kapitza Club, which he joined in the fall of the same year. In both clubs membership was limited and was decided by election, and meetings took place in the college rooms, often in Dirac's room in St. John's. The $\nabla^2 V$ Club was mainly for mathematical physicists, who presented their own work at the meetings. Most Cambridge theorists were members of this club, which in 1924 included Eddington, Jeffreys, Milne, Chadwick, Hartree, Blackett, Fowler, Stoner, Kapitza, and Dirac. The Kapitza Club, an informal discussion club where papers on recent developments in physics were read and discussed at Trinity on Tuesdays, was started in 1922 by the colorful Soviet physicist Peter Kapitza, then a research student under Rutherford at the Cavendish. Experimental physics had predominance in the Kapitza Club, contrary to the theoretical orientation of the $\nabla^2 V$ Club. After his election to the Kapitza Club in the fall of 1924, Dirac listened to lectures by distinguished foreign guests such as James Franck (October 1924) and Niels Bohr (May 1925). The meetings of the club continued with Kapitza in charge until the summer of 1934, when Kapitza was unable to return to Cambridge from a visit to his homeland (see also Chapter 7). By that time, the club had held 377 meetings, many of them with Dirac as a participant. He remained an active member of both clubs until the war, and in the fall term of 1930 he served as president of the $\nabla^2 V$ Club.[28]

Paul Dirac lived a quiet life in Cambridge, totally absorbed in studies and research. Theoretical physics belonged to the Mathematics Faculty, which did not have its own building. There was no tradition of social or professional contact between the few students of theoretical physics, who usually sat alone in their college rooms or in the small library – which also served as a tea room – at the Cavendish Laboratory. It was "a terribly isolated business" to be a physics student at Cambridge, Nevill Mott recalled.[29] Yet Dirac did not find the isolation terrible at all. Had he wanted to, he could have taken part in what little extramural student life there was; but he did not want to. He deliberately kept away from external activities – whether politics, sports, or girls – that might disturb his studies. According to his recollections:[30]

At that time, I was just a research student with no duties apart from research, and I concentrated all my energy in trying to get a better understanding of the problems facing physicists at that time. I was not interested at all in politics, like most students nowadays. I confined myself entirely to the scientific work, and continued at it pretty well day after day, except on Sundays when I relaxed and, if the weather was fine, I took a long solitary walk out in the country. The intention was to have a rest from the intense studies of the week, and perhaps to try and get a new outlook with which to approach the problem the following Monday. But the intention of these walks was mainly to relax, and I had just the problems maybe floating about in the back of my mind without consciously bringing them up. That was the kind of life that I was leading.

Within an astonishingly short time, Dirac managed to transform himself from a student into a full-fledged scientist. After only half a year at Cambridge, in March 1924 he was able to submit his first scientific paper to the *Proceedings of the Cambridge Philosophical Society,* a local but internationally recognized periodical.[31] This paper dealt with a problem of statistical mechanics suggested by Fowler, his supervisor. Neither the problem nor the paper was of particular significance. It was merely an exercise, as debut papers often are. Dirac was then determined to become a research physicist and knew that he was good enough to contribute to the advancement of science, but he still did not have any definite ideas about which subject to specialize in. He had a preference for the fundamental and general problems of physics but only had vague ideas about how to deal with these problems in a new way. As a result, his first papers dealt with a rather scattered field of specific problems, mostly in relativity, quantum theory, and statistical mechanics (see the bibliography in Appendix II).

Dirac was very productive, publishing seven papers within two years, and succeeded making himself known to the small community of British theoretical physicists. Dirac's ability to solve difficult theoretical problems was soon noticed, both inside and outside Cambridge. Charles Galton Darwin, professor of natural philosophy (physics) at the University of Edinburgh and grandson of the famous naturalist, was told about the bright student by Fowler. Darwin asked Dirac to solve a mathematical-physics problem with which he had occupied himself, namely, proving that quantizing a dynamical system results in the same answer no matter what coordinates are used. This was just the problem to suit Dirac's taste.[32]

Dirac's early works appeared in the most recognized British journals and were communicated by Fowler, Milne, Eddington, and Rutherford. His approach in the papers was to take an already known result, based on established theory, and to criticize it in order to reach a better understanding. If possible, Dirac used relativistic arguments to discuss the results and make them more general:[33]

There was a sort of general problem which one could take, whenever one saw a bit of physics expressed in a nonrelativistic form, to transcribe it to make it fit in with special relativity. It was rather like a game, which I indulged in at every opportunity, and sometimes the result was sufficiently interesting for me to be able to write up a little paper about it.

One of these little papers dealt with an astrophysical problem: how to calculate the red-shift of solar lines on the assumption that the radiation emitted from the interior of the sun is Compton scattered in the atmosphere of the sun. This problem was suggested by the mathematician and astronomer Edward A. Milne, who, in the first months of 1925, became Dirac's supervisor while Fowler was on leave in Copenhagen. Dirac was not particularly interested in astrophysics, but he had followed Milne's course of lectures on the physics of stellar atmospheres and had obtained a good knowledge of the field.[34] He solved the problem suggested by Milne, concluding that the suggested mechanism could not account for the observed red-shift. This result ran against the expectations of Milne. Dirac did not deal with astrophysics again, but his later research covered topics also cultivated by Milne (see Chapter 11).

Another of the early papers dealt with a problem in the theory of relativity concerning the definition of velocity.[35] The problem had been stated by Eddington in *The Mathematical Theory of Relativity,* and he took a keen interest in Dirac's paper before its publication. Eddington suggested various alterations, mainly of an editorial kind, which Dirac was glad to accept. Before communicating Dirac's paper, Eddington commented on the manuscript: "[The paper needs] an introductory paragraph . . . to run something like this . . . you will no doubt reword this . . . look at these points and let me have it back."[36]

To make a long story short, Dirac's situation in the summer of 1925 was as follows: He had proved to be a talented physicist with a flair for complex theoretical problems and the use of mathematical methods. He had earned himself a name in Cambridge as a promising theorist, but outside Britain he was unknown. His contributions were interesting, but not remarkably so, and not of striking originality. In retrospect, his first seven publications can be seen to have been groundwork for more complex problems, the nature of which was then still unknown to Dirac. He vaguely felt that he was ready for bigger prey, but it was only after Heisenberg's pioneering discovery of quantum mechanics that Dirac knew his true hunting ground. Then things happened very quickly and he metamorphosed from a rather ordinary physicist into a *natural philosopher* whose name could rightly be placed alongside those of Maxwell and Newton. Only ten years after he entered Cambridge University, Dirac received the Nobel Prize in physics.

What was Dirac's life like before he found quantum mechanics? As mentioned, he lived a modest and undramatic life filled with physics and little else. His introvert character did not change much. Although he was in contact with several of the Cambridge physicists, and Fowler in particular, these contacts did not evolve into friendships. His contacts with other students at Cambridge were almost nil. Dirac spent much of his time alone in libraries and relaxed only on his solitary Sunday walks. "I did my work mostly in the morning," he wrote. "Mornings I believe are the times when one's brain power is at its maximum, and towards the end of the day I was more or less dull, especially after dinner."[37]

At an early stage of his career, Dirac developed the concise style that was to characterize all of his writings. Conceptual clarity, directness, technical accuracy, and logical presentation were virtues he cultivated from an early age. When writing a manuscript for a paper, he would first try to draw up the whole work in his mind. Only then would he write it down on paper in his meticulous handwriting, and this first draft would need few if any corrections. Niels Bohr, whose working habits and mental constitution were very different from Dirac's, once remarked: "Whenever Dirac sends me a manuscript, the writing is so neat and free of corrections that merely looking at it is an aesthetic pleasure. If I suggest even minor changes, Paul becomes terribly unhappy and generally changes nothing at all."[38] In the same vein, Igor Tamm recounted an exchange that took place when Bohr read the proofs of one of Dirac's papers:[39]

Bohr: "Dirac, why have you only corrected few misprints, and added nothing new to the text? So much time has passed since you wrote it! Haven't you had any new ideas since then?"
Dirac: "My mother used to say: think first, then write."

CHAPTER 2

DISCOVERY OF
QUANTUM MECHANICS

IRAC'S scientific life took a dramatic turn in the early fall of 1925, when he became acquainted with the work of Werner Heisenberg in which the fundamental ideas of quantum mechanics were first stated. On July 28, Heisenberg, Dirac's senior by only eight months, delivered a lecture in Cambridge at a meeting of the Kapitza Club. His subject was "Term-zoology and Zeeman-botany," that is, theoretical spectroscopy within the framework of the then existing "old" quantum theory of Bohr and Somerfeld. In the lecture Heisenberg did not refer to the new, still unpublished theoretical scheme he had just discovered. Presumably, Dirac was not in Cambridge at the time and thus missed the opportunity to attend Heisenberg's lecture.[1] However, Fowler was present, and he understood, perhaps from informal discussions with Heisenberg following his lecture, that the young German physicist had recently been able to derive some of the spectroscopic rules in a completely new way. In August, Fowler received the proof-sheets of Heisenberg's new paper. He ran through them and sent them on to Dirac, requesting him to study the work closely. At that time, the end of August, Dirac was in Bristol with his parents.

Heisenberg's aim in his historic paper was to establish a quantum kinematics that was in close accordance with Bohr's correspondence principle but that involved only observable quantities.[2] For this purpose he considered the classical Fourier expansion of an electron's position coordinate, for an electron being in its n'th stationary state

$$x(n) = \sum_{\alpha=-\infty}^{+\infty} x(n,\alpha)e^{2\pi i \nu(n,\alpha)t} \tag{2.1}$$

Here $\nu(n,\alpha) = \alpha\nu(n)$ and $x(n,\alpha) = x(n,-\alpha)$ denote the Fourier frequencies and amplitudes, respectively, where the condition on the latter guarantees that $x(n)$ is real. Since $x(n)$ is not directly observable, Heisenberg wanted to replace it with an expression that could be given a more satisfactory quantum theoretical interpretation. He suggested that $v(n,\alpha)$ be replaced with $v(n,n - \alpha)$, where the latter expression signifies the frequency corresponding to the quantum transition $n \to n - \alpha$. The Fourier coefficients $x(n,\alpha)$ were similarly replaced with $x(n,n - \alpha)$, interpreted to be transition amplitudes. Heisenberg argued that only the individual terms – not the summation – in equation (2.1) can be taken over into the quantum domain. These terms are then of a form

$$x(n,n - \alpha)e^{2\pi i v(n,n-\alpha)t}$$

which can be arranged in a two-dimensional array (a "Heisenberg array" or, as was recognized a few months later, a matrix).

On extending his analysis to the case of the anharmonic oscillator, Heisenberg was faced with the problem of how to represent a quantity like x^2, whose classical expression is

$$x^2(n) = \left(\sum_\alpha x(n,\alpha)e^{2\pi i v(n,\alpha)t} \right) \left(\sum_\beta x(n,\beta)e^{2\pi i v(n,\beta)t} \right)$$

He showed that for a single $x^2(n)$ term, the corresponding term in quantum theory can be written as

$$x^2(n,m)e^{2\pi i v(n,m)t}$$

The complete $x^2(n)$ expression can again be written in an array, each of the terms being related to the $x(n)$ terms by

$$x^2(n,m)e^{2\pi i v(n,m)t} = \sum_k x(n,k)x(k,m)e^{2\pi i v(n,m)t}$$

For the amplitude factors, this yields

$$x^2(n,m) = \sum_k x(n,k)x(k,m)$$

In the case of a product of two different quantities, x and y, the elements are formed similarly:

$$xy(n,m) = \sum_k x(n,k)y(k,m) \tag{2.2}$$

This is Heisenberg's famous law of multiplication. The fact that in general it is not commutative was noticed by Heisenberg. He found this most disturbing and at first considered it to be a flaw in his theory.

When Dirac first read Fowler's copy of Heisenberg's paper, he did not find it very interesting. In his early work in quantum theory, Dirac had stuck to the research program of the Bohr–Sommerfeld theory, which based the theory of atoms on Hamiltonian methods by extensive use of the angle and action variable technique known from classical mechanics. But, as he later realized, this approach was too restricted and not suited to an appreciation of Heisenberg's work:[3]

I was very much impressed by action and angle variables. Far too much of the scope of my work was really there; it was much too limited. I see now that it was a mistake; just thinking of action and angle variables one would never have gotten on to the new mechanics. So without Heisenberg and Schrödinger I would never have done it by myself.

It was only when Dirac again studied the proof-sheets, a week or so after he first read them, that he realized that Heisenberg had initiated a revolutionary approach to the study of atoms. Dirac now occupied himself intensely with Heisenberg's ideas, trying to master them and also to improve them. He found Heisenberg's formulation complicated and unclear and was also dissatisfied because it did not take relativity into account. He felt that it should be possible to state the quintessence of Heisenberg's theory in a Hamiltonian scheme that would conform with the theory of relativity.

After the summer vacation ended, Dirac returned to Cambridge, thinking deeply about Heisenberg's paper and the strange appearance in it of noncommuting dynamical variables. In order to proceed with his plan for setting up a Hamiltonian version of the new mechanics, he would have to have a classical expression to correspond with the quantity $xy - yx$ appearing implicitly in Heisenberg's theory. However, Dirac's first attempt to develop the theory went in another direction, that of extending Heisenberg's mechanics to systems involving rapidly moving electrons. In an unfinished manuscript, written in early October, Dirac argued that the Heisenberg variables [such as $x(n,m)$] referred not only to two energy levels but also to the two associated momenta.[4] In this case, the variables are connected with what he called "the theory of the uni-directional emission of radiation," that is, with light moving in a particular direction. Elaborating on this idea, Dirac proposed that, for relativistic velocities, the Heisenberg variables $x(n, n - \alpha)$ should be generalized by replacing t by $t - z/r$, where z is the direction of the light and r its distance from the source. However, Dirac soon sensed that he was on the wrong track and left his paper incomplete. Referring to this episode, he later recalled:

"There was a definite idea which I could work on, and I proceeded to write it up, but I never got very far with it."[5] In this first, abortive work on the new quantum mechanics, Dirac made use of some of his earlier works, in particular a short paper of 1924 in which he had proved the relativistic invariance of Bohr's frequency condition.[6]

During his unsuccessful attempt to introduce relativistic arguments into Heisenberg's theory, Dirac continued to ponder the puzzling non-commutativity. He later told how he discovered what became his own key for unlocking the quantum mysteries:[7]

I went back to Cambridge at the beginning of October, 1925, and resumed my previous style of life, intense thinking about these problems during the week and relaxing on Sunday, going for a long walk in the country alone. The main purpose of these long walks was to have a rest so that I would start refreshed on the following Monday. . . . It was during one of the Sunday walks in October, 1925, when I was thinking very much about this $uv - vu$, in spite of my intention to relax, that I thought about Poisson brackets. I remembered something which I had read up previously in advanced books of dynamics about these strange quantities, Poisson brackets, and from what I could remember, there seemed to be a close similarity between a Poisson bracket of two quantities, u and v, and the commutator $uv - vu$. The idea first came in a flash, I suppose, and provided of course some excitement, and then of course came the reaction "No, this is probably wrong." I did not remember very well the precise formula for a Poisson bracket, and only had some vague recollections. But there were exciting possibilities there, and I thought that I might be getting to some big new idea. It was really a very disturbing situation, and it became imperative for me to brush up my knowledge of Poisson brackets and in particular to find out just what is the definition of a Poisson bracket. Of course, I could not do that when I was right out in the country. I just had to hurry home and see what I could then find about Poisson brackets. I looked through my notes, the notes that I had taken at various lectures, and there was no reference there anywhere to Poisson brackets. The textbooks which I had at home were all too elementary to mention them. There was just nothing I could do, because it was a Sunday evening then and the libraries were all closed. I just had to wait impatiently through that night without knowing whether this idea was really any good or not, but still I think that my confidence gradually grew during the course of the night. The next morning I hurried along to one of the libraries as soon as it was open, and then I looked up Poisson brackets in Whittaker's *Analytical Dynamics,* and I found that they were just what I needed.

The quantity which Dirac looked up after his sleepless night was first introduced by the French mathematical physicist Siméon Poisson in 1809. It is defined as

$$[x,y] = \sum_j \left(\frac{\partial x}{\partial q_j} \frac{\partial y}{\partial p_j} - \frac{\partial x}{\partial p_j} \frac{\partial y}{\partial q_j} \right) \tag{2.3}$$

where p and q represent any two canonical variables for the system in question, and the summation is over the number of degrees of freedom of the system. Although the idea that Poisson brackets were relevant came to Dirac "rather out of the blue,"[8] it obviously stemmed from the fact that Hamiltonian dynamics can be formulated by means of the non-commuting Poisson bracket algebra. In particular one has that

$$[q_j,q_k] = 0, \, [p_j,p_k] = 0, \, [p_j,q_j] = \delta_{jk} \tag{2.4}$$

where δ_{jk}, the Kronecker delta, has a value of one for $j = k$ and a value of zero otherwise. The connection between Poisson brackets and Heisenberg products conjectured by Dirac that Monday morning in October was the following:

$$(xy - yx) \equiv \frac{ih}{2\pi} [x,y] \tag{2.5}$$

Armed with this idea, Dirac began to write his paper "The Fundamental Equations of Quantum Mechanics," which became one of the classics of modern physics. The paper received quick publication in *Proceedings of the Royal Society,* no doubt because of Fowler, who recognized its importance. Only three weeks intervened between the receipt of Dirac's paper by the Royal Society and its appearance in print.

Dirac did not introduce the Poisson bracket formulation at once in his paper; he did so only after deriving the rules of quantum differentiation. In most cases the structure of Dirac's research publications reflects fairly well the order in which the ideas occurred to him; that is, the "context of justification" roughly agrees with the "context of discovery." But in this case, Dirac "preferred to set up the theory on this basis where there was some kind of logical justification for the various steps which one made."[9] Let us briefly look at the main results of the paper.

Dirac's primary aim was to construct algebraic operations of the quantum variables in agreement with Heisenberg's theory. In particular, he looked for a process of quantum differentiation, which could give meaning to quantities like dx/dv, where x and v are quantum variables corresponding to Heisenberg's quantum amplitudes (matrices). Dirac found the result

$$\left(\frac{dx}{dv}\right)(nm) = \sum_k [x(nk)a(km) - a(nk)x(km)]$$

where the a coefficients represent another quantum variable. In condensed notation the formula was just written as

$$\frac{dx}{dv} = xa - ax \qquad (2.6)$$

Quantum differentiation of a quantity x, according to Dirac, was then equivalent with "taking the difference of its Heisenberg products with some other quantum variable." What does equation (2.6) correspond to classically? By means of an argument based on Bohr's correspondence principle, Dirac proved relation (2.5) and explained, "We make the fundamental assumption that *the difference between the Heisenberg products of two quantum quantities is equal to ih/2π times their Poisson bracket expression.*"[10]

It is remarkable that Dirac's deduction of equation (2.6) relied heavily on the correspondence principle. This principle played a crucial role in Heisenberg's road to quantum mechanics, but in general Dirac did not appreciate correspondence arguments; unlike his colleagues in Germany and Denmark, he made almost no use of them. Although, in principle, quantum mechanics made Bohr's correspondence principle obsolete, or at least far less important, many physicists continued to apply correspondence arguments after 1925.

With relation (2.5) at his disposal, Dirac could now proceed to formulate the fundamental laws of quantum mechanics by simply taking them over from classical mechanics in its Poisson bracket formulation. He no longer had need of the correspondence principle which was, so to speak, once and for all subsumed in relation (2.5). The quantum mechanical commutation relations follow from equations (2.4), yielding

$$q_j q_k - q_k q_j = p_j p_k - p_k p_j = 0$$

$$\text{and} \quad q_j p_k - p_k q_j = \frac{ih}{2\pi} \delta_{jk} \qquad (2.7)$$

From classical theory he further obtained the relation $dx/dt = [x,H]$, where H is the Hamiltonian and x is any dynamical variable of p and q ($[x,H]$ denotes the Poisson bracket, not the quantum mechanical commutator). Translating this into quantum mechanics, he obtained the fundamental equation of motion

$$\frac{dx}{dt} = \frac{2\pi}{iH}(xH - Hx) \qquad (2.8)$$

and from this he concluded that if $AH - HA = 0$, the quantum variable A must be a constant of motion. This result included the law of energy conservation: if $x = H$ in equation (2.8), then $dH/dt = 0$; i.e., H is constant.

Having thus set up the general scheme of quantum algebra, Dirac showed that it could be used to give a satisfactory definition of stationary states that agreed with that of the old quantum theory. For such states he derived Bohr's frequency relation of 1913, $E_m - E_n = h\nu$. Since in the old quantum theory the frequency relation, as well as the notion of stationary states, both had the status of postulates, it was most satisfying to Dirac that he could now deduce them from his new theory.

The general commutation relations (2.7) were discovered by several physicists in the fall of 1925. Apart from Heisenberg and Dirac, Wolfgang Pauli and Hermann Weyl also proposed, but did not publish, the relations, and they also figured prominently in an important paper by Max Born and Pascual Jordan.[11]

After Dirac completed his paper, he sent a handwritten copy to Heisenberg, who congratulated him for the "extraordinarily beautiful paper on quantum mechanics." In particular, Heisenberg was impressed by its representation of the energy conservation law and Bohr's frequency condition. In his letter Heisenberg reported to Dirac the rather disappointing news that most of his results had however already been found in Germany:[12]

Now I hope you are not disturbed by the fact that indeed parts of your results have already been found here some time ago and are published independently here in two papers – one by Born and Jordan, the other by Born, Jordan, and me – in *Zeitschrift für Physik*. However, because of this your results by no means have become less important [*unrichtiger*]; on the one hand, your results, especially concerning the general definition of the differential quotient and the connection of the quantum conditions with the Poisson brackets, go considerably further than the just mentioned work; on the other hand, your paper is also written really better and more concisely than our formulations given here.

It must have been disappointing to Dirac to hear about the work of Jordan and Born in which most of his results had been derived – and more than a month earlier at that.[13] In their paper, Born and Jordan for the first time used matrices representing quantum mechanical variables. On this basis they proved the equivalent of Dirac's equation (2.7), written in matrix notation as

$$\mathbf{pq} - \mathbf{qp} = \frac{h}{2\pi i}\mathbf{1}$$

where **1** is the unit matrix. They also proved the frequency condition and the energy conservation law, both of which figured in Dirac's paper. But they did not make the Poisson bracket connection.

During the fall of 1925 and the following winter, the formulation of the new quantum mechanics initiated by Heisenberg's paper was attended by stiff competition, primarily between the German physicists (Heisenberg, Jordan, Born, and Pauli) and Dirac in England. The Germans had the great advantage of formal and informal collaboration, while Dirac worked on his own. Even had he wanted to (which he did not), there were no other British physicists with whom he could collaborate on an equal footing. That he lost the competition under such circumstances is no wonder. However, Dirac was satisfied to know that it was possible to develop quantum mechanics independently in accord with his ideas. He was confident that the theory was correct and his method appropriate for further development.

Though handicapped relative to his German colleagues, Dirac, having quick access to the results obtained on the Continent, was better off than most American physicists. The competition in quantum mechanics at the time was given expression by John Slater, who, in a letter to Bohr of May 1926, told somewhat bitterly of his frustration at being beaten in the publication race: "It is very difficult to work here in America on things that are changing so fast as this [quantum mechanics] is, because it takes us longer to hear what is being done, and by the time we can get at it, probably somebody in Europe has already done the same thing." As an example of this experience, Slater mentioned that he had independently duplicated most of Dirac's results: "I had all the results of Dirac, the interpretation of the expressions ($pq - qp$) in terms of Poisson's bracket expressions, with applications of that, before his paper came, and was almost ready to send off my paper when his appeared."[14] Born, who visited MIT from November 1925 to January 1926, brought with him a copy of the still unpublished Born–Jordan paper, which he showed to Slater. The manuscript to which Slater referred in his letter to Bohr was written at the end of December. Entitled "A Theorem in the Correspondence Principle," it contained a full account of the Poisson formalism in quantum theory. However, at that time Dirac's work had already appeared in Europe.[15] Independently of Dirac, and almost at the same time, the Dutch physicist Hendrik Kramers observed the algebraic identity between Poisson brackets and the quantum mechanical commutators, but he did not realize that this identity was of particular significance and merely used it to confirm his belief that quantum mechanical problems always have a classical counterpart.[16] Dirac's conclusion was completely different and immensely more fruitful.

At that time Dirac was twenty-three years old. He was still a student, barely known to the Continental pioneers of quantum theory. The German physicists were surprised to learn about their colleague and rival in Cambridge. "The name Dirac was completely unknown to me," recalled Born. "The author appeared to be a youngster, yet everything was perfect in its way and admirable."[17] A few days after receiving Dirac's paper, Heisenberg mentioned to Pauli:[18]

An Englishman working with Fowler, Dirac, has independently re-done the mathematics for my work (essentially the same as in Part I of Born–Jordan). Born and Jordan will probably be a bit depressed about that, but at any rate they did it first, and now we really know that the theory is correct.

Dirac's reputation in the physics community was soon to change. While in the fall of 1925 he was referred to as just "an Englishman," within a year he would rise to become a star in the firmament of physics. In Cambridge Dirac quickly established himself as the local expert in the new quantum theory, lecturing frequently to the Kapitza Club on various aspects of the subject, including his own works.[19]

At the end of 1925, things were evolving very rapidly in quantum theory. Heisenberg's theory was established on a firm basis with the famous *Dreimännerarbeit* of Heisenberg, Born, and Jordan, and the new theory was now often referred to as "matrix mechanics" or the "Göttingen theory." But in spite of, or perhaps because of, the rapid development of the theory, many physicists felt uneasy about it; they wanted to see if it was also empirically fruitful and not merely a strange game with mathematical symbols. As recalled by Van Vleck: "I eagerly waited to see if someone would show that the hydrogen atom would come out with the same energy levels as in Bohr's original theory, for otherwise the new theory would be a delusion."[20]

In his next contribution to quantum mechanics, Dirac attacked the problem mentioned by Van Vleck.[21] Using an elaborate scheme of action and angle variables, he was able to prove that the transition frequencies for the hydrogen atom are given by the expression

$$\omega_n = \frac{\pi m e^4}{h}\left(\frac{1}{P^2} - \frac{1}{(P + nh/2\pi)^2}\right)$$

where n is an integer. As Dirac noticed, provided the quantity P is an integral multiple of $h/2\pi$, this is the same result obtained in the experimentally confirmed Bohr theory of 1913. However, since he was unable to prove that P is in fact an integral multiple of $h/2\pi$, Dirac's derivation of the hydrogen spectrum was not complete.

At this place it is appropriate to introduce the symbol $\hbar$ as an abbreviation for $h/2\pi$, where h denotes the usual Planck constant. Dirac first used the symbol in 1930, although in his paper on the hydrogen spectrum and in subsequent work up to 1930 he decided to let the symbol h ("Dirac's h") denote the quantity $h/2\pi$.[22] In what follows, h will mean the usual Planck constant.

By January 1926, Dirac had known for some time that a derivation of the hydrogen spectrum had already been obtained by Pauli. He had been so informed by Heisenberg in his letter of November 20 and had received proofs of Pauli's paper before publication. Although Pauli solved the hydrogen spectrum before Dirac, in fact Dirac's paper appeared in print before Pauli's. This was a result not only of fast publication of the *Proceedings of the Royal Society* but also of the fact that Pauli's paper was subject to considerable delay.[23] Dirac said that he "was really competing with him [Pauli] at this time."[24] The fact that Pauli had priority did not matter too much to Dirac, whose primary aim was to test his own scheme of quantum mechanics. Furthermore, Dirac's derivation was completely different from and much more general than that of Pauli, who made use of a rather special method.

Heisenberg praised Dirac's work on the hydrogen atom:[25]

I congratulate you. I was quite excited as I read the work. Your division of the problem into two parts, calculation with q-numbers on the one side, physical interpretation of q-numbers on the other side, seems to me completely to correspond to the reality of the mathematical problem. With your treatment of the hydrogen atom, there seems to me a small step towards the calculation of the transition probabilities, to which you have certainly approached in the meantime. Now one can hope that everything is in the best order, and, if Thomas is correct with the factor 2, one will soon be able to deal with all atom models.

Although Pauli's work on the hydrogen atom preceded the work of Dirac, Pauli realized that Dirac's treatment included a general treatment of action and angle variables which he (Pauli) had not yet obtained. In connection with his efforts to establish a more complete (relativistic) matrix mechanical theory of the hydrogen atom, Pauli had worked hard to formulate a method for treating action and angle variables. He was therefore a bit disappointed to see that he was, in this respect, superseded by Dirac. In March, after studying Dirac's paper, he wrote to Kramers: "In the March volume of the Proceedings of the Royal Society there is a very fine work by Dirac, which includes all of the results that I have thought out since Christmas about the extension of matrix calculus to non-periodic quantities (such as polar angles). I am sorry to have lost so much time working on this, when I could have been doing something else!"[26]

Dirac's (and Pauli's) work on the hydrogen spectrum was further developed by the Munich physicist Gregor Wentzel, who also treated the relativistic case.[27] Wentzel's approach was rather close to that of Dirac, but although Wentzel knew of Dirac's paper, he had obtained his main results independently. Still another, and completely different, theory of the hydrogen atom was worked out in Zürich by Erwin Schrödinger on the basis of his new wave mechanics. Schrödinger was not impressed by the works founded on matrix or q-number mechanics. In June, he wrote to Lorentz: "Dirac (Proc.Roy.Soc.) and Wentzel (Z.f.Phys.) calculate for pages and pages on the hydrogen atom (Wentzel relativistically, too), and finally the only thing missing in the end result is just what one is really interested in, namely, whether the quantization is in 'half integers' or 'integers'!"[28]

In his work on the hydrogen atom, Dirac did not consider the problem of how to incorporate spin and relativistic corrections, a problem to which he would give a complete solution less than two years later (see Chapter 3). At the beginning of 1926, it was known that in order to find these corrections one would have to calculate the quantum mechanical mean values of $1/r^2$ and $1/r^3$; this was a difficult and as yet unsolved mathematical problem. When the young American physicist John H. Van Vleck read Dirac's paper in the spring of 1926, he realized that the q-number formalism furnished a means for the calculations. Van Vleck was one of the few physicists who adopted Dirac's q-number technique for practical calculations. When he arrived in Copenhagen a few weeks later, he had found the corrections only to learn that the results, based on the methods of matrix mechanics, had just been published by Heisenberg and Jordan.[29]

Dirac wanted to establish an algebra for quantum variables, or, as he now termed them, q-numbers (q for "quantum" or, it was said, perhaps for "queer"). He wanted his q-number algebra to be a general and purely mathematical theory that could then be applied to problems of physics. Although it soon turned out that q-number algebra was equivalent to matrix mechanics, in 1926 Dirac's theory was developed as an original alternative to both wave mechanics and matrix mechanics. It was very much Dirac's own theory, and he stuck to it without paying much attention to what went on in matrix mechanics. In contrast to his colleagues in Germany, who collaborated fruitfully and also benefited from close connections with local mathematicians (such as Hilbert, Weyl, and Courant), Dirac worked in isolation. He probably discussed his work with Fowler, when he was available, but collaborated neither with him nor with other British physicists. He relied on a few standard textbooks, in particular Whittaker's *Analytical Dynamics* and Baker's *Principles of Geometry*, but did not seek the assistance of the Cambridge mathematicians. Baker's book proved particularly valuable in connection with the q-number algebra.[30]

Q-numbers are quantum variables that do not satisfy the usual commutation law for ordinary numbers, or, as Dirac called them, c-numbers (c for "classical"). If q-numbers represent observable physical quantities, then "in order to be able to get results comparable with experiment from our theory, we must have some way of representing q-numbers by means of c-numbers, so that we can compare these c-numbers with experimental values."[31] Dirac showed that q-numbers satisfy Heisenberg's law of multiplication [equation (2.2)]; that is, they can be represented by matrices. However, in the paper of January 1926 he did not say so explicitly and did not, in fact, mention the word "matrix" at all. At the time, he knew, of course, of the Göttingen matrix mechanics, but he seems not yet to have realized that q-numbers are equivalent to matrices. Dirac was not much impressed by the matrix formulation and believed that his scheme of quantum mechanics was superior in clarity as well as in generality. "It took me quite some time," he wrote, "to get reconciled to the view that my q-numbers were not really more general than matrices, and had to have the same limitations that one could prove mathematically in the case of matrices."[32]

In the summer of 1926, Dirac published a new and very general version of q-number algebra, this time presented as a purely mathematical theory.[33] In this paper he did not refer to physics at all. In his attempt to state a general and autonomous theory, he even went so far as not to include Planck's constant explicitly (that is, he put $ih/2\pi = 1$). The work had little impact on the physics community but seems to have been appreciated by those who cultivated the mathematical aspects of quantum physics. Jordan, who was such a connoisseur, wrote, "I find this paper . . . very beautiful; for to me the mathematics is just as interesting as the physics!"[34] The following are a few of the formulae obtained by Dirac in his quantum algebra.

If q and p are canonically conjugate, any other set of canonical variables Q and P can be written by means of the transformations

$$Q_j = bq_jb^{-1} \quad \text{and} \quad P_j = bp_jb^{-1} \tag{2.9}$$

where b is another q-number and b^{-1} is the quantity defined by $bb^{-1} = 1$. Similar transformation formulae were given in the Born–Jordan paper in matrix formulation, and they played an important role in the *Dreimännerarbeit*. However, when Dirac first stated them, he did not fully recognize their importance. "These formulae," he wrote, "do not appear to be of great practical value."[35] Equation (2.9) implies that Q and P are canonical variables if $bb^{-1} = 1$:

$$Q_jP_k - P_kQ_j = \frac{ih}{2\pi}\delta_{jk} \quad \text{and} \quad Q_jQ_k - Q_kQ_j = P_jP_k - P_kP_j = 0$$

just as stated in his first paper on quantum mechanics.

Functions of q-numbers may be differentiated not only with respect to the time but with respect to any q-number. The general definition of q-number differentiation, as given by Dirac, was as follows: Let the q-number q be conjugate to p, so that $qp - pq = ih/2\pi$; if $Q = Q(q)$, then dQ/dq is defined as

$$\frac{dQ}{dq} = Qp - pQ$$

As to the q-numbers representing angular momentum, Dirac showed that they satisfy the commutation relations[36]

$$L_zx - xL_z = \frac{ih}{2\pi}y$$

$$L_zp_x - p_xL_z = \frac{ih}{2\pi}p_y \quad \text{(and cyclic permutations)}$$

$$L_xL_y - L_yL_x = \frac{ih}{2\pi}L_z$$

and

$$L^2L_z - L_zL^2 = 0, \quad \text{etc.}$$

These relations, too, had been obtained earlier in matrix formulation by Heisenberg, Born, and Jordan.

In the summer of 1926, q-number algebra was one of several, competing schemes of quantum mechanics; the other versions were matrix mechanics (Heisenberg, Born, Jordan), wave mechanics (Schrödinger), and operator calculus (Born, Weiner). Physicists increasingly turned to wave mechanics when calculations had to be made, while q-number algebra remained almost exclusively Dirac's personal method. At this time, it was felt that what was needed was a general and unified quantum mechanical formalism – a feeling that Dirac shared. Before dealing with his contributions to this end, we shall briefly survey some other results he obtained in 1926.

Dirac was not, like the young German quantum theorists, raised in the spectroscopic tradition of the old quantum theory. This tradition was very much a Continental one and never received focal interest in England. But as Fowler's student, Dirac was acquainted with the literature and well aware of the connections between the new quantum mechanics and the various spectroscopic subtleties. In continuation of his work on the hydrogen atom, he used his method to throw light on some of the spectroscopic problems that had haunted the old quantum theory.[37]

In the old quantum theory the magnitude of the angular momentum of a single-electron atom, in units of $h/2\pi$, was assumed to be equal to the action variable k. Dirac showed that this is not so in quantum mechanics. If m is the magnitude of the angular momentum, the correct result is

$$m^2 = \left(k + \frac{1}{2} \frac{h}{2\pi} \right) \left(k - \frac{1}{2} \frac{h}{2\pi} \right)$$

For the total angular momentum of many-electron atoms, he found similar relations. Having established the general formulae for obtaining action and angle variables, Dirac turned to spectroscopic applications. His program was:[38]

To obtain physical results from the present theory one must substitute for the action variables a set of c-numbers which may be regarded as fixing a stationary state. The different c-numbers which a particular action variable may take form an arithmetical progression with constant difference h, which must usually be bounded, in one direction at least, in order that the system may have a normal state.

For a single-electron atom, he proved that for a given k the z-component of the angular momentum takes values ranging from $|k| - \frac{1}{2}h/2\pi$ to $-|k| + \frac{1}{2}h/2\pi$. Furthermore,[39]

k takes half integral quantum values ... and thus has the values $\pm \frac{1}{2}h$, $\pm \frac{3}{2}h$, $\pm \frac{5}{2}h$, ..., corresponding to the $S, P, D, ...$ terms of spectroscopy. There will thus be 1, 3, 5, ... stationary states for $S, P, D, ...$ terms when the system has been made non-degenerate by a magnetic field, in agreement with observation for singlet spectra.

He also proved the selection rules for k and m_z, and in particular that $S \rightarrow S$ transitions (i.e., from $k = \frac{1}{2}$ to $k = -\frac{1}{2}$) are forbidden. Then he proceeded to consider the anomalous Zeeman effect, one of the riddles of the old quantum theory. He showed that q-number theory is able to reproduce the correct *g*-factor for the energy of the stationary states in a weak magnetic field,[40] and also derived formulae for the relative intensities of multiplet lines that agreed with the formulae obtained by using the old quantum theory. Most of the results obtained by Dirac in his paper "On the Elimination of the Nodes in Quantum Mechanics" had been found earlier by the German theorists using the method of matrix mechanics; but Dirac was able to improve on some of the results and deduce them from his own system of quantum mechanics.

Dirac did not introduce the electron's spin in his treatment of the spectroscopic phenomena. He therefore had to rely on the largely ad hoc assumption of the old quantum theory that the gyromagnetic ratio of the atomic core is twice its classical value. Apart from this, his results did not depend on special assumptions concerning the structure of atoms.

A month later, at the end of April, Dirac considered another empirically well-established subject, Compton scattering, and showed that it too followed from his theory.[41] In doing so, Dirac extended his formalism to cover relativistic motions, making use of some of his ideas from his unpublished manuscript of October 1925 (see also Chapter 3). As is well known, the basic features of the scattering of high-frequency radiation (e.g., X-rays and gamma rays) on matter were explained in 1923 by Arthur H. Compton with the assumption of light-quanta, or, as they were called by Gilbert Lewis three years later, photons. Compton's theory convinced physicists of the reality of light quanta, although some, most notably Bohr, continued to consider the concept controversial.

For the sake of argument, Dirac accepted the light quantum hypothesis; but he was not particularly interested in whether electromagnetic radiation was "really" made up of corpuscles or waves. Dirac was content to get the correct formulae. For the change in wavelength of the radiation, he managed to reproduce Compton's formula, which expresses conservation of energy and momentum. As to the intensity of the scattered radiation, he obtained a result very close to that found by Compton in 1923 but not quite identical with it. "This is the first physical result obtained from the new mechanics that had not been previously known," Dirac proudly declared.[42]

Dirac's treatment of the Compton effect was recognized to be a work of prime importance. In the period 1926–9, the paper was cited at least 33 times and thus became the first of his papers to have a considerable impact on the physics community (see Appendix I). Dirac's work was generally considered to be very difficult. "We saw a paper by Dirac [on the Compton effect] which was very hard to understand," Oskar Klein recalled. "I never understood how he did it, but I've always admired the fact that he did it because he got the right result."[43] The work was discussed in Copenhagen before publication. In March, Sommerfeld visited

Cambridge, where he stayed with Eddington. When Sommerfeld was told about the still unfinished calculations by Dirac, he was much surprised. On Eddington's initiative, a meeting over a cup of tea with Dirac was arranged.[44]

Dirac was pleased with his work and felt that he had finally obtained something new. He discussed carefully how his results compared with experiment. In earlier as well as in later papers, Dirac showed little interest in experimental tests and preferred to emphasize the theoretical significance of his results. This time he was eager to show that his quantum algebra produced a result that fit the data better than earlier theories. He even went to the extreme of illustrating the fit with a diagram.[45] When he observed that Compton's experimental data were all a little below those predicted by his theory, he did not conclude that the theory was incomplete or faulty; no, he concluded that the discrepancy "suggests that in absolute magnitude Compton's values are 25 per cent cent too small."[46] Dirac had complete confidence in his theory.

When Compton read Dirac's paper, he was impressed and wrote to Dirac that physicists at the University of Chicago had performed X-ray measurements that nicely confirmed the new theory:[47]

Mr. P. A. M. Dirac:

You will be interested to know that Messrs. Barrett and Bearden, working here, have completed a set of measurements of the angle of maximum polarization for X-rays of effective wave-length of about .3A, .2A and .18A, finding in every case an angle of maximum polarization within 1 degree of 90°, in good accord with your theory.

Yours sincerely,

Arthur H. Compton

Later that year Dirac returned to the Compton effect, which he next treated by making extensive use of wave mechanics.[48] With the new method, he derived exactly the same expressions that he had found in his first paper on the subject.

In this period of hectic research activity, Dirac found time to write his Ph.D. dissertation, which was completed in May. It consisted mainly of a survey of work he had already published or was about to publish.[49] Dirac was completely absorbed in physics and spent most of his time alone in his study room at Cambridge. He had neither time nor desire to become involved in social or other extrascientific activities. In these months, there was much political and social unrest in England, which culminated in the General Strike declared on May 3. The conservative government had established an emergency plan that included a flood of antistrike volunteers. Many of Dirac's fellow students left their studies for a time to act as antistrike volunteers. One of them was Nevill Mott, who was at the time preparing for the tripos examination.[50] Dirac did not want the strike, or anything else, to interfere with his scientific work and did not join the volunteers.[51]

While still working on his thesis, Dirac was assigned by Fowler to lecture on quantum theory to the few students of theoretical physics at Cambridge. The title of Dirac's course was first announced as "Quantum Theory of Specific Heats" but was changed to "Quantum Theory (Recent Developments)." It was the first course on quantum mechanics ever taught at a British university. The students included A. H. Wilson, B. Swirles, J. A. Gaunt, N. F. Mott, and the American J. R. Oppenheimer. Fowler, who gave another course on quantum theory at the same time as Dirac, attended with his entire class. "Dirac gave us what he himself had recently done, some of it already published, some, I think, not," recalled one of the attendants of the lectures. "We did not, it is true, form a very sociable group, but for anyone who was there it is impossible to forget the sense of excitement at the new work. I stood in some awe of Dirac, but if I did pluck up courage to ask him a question I always got a direct and helpful answer, with no beating about the bush if I was getting things wrong."[52] Beginning in 1927, Dirac gave a regular course on quantum mechanics and was also assigned other teaching duties. As a teacher and supervisor, Dirac was "unapproachable," according to Alan Wilson, who was a research student in 1927.[53] The slightly younger Mott related the following episode from November 1927, when he had worked out some results that he wanted Dirac and Fowler to see:[54]

Dirac was there, and Fowler called him and Dirac said timidly that it was all nonsense, and referred me to one of his papers – which is about something quite different. Dirac said that the general theory allowed us to assume. . . . I asked him how he knew, and because I thought that the great man was being stupid, I may have summoned up courage to hector the great. Then I suddenly realised that the great man was timid and that I was being a bully! Funny moment. Fowler suggested that I should write it all nicely and that Dirac should read it and Dirac said he would – I hope he won't hate me too much!

Unknown to other physicists, since December 1925, Erwin Schrödinger in Zürich had worked on a completely new atomic theory in which quantum phenomena were seen as a kind of wave phenomenon. Schrödinger's "wave mechanics" developed ideas previously suggested by Louis de Broglie in Paris. The theory made its entry in March 1926, when Schrödinger published the first of a series of monumental papers on quantum mechanics.[55] The core of the theory was a differential equation, soon known as the Schrödinger equation:

$$\nabla^2\psi + \frac{8\pi^2 m}{h^2}(E - E_{\text{pot}})\psi = 0$$

Schrödinger's theory at once aroused intense interest, and it almost divided the physics community into two camps. Heisenberg and his circle criticized the theory for being inconsistent and conceptually regressive. Schrödinger, on his side, expressed a dislike for matrix mechanics' *Unanschaulichkeit* and "transcendental algebra," a dislike he presumably also held with respect to Dirac's formulation.

At an early stage, Dirac had studied the quantum statistics of Bose and Einstein and also de Broglie's approach to radiation phenomena. In the summer of 1925, when he gave a talk on the subject to the Kapitza Club, he was sympathetic to de Broglie's wave theory of matter; he argued that it was equivalent to the light quantum theory of Bose and Einstein.[56] But Dirac became occupied with Heisenberg's new theory and did not think of developing de Broglie's ideas into a quantum mechanics. Dirac probably first heard about Schrödinger's theory in mid-March, when Sommerfeld visited Cambridge. About a month later, Heisenberg wrote Dirac, wanting to know how his treatment of the hydrogen atom was related to Schrödinger's method:[57]

A few weeks ago an article by Schrödinger appeared . . . whose contents to my mind should be mathematically closely connected with quantum mechanics. Have you considered how far Schrödinger's treatment of the hydrogen atom is connected with the quantum mechanical one? This mathematical problem interests me especially because I believe that one can win from it a great deal for the physical significance of the theory.

But Dirac was much too absorbed in his own theory to consider Schrödinger's theory worthy of careful study:[58]

I felt at first a bit hostile towards it [Schrödinger's theory]. . . . Why should one go back to the pre-Heisenberg stage when we did not have a quantum mechanics and try to build it up anew? I rather resented this idea of having to go back and perhaps give up all the progress that had been made recently on the basis of the new mechanics and start afresh. I definitely had a hostility to Schrödinger's ideas to begin which, with persisted for quite a while.

It was only somewhat later, as a result of another letter from Heisenberg, that Dirac's attitude changed. Right after the publication of Schrödinger's first paper, many physicists wondered if wave mechanics and matrix mechanics, two theories that were different in style and content yet covered the same area of physics, were in fact deeply connected. Schrödinger was the first to prove the formal equivalence between the two theories.

But some time before Schrödinger's paper appeared, the result was known to the German physicists, thanks to an independent proof by Pauli, whose calculations were not published but were circulated quickly to the insiders.[59] In a letter to Dirac of May 26, Heisenberg reproduced Pauli's demonstration of the connection between wave mechanics and matrix mechanics; furthermore, he reported that when relativity was incorporated, the Schrödinger equation would become

$$\nabla^2\psi - \frac{1}{E^2}[\,(E - E_{pot})^2 - m_0^2 c^4]\,\frac{\partial^2\psi}{\partial t^2} = 0 \qquad (2.10)$$

where E is the total energy, including the rest mass $m_0 c^2$. As to his general opinion regarding Schrödinger's theory, Heisenberg was negative:[60]

I agree quite with your criticism of Schrödinger's paper with regard to a wave theory of matter. This theory must be inconsistent as just like the wave theory of light. I see the real progress made by Schrödinger's theory in this: that the same mathematical equations can be interpreted as point mechanics in a non-classical kinematics *and* as wave-theory accor. w. Schröd. I always hope that the solution of the paradoxes in quantum theory later could be found in this way. I should very much like to hear more exactly what you have done with the Compton-effect. We all here in Cophenhagen have discussed this problem so much that we are very interested in its quantum mechanical treatment. . . . I hope very much to see you in Cambridge in July or August. My best regards to Mr. Fowler.

The fact that Schrödinger's wave mechanics turned out to be mathematically equivalent to quantum mechanics caused Dirac's hostility to vanish. He realized that, computationally, wave mechanics is in many cases preferable. As to the physical interpretation, not to mention the ontology, associated with Schrödinger's theory, Dirac did not care much: "The question as to whether the [ψ] waves are real or not would not be a question which would bother me because I would think upon that as metaphysics."[61] Dirac was interested in formulae and simply found wave mechanics suitable for this purpose. Consequently, he began an intense study of the theory, which he soon mastered. This was difficult since the mathematics of wave mechanics, such as the theory of eigenvalues and eigenfunctions, was not part of Dirac's education and was little known in Cambridge. Dirac's view of the different formulations of quantum mechanics was essentially pragmatic. He never became a "wave theorist" in the sense of Schrödinger or de Broglie but freely used wave mechanics when he found it useful. Often he mixed it with his own q-number algebra.

Schrödinger recognized the formal beauty of Dirac's version of quantum mechanics but preferred to translate its results into the language of wave mechanics and did not, at that time, find himself congenial to Dirac's way of doing physics, which he found strange and difficult to understand.[62] Sometime during 1926, Schrödinger requested one of his students to give a review of one of Dirac's papers for a seminar in Zürich. The student, Alex von Muralt, tried hard but in vain to understand the paper, and Schrödinger had to give the review himself. He confessed to his depressed student that Dirac's paper had also caused him great difficulty.[63]

The first work in which Dirac considered Schrödinger's theory was the important paper "On the Theory of Quantum Mechanics," which was completed in late August. While finishing this article, he first met Van Vleck in Cambridge. Van Vleck had participated in a meeting of the British Association for the Advancement of Science in Oxford during August 4–11. Dirac told him about his new work in quantum mechanics, but Van Vleck, who had not yet studied Schrödinger's theory, found Dirac's ideas very difficult to understand.

Dirac started out from the wave equation, which he considered "from a slightly more general point of view," writing it as

$$\left\{ H\left(q_r, \frac{ih}{2\pi} \frac{\partial}{\partial q_r} \right) - W \right\} \psi = 0 \qquad (2.11)$$

If ψ is a solution to (2.11), it can be written as

$$\psi = \sum_n c_n \psi_n$$

where the coefficients are arbitrary constants and ψ_n are the eigenfunctions.[64] Dirac interpreted $|c_n|^2$ as the number of atoms in the n'th quantum state. The eigenfunctions ψ_n satisfy the equation $a\psi_n = a_n\psi_n$, where a is a q-number and a_n is a c-number. According to Dirac, ψ_n represents a state in which a has a definite numerical value, a_n. In the case of a system disturbed by the time-dependent perturbation energy $A(p,q,t)$, the wave equation was written as

$$(H - W - A)\psi = 0 \qquad (2.12)$$

for which the solution is

$$\psi = \sum_n a_n \psi_n$$

where a_n now depends on the time and $|a_n|^2$ denotes the number of atoms in state n at time t. The matrix elements corresponding to A are the coefficients of the expansion

$$A\psi_n = \sum_m A_{mn}\psi_n$$

Dirac used this expression to derive a general expression for time-dependent perturbations, namely

$$\frac{ih}{2\pi}\frac{d}{dt}|a_m|^2 = \sum_n (a_n A_{mn} a^*_m - a^*_n A_{nm} a_m)$$

which gives the rate of change of the number of atoms in state m. As an important application of the perturbation theory, Dirac considered the emission and absorption of radiation, a subject to which we shall return in Chapter 6. At the time when Dirac wrote his paper, time-dependent perturbation theory had already been developed by Schrödinger, but Dirac was unaware of this.[66]

In another part of this paper, Dirac examined what subsequently became known as Fermi–Dirac quantum statistics. His point of departure was Heisenberg's positivistic credo that a fundamental physical theory should enable one to calculate only those quantities that can be measured experimentally. "We should expect this very satisfactory characteristic to persist in all future developments of the theory," Dirac wrote.[66] Dirac adapted Heisenberg's philosophy to an atom with two electrons in states m and n, respectively, asking if the systems (mn) and (nm) should be counted as one state or two. Since the states are empirically indistinguishable, then "in order to keep the essential characteristic of the theory that it shall enable one to calculate only observable quantities, one must adopt the second alternative that (mn) and (nm) count as only one state."[67] Dirac argued that this restricts the set of eigenfunctions for the whole atom (neglecting electron–electron interactions) to the form

$$\psi_{mn} = a_{mn}\psi_m(1)\psi_n(2) + b_{mn}\psi_m(2)\psi_n(1) \tag{2.13}$$

where a_{mn} and b_{mn} are constants, and $\psi_m(1)$ is the eigenfunction of electron number 1, being in state m, etc. He then proved that equation (2.13) is a complete solution only if $a_{mn} = b_{mn}$ or $a_{mn} = -b_{mn}$. In the first case the wave function is symmetrical in the two electrons, i.e., $\psi_{mn}(1,2) = \psi_{nm}(2,1)$; in the latter case it is antisymmetrical, i.e., $\psi_{mn}(1,2) = -\psi_{nm}(2,1)$. Quantum mechanics does not predict which of the two cases is the correct one for electrons, but with the help of Pauli's exclusion principle, Dirac concluded that it must be the antisymmetrical solution, because then the wave function is of the form

$$\psi_m(1)\psi_n(2) - \psi_m(2)\psi_n(1)$$

implying that if two electrons are in the same state ($n = m$), then $\psi_{mn} = 0$, which means that there can be no such state. This agrees with the Pauli principle, which was known to hold for electrons, while the other possibility – the symmetrical case – does not rule out states with $n = m$.

Dirac further showed that light quanta are described by symmetrical wave functions which thus must be associated with Bose–Einstein statistics. By a curious (and erroneous) argument he assumed that gas molecules are governed by antisymmetrical eigenfunctions (because "one would expect molecules to resemble electrons more closely than light quanta"). However, the belief that gas molecules satisfy the same statistics as electrons was not peculiar to Dirac; it was rather generally assumed in 1926 and was, for example, also a part of Fermi's early work on quantum statistics. Using standard statistical methods, Dirac went on to find the energy distribution, the so-called statistics, of molecules. If A_s denotes the number of states with energy E_s, the number of molecules in state s he found to be

$$N_s = \frac{A_s}{\exp(\alpha + E_s/kT) + 1}$$

where k is Boltzmann's constant, α is another constant, and T is the temperature. This expression is the basic distribution law for particles obeying Fermi–Dirac statistics, such as electrons.

Dirac knew that Heisenberg had also applied quantum mechanics to many-particle systems, especially to the helium atom.[68] In a letter of April 9, Heisenberg had informed him as follows:[69]

Since I am in Copenhagen I tried to treat the heliumproblem on the basis of quantum mechanics. There was an essential difficulty for the explanation of the large distance between Singlet- and Tripletsystem, because this distance could *not* be explained by interaction of two magnets only. But now I think we have in the helium to deal with a resonance effect of a typical quantum mechanical feature. Really one gets in this way a qualitative explanation of the spectrum with regard as well to the frequencies as to the intensities. And I hope, the quantitative agreement is only a question of long numerical work.

Heisenberg's paper appeared a little before Dirac's and contained the same distinction between symmetrical and antisymmetrical eigenfunctions, including the connection to the exclusion principle. The other part of Dirac's result, the quantum statistics of gas molecules, had also been obtained earlier, by Enrico Fermi in a paper from the spring of 1926.[70] When Fermi saw Dirac's article, he was naturally disturbed that there was no reference to his own work. He wrote at once to Cambridge:[71]

In your interesting paper "On the theory of Quantum Mechanics" (Proc.Roy.Soc. *112*, 661, 1926) you have put forward a theory of the Ideal Gas based on Pauli's exclusion Principle. Now a theory of the ideal gas that is practical identical to yours was published by me at the beginning of 1926 (Zs.f.Phys. *36*, p. 902; Lincei Rend., February 1926). Since I suppose that you have not seen my paper, I beg to attract your attention on it.

The situation was embarrassing to Dirac, who hurried to write a letter of apology to Fermi. Much later, Dirac recalled the situation as follows:[72]

When I looked through Fermi's paper, I remembered that I had seen it previously, but I had completely forgotten it. I am afraid it is a failing of mine that my memory is not very good and something is likely to slip out of my mind completely, if at the time I do not see its importance. At the time that I read Fermi's paper, I did not see how it could be important for any of the basic problems of quantum theory; it was so much a detached piece of work. It had completely slipped out of my mind, and when I wrote up my work on the antisymmetric wave functions, I had no recollection of it at all.

Although virtually all of Dirac's results in "On the Theory of Quantum Mechanics" were thus obtained independently and earlier by other physicists, the work is justly seen as a major contribution to quantum theory. The new statistics was soon known under the joint names of Fermi and Dirac, although Fermi's priority was recognized (occasionally the statistics was referred to as Pauli–Fermi statistics). Incidentally, years later Dirac invented the names *fermions* and *bosons* for particles that obey Fermi–Dirac and Bose–Einstein statistics, respectively. The names date from 1945 and are today a part of physicists' general vocabulary.[73] Following the publication of Dirac's paper, the new statistics was eagerly taken up and applied to a variety of problems. The first application was made by Dirac's former teacher, Fowler; as an expert in statistical physics, he was greatly interested in the Fermi–Dirac result. Fowler studied a Fermi–Dirac gas under very high pressure, thus beginning a chapter in astrophysics that, a few years later, would be developed into the celebrated theory of white dwarfs by his student Chandrasekhar.[74] In Germany, Pauli and Sommerfeld made other important applications of the new quantum statistics, with which they laid the foundation for the quantum theory of metals in 1927.[75]

"On the Theory of Quantum Mechanics" became the most cited of Dirac's early papers and was studied with interest by both matrix and wave theorists. Although the paper was recognized as an important work, many physicists felt that it was difficult to understand and even cryptic. Schrödinger may be representative in this respect (see also the views of Einstein and Ehrenfest, quoted below). In October, when Dirac was in Copenhagen, Schrödinger told Bohr about his troubles in reading Dirac:[76]

I found Dirac's work extremely valuable, because it translates his interesting set of ideas at least partly into a language one can understand. To be sure, there is still a lot in this paper which I find obscure, . . . Dirac has a completely original and unique method of thinking, which – precisely for this reason – will yield the most valuable results, hidden to the rest of us. But he has no idea how *difficult* his papers are for the normal human being.

After completing his dissertation, Dirac wanted to go abroad to study with some of his Continental peers. At the time, the spring of 1926, he was well acquainted with Heisenberg, and hence it was natural for him to prefer Göttingen as his first destination. Göttingen was the center and birthplace of quantum mechanics, and its physics institute included not only Heisenberg but also Born and Jordan, as well as a number of other talented young physicists. However, on Fowler's advice Dirac decided first to spend a term at Bohr's institute in Copenhagen. This was a wise decision, for although Bohr no longer published on the technical aspects of quantum mechanics, he was very active as an organizer and source of inspiration; his flourishing institute was no less a center of quantum physics than was Göttingen. Bohr had close contacts to Germany, and German physicists often stayed in Copenhagen. Heisenberg spent much of the period from May 1926 to June 1927 with Bohr, during which time Pauli too was a frequent visitor. Bohr was happy to accept Fowler's request, and Dirac arrived in September.[77] In Copenhagen he met with Heisenberg, Friedrich Hund, Klein, Ehrenfest, and Pauli, and of course with Bohr. "I learned to become closely acquainted with Bohr, and we had long talks together, long talks on which Bohr did practically all of the talking."[78]

Although Dirac was now part of an intense intellectual environment in which cooperation and group discussions were much valued, he largely kept to his Cambridge habits of working alone. Not even the friendly atmosphere at Bohr's institute could break his deep-rooted preference for isolation. According to the Danish physicist Christian Møller, then a young student:[79]

[Dirac] appeared as almost mysterious. I still remember the excitement with which we [the young students] in those years looked into each new issue of

Proc.Roy.Soc. to see if it would include a work of Dirac. . . . Often he sat alone in the innermost room of the library in a most uncomfortable position and was so absorbed in his thoughts that we hardly dared to creep into the room, afraid as we were to disturb him. He could spend a whole day in the same position, writing an entire article, slowly and without ever crossing anything out.

Dirac was impressed by Bohr's personality. He later said that "he [Bohr] seemed to be the deepest thinker that I ever met."[80] Although Bohr's way of thinking was strikingly different from his own, the taciturn Dirac did not avoid being influenced by the thoughts of the talking Bohr. He was certainly influenced by the discussions at the institute, which, in the fall of 1926, concentrated on the physical interpretation of quantum mechanics. Dirac arrived in Copenhagen shortly after Schrödinger had left. Schrödinger's meeting with Bohr had resulted in heated discussions about the foundational problems of quantum theory, discussions that continued during the following months. But Dirac was unwilling to participate in the lofty, epistemological debate. He stuck to his equations.

In September 1926, a conceptual clarification of quantum mechanics was felt to be a pressing need. After Born's probabilistic interpretation of Schrödinger's theory, the question of how to generalize the probability interpretation and relate it to matrix mechanics came to the forefront. The essential step in this process, leading to a completely general and unified formalism of quantum mechanics, was the transformation theory. This theory was fully developed by Dirac and Jordan at the end of the year. A generalized quantum mechanics was in the air at the time and had already been developed to some extent by Fritz London.[81] Pauli, in close contact with Heisenberg in Copenhagen, suggested a probabilistic interpretation also holding in momentum space and related it to the diagonal elements of the matrices; but he was not able to go further.[82] The problem occupied Dirac, who thought much about how to interpret wave mechanical quantities in the more general language of quantum mechanics. At the end of October, Heisenberg reported to Pauli about Dirac's still immature ideas:[83]

In order to clarify Schrödinger's electrical density, Dirac has reflected on it in a very funny way. Question: "What is the quantum mech[anical] matrix of the electrical density?" Definition of density: I. It is zero everywhere where there is no electron. In equations:

$$\rho(x_0, y_0, z_0, t) \, (x_0 - x(t)) = 0$$
$$\rho(\cdots \quad \quad) \, (y_0 - y(t)) = 0$$

. . . .

Furthermore, the total charge is e:

$$\int \rho \,(x_0,y_0,z_0,t)dx_0dy_0dz_0 = e$$

The solution is (as can rather easily be proved):

$$\rho_{nm}(x_0y_0z_0t) = e\psi_n\psi_m^*(x_0y_0z_0t)$$

where ψ_n and ψ_m are Schröd[inger]'s normalized functions. This formulation seems really quite attractive to me.

Dirac had his transformation theory ready a month later, inspired by recent works of Heisenberg and Cornelius Lanczos.[84] Whereas many of Dirac's great works were based on ideas that came unexpectedly to him – from "out of the blue" – this theory was the result of a more direct and logical procedure. To follow this procedure appealed to Dirac's intellect. "This work gave me more pleasure in carrying it through than any of the other papers which I have written on quantum mechanics either before or after," he later wrote.[85]

In Copenhagen, Dirac gave a seminar on his theory sometime before he submitted the paper for publication. Oskar Klein recalled the difficulty of following Dirac's thoughts as expressed at the seminar: "It took us some time to understand the things because he gave some lectures and he wrote all the figures on the blackboard very nicely and he said a few words to them, but they were very, very hard to get."[86] Heisenberg at once reported to Pauli about Dirac's "extraordinarily grandiose generalization" of the transformaton theory. He was clearly impressed by the generality and logical rigor of Dirac's work, which he judged to be "an extraordinary advance."[87] Let us turn to this extraordinary paper.

Dirac addressed the fundamental problem of what questions can be given an unambiguous answer in quantum theory. "In the present paper a general theory of such questions and the way the answers are to be obtained will be worked out. This will show all the physical information that one can hope to get from the quantum dynamics, and will provide a general method for obtaining it, which can replace all the special assumptions previously used and perhaps go further."[88] In the Göttingen matrix mechanics, canonical transformations of the form

$$G = bgb^{-1} \tag{2.14}$$

where g is some dynamical variable and b is a transformation matrix, played an important role. In particular, they served to diagonalize the

Hamiltonian. Dirac asked about the significance of the transformation matrix and its relation to Schrödinger's wave function ψ. In order to state the result as generally as possible, he considered continuous matrices, i.e., matrices in which the parameters that label the rows and columns may vary continuously.

If $g(\alpha'\alpha'')$ denotes the dynamical variable g expressed in the matrix scheme labeled by the c-numbers α' and α'', the transformed matrix G is the variable labeled by another set of indices, say λ' and λ'': $G = g(\lambda'\lambda'')$. Considering transformations between any two matrix schemes (α) and (λ), Dirac showed that the canonical transformation (2.14) can be written as the integral

$$g(\lambda'\lambda'') = \int \int (\lambda'/\alpha')d\alpha' g(\alpha'\alpha'')(\alpha''/\lambda'')d\alpha''$$

In this expression the transformation functions appear as the symbols (λ'/α') and (α''/λ''), which correspond to $b(\lambda'\alpha')$ and $b^{-1}(\alpha''\alpha'')$, respectively. Any function $F(\lambda,\sigma)$ of conjugate variables λ and σ was now shown to be transformable into a matrix scheme (α) in which F is diagonal. Dirac proved that the transformation is given by

$$F\left(\lambda', -\frac{ih}{2\pi}\frac{\partial}{\partial\lambda'}\right)(\lambda'/\alpha') = F(\alpha')\,(\lambda'/\alpha')$$

where $F(\alpha')$ are the diagonal elements of the diagonal matrix that represents F. This equation is a differential equation in which F on the left side is a differential operator operating on the transformation function (λ'/α'). If, as a special case, λ' is taken to be equal to q, and F is the Hamiltonian, the equation reads

$$H\left(q, -\frac{ih}{2\pi}\frac{\partial}{\partial\lambda'}\right)(\lambda'/\alpha') = E(\lambda'/\alpha')$$

This is Schrödinger's wave equation, with the transformation function appearing instead of the usual ψ-function. Hence Dirac's conclusion: "The eigenfunctions of Schrödinger's wave equation are just the transformation functions . . . that enable one to transform from the (q) scheme of matrix representation to a scheme in which the Hamiltonian is a diagonal matrix."[89]

In his work with continuous matrices Dirac introduced an important formal innovation, the famous δ-function. He took its defining properties to be

$$\delta(x) = 0 \quad \text{for} \quad x \neq 0$$

$$\text{and} \quad \int_{-\infty}^{+\infty} \delta(x)dx = 1$$

and derived some other properties of $\delta(x)$. With the help of this function, or improper function, he showed, for example, that if a dynamical variable $\lambda(\lambda'\lambda'')$ is expressed in some matrix scheme, then its canonical conjugate in the same scheme is

$$\eta(\lambda'\lambda'') = -i\hbar\delta'(\lambda' - \lambda'')$$

where δ' means the differential quotient of δ. The commutation relation can then be written as

$$(\lambda\eta - \eta\lambda)(\lambda'\lambda'') = i\hbar\delta(\lambda' - \lambda'')$$

The δ-function had a long prehistory in 1926.[90] It was not really invented by Dirac but he introduced it independently, and it was only with his work that it became a powerful tool in physics. Perhaps, as Dirac later stated, the idea of the δ-function grew out of his early engineering training.[91] At any rate, "Dirac's δ-function" soon became a standard tool in physics. Originally considered to be merely an elegant and useful notation, it has proved to be of extreme importance in virtually all branches of physics. In the realm of pure mathematics, it may be seen as a predecessor of the theory of distributions created in 1945 by the Swiss mathematician Laurent Schwartz, whom Dirac later met, in August 1949, when they both lectured at a seminar arranged by the Canadian Mathematical Congress in Vancouver.[92]

It was not difficult for Dirac to show that his theory comprised Born's probabilistic interpretation of wave mechanics, according to which $|\psi|^2$ determines a probability density. But Dirac's interpretation was much more general, because it rested on the more general transformation functions. According to this interpretation, if the dynamical variable α has the initial value α_0, then the probability that the system at the time t is in a state between α'_t and $\alpha'_t + d\alpha'_t$ is given by

$$|(\alpha'_t/\alpha'_0)|^2 d\alpha'_t$$

In his work with the transformation theory, Dirac came close to formulating the indeterminacy principle, later given by Heisenberg. Dirac wrote: "One cannot answer any question on the quantum theory which refers to numerical values for both the q_{r0} and the p_{r0}." And again in his conclusion he said:[93]

One can suppose that the initial state of a system determines definitely the state of the system at any subsequent time. If, however, one describes the state of the system at an arbitrary time by giving numerical values to the co-ordinates and momenta, then one cannot actually set up a one-one correspondence between the values of these co-ordinates and momenta initially and their values at a subsequent time. All the same one can obtain a good deal of information (of the nature of averages) about the values at the subsequent time considered as functions of the initial values.

However, in some contrast to Born and Jordan, Dirac did not consider the probabilistic interpretation as something inherent in the quantum mechanical formalism, but rather as something that relied on assumptions that could be criticized. He ended his paper: "The notion of probabilities does not enter into the ultimate description of mechanical processes; only when one is given some information that involves a probability . . . can one deduce results that involve probabilities."[94] This view appealed to Heisenberg, who preferred it to the more deeply entrenched probabilistic physicalism of Born and Jordan. In February, Heisenberg wrote to Pauli: "One can say, as Jordan does, that the laws of nature are statistical. But one can also say, with Dirac (and this seems to me substantially more profound), that all statistics are produced by our experiments."[95] Pauli agreed. He felt that statistical notions should not enter the fundamental equations of a really satisfactory physical theory.[96]

Almost simultaneously with Dirac, Pascual Jordan published another transformation theory that matched Dirac's in generality and scope and essentially contained the same results.[97] Jordan's detailed and mathematically intricate work was independent of Dirac's and was based on a strictly probabilistic view. Since at least in a formal sense the two theories were equivalent, physicists soon talked of "the Dirac–Jordan transformation theory." Dirac knew about Jordan's work before it was published but not before he submitted his own paper. On Christmas Eve in 1926, he outlined the essence of his tranformation theory in a letter to Jordan and argued that the two theories were equivalent:[98]

Dr Heisenberg has shown me the work you sent him, and as far as I can see it is equivalent to my own work in all essential points. The way of obtaining the results may be rather different though. . . . In your work you consider transformations from one set of dynamical variables to another, instead of a transformation from one scheme of matrices representing the dynamical variables to another scheme representing the same dynamical variables, which is the point of view adopted throughout my paper. The mathematics would appear to be the same in the two cases however.

In the letter, Dirac did not introduce the δ-function explicitly but mentioned that the quantity $\int(\lambda'/\alpha')d\alpha(\alpha'/\lambda'')$ would equal zero when $\lambda' \neq \lambda''$ and equal "a certain kind of infinity" when $\lambda' = \lambda''$.

At the beginning of February 1927, Dirac took leave of Bohr and parted for Göttingen by train. On his way he stopped in Hamburg, where a local meeting of the German Physical Society took place on February 5–6. In Hamburg he met with many of Germany's best physicists. The day before the meeting began, Dirac chaired an informal seminar on quantum mechanics.[99] From Hamburg he went on to Göttingen in the company of other physicists who had joined the Hamburg meeting.

Dirac's months in Göttingen were divided between research and further education in some of the mathematical methods of physics with which he had not become well acquainted during his studies in Cambridge. Among other things, Dirac went to a course of lectures on group theory given by the distinguished mathematician Hermann Weyl. Group theory, extensively introduced in quantum mechanics by Weyl and Eugene Wigner during 1927–9, was then a very new subject in theoretical physics, but Dirac did not find it particularly appealing. Many physicists objected to the *Gruppenpest* fashion because of the abstract and unfamiliar character of group theory.[100] Dirac was not frightened by abstract mathematics but felt that group theory was largely unnecessary for physical applications. He always preferred to do without group theoretical methods and thought that, instead of adapting quantum mechanics to the mathematical structure of group theory, one should consider groups as merely a part of ordinary quantum mechanics. In January 1929, Dirac gave a talk to the Kapitza Club in Cambridge entitled "Quantum Mechanics without Group Theory," and slightly later he had this to say about the subject:[101]

Group theory is just a theory of certain quantities that do not satisfy the commutative law of multiplication, and should thus form a part of quantum mechanics, which is the general theory of all quantities that do not satisfy the commutative law of multiplication. It should therefore be possible to translate the methods and results of group theory into the language of quantum mechanics and so obtain a treatment of the exchange phenomena which does not presuppose any knowledge of groups on the part of the reader.

An important part of Dirac's stay abroad, in Göttingen as in Copenhagen, was the people he met and the contacts he made. In Göttingen were, of course, Born, Weyl, Jordan, and Heisenberg, although Heisenberg spent much of his time with Bohr in Copenhagen. Dirac seems to have impressed the people in Göttingen, the Germans as well as the visiting foreigners. "Dirac is at Göttingen and is the real master of the situation," wrote the American physicist Raymond T. Birge with regard to the experiences of his former student Edward Condon, who visited Göttingen during the same period as Dirac. "When he [Dirac] talks, Born just sits and listens to him open-mouthed. That Dirac thinks of absolutely nothing but physics."[102] Dirac also met the American John Robert Oppenheimer, with whom he established a lasting friendship. Oppenheimer, Dirac's junior by two years, had studied at the Cavendish and then moved on to Göttingen in order to prepare his doctoral dissertation under Born. He knew Dirac from Cambridge, but it was only in Göttingen that their relationship evolved into a friendship. Together they took walks in the surroundings of Göttingen, including an expedition to the Harz Mountains.

Scientifically, Dirac's stay in Göttingen was very fruitful. It resulted, in particular, in important work on the quantum theory of radiation, to which we shall return in Chapter 6. Another very important development during the period was Heisenberg's new theory of the physical interpretation of quantum mechanics, the core of which was the famous indeterminacy relations.[103] Dirac was not directly involved in this theory, which was mainly worked out during Heisenberg's stay in Copenhagen; but Dirac's work on the transformation theory and his discussions with Heisenberg in Copenhagen were instrumental in forming Heisenberg's views. As far as the quantitative aspects are concerned, the indeterminacy principle grew out of the Dirac–Jordan transformation theory; with regard to the no less important qualitative and philosophical aspects, it was indebted to Heisenberg's long discussions with Bohr. As mentioned, Dirac was technically close to the indeterminacy relations in December 1926, but he felt no pressing need to formulate the insight of his transformation theory in a principle of coordinated measurement. Perhaps Dirac was, at any rate, of too unphilosophical a mind to proceed in the direction followed by Heisenberg.

Although Dirac's primary occupation was not with the measurement problem, he was, of course, interested in the new principle of indeterminacy. In May, he wrote to Heisenberg, raising by means of *Gedankenexperimente* various objections to the limited accuracy of coordinated measurements in quantum mechanics.[104] Apparently he was not at that time convinced of the universal validity of the indeterminacy relations. Heisenberg had no difficulty in countering Dirac's objections and explained to him how he and Bohr had analyzed the impossibility of measuring both the position and velocity of an electron at the same time by means of the so-called gamma-microscope.

Dirac left Göttingen at the end of June, spending a few days in Holland on his way back to England. Invited by Paul Ehrenfest, he went to Leiden together with Oppenheimer. At that time, Holland had a strong tradition in theoretical physics, as witnessed by names such as Kramers, Uhlenbeck, Fokker, and Ehrenfest. The dean of Dutch physics was seventy-three–year–old Hendrik Lorentz, who was still remarkably active and followed the new developments in physics with interest. Lorentz was impressed by Dirac's work and wanted him to come to Leiden for the two terms of 1927–8: "I think I have well understood your general trend of thought, admiring the beauty of your method and your remarkable deduction of Schrödinger's wave equation," he wrote to Dirac in June 1927.[105]

Dirac at the Kammerlingh-Onnes Laboratory in Leiden, 1927. In the front row, left to right: G. Uhlenbeck, H. Hönl, F. Florin, unidentified, A. D. Fokker, H. A. Kramers, and S. A. Goudsmit. Behind them, left to right: K. F. Niessen, P. A. M. Dirac, J. R. Oppenheimer, L. Polak, T. Ehrenfest, P. Ehrenfest, and H. R. Woltjer. The two persons in the third row are unidentified. Reproduced with permission of AIP Niels Bohr Library.

Dirac did not accept Lorentz's offer, but he did stay in Leiden for a few days and also spent a day in Utrecht visiting Kramers. For some time, Ehrenfest had recognized the originality of Dirac's physics, which he admired but found hard to understand. He had wanted Dirac to come to Leiden earlier, in October 1926, and was now happy to get the opportunity to meet the young Englishman.[106] After Dirac had agreed to visit Leiden, Ehrenfest wrote to him that the Dutch physicists had discussed "the last three Diracian crossword puzzles: Physical interpretation, emission and absorption, dispersion. We spent many, many hours going over a few pages of your work before we understood them! And many points are still as dark to us as the most moonless night!"[107] One of the features in Dirac's scheme of quantum mechanics, to which Ehrenfest objected, was what he called the "time illness." After Dirac's stay in Leiden in the summer of 1927, Ehrenfest wrote to Uhlenbeck about his opinion of the current status of quantum mechanics: "In fact, the Dirac theory works *not at all with the time,* although indeed with a q and p which are algebraically ill," he wrote. "Dirac denies that one should ask *when* something happens. One should be content knowing with what probability something happens. . . . [Dirac's] apparatus is exaggeratedly blind."[108] Evidently Ehrenfest studied Dirac's work very hard. In correspondence with Max Planck, he offered his view on Dirac's quantum theory, suggesting that it might be clarified by making an analogy between q-number algebra and the more familiar tensor algebra. Planck found Ehrenfest's interpretation helpful and decided to study Dirac once again during his summer vacation.[109]

After returning to Cambridge following his journey to the Continent, Dirac was faced with a problem of a more mundane nature, that of remaining within the academic institution. His fellowship was running out, but in November 1927 he received a new fellowship at St. John's College. Apparently he had told Oppenheimer that he would take a rest from quantum mechanics during the summer vacation, a promise that he was not quite able to keep. Oppenheimer, then back in the United States, wrote to Dirac in November: "I have just heard that you received the fellowship. My very best felicitations. There has been no direct news of what you have been doing for a long time. Did you keep your promise to stop quantum mechanics over the summer? I should very much appreciate it if you would let me know what you have got."[110]

In the fall of 1927, the so-called Copenhagen interpretation of quantum mechanics had emerged as a powerful paradigm in physics. It was based on Heisenberg's indeterminacy relations, Bohr's complementarity principle, and a strictly acausal and probabilistic interpretation of the subatomic domain. On the formal side, the Copenhagen interpretation became based on von Neumann's axiomatic Hilbert space theory, a mathematically advanced development of the Dirac–Jordan transformation theory. As is well known, leading physicists such as Schrödinger and Einstein did not accept the Copenhagen doctrines. At the Solvay Conference of October 1927, the discussion centered on the interpretation of quantum mechanics, and the highlight was Bohr's successful defense of the Copenhagen interpretation in face of the objections raised by Einstein and others. Dirac was invited to participate in this Solvay Conference, which indicates that he was then recognized as one of the world's top physicists. We shall return to the 1927 Solvay Conference, and Dirac's stand in the long and lasting debate on the interpretation of quantum mechanics, in Chapters 4 and 13.

CHAPTER 3

RELATIVITY AND SPINNING ELECTRONS

W HEN Dirac went to the Solvay Conference in October 1927, he could look back at two years of successful involvement in quantum mechanics. Twenty-five years old, he was a physicist of international repute. But in spite of all his productivity and originality, most of his results had also been obtained by other physicists who, more often than not, had published their works a little before Dirac. He felt that he still lived in the shadow of Heisenberg and the other German theorists and that he had not yet produced a deep and really novel theory, a theory nobody else had thought of. These ambitions were fulfilled by the end of the year with Dirac's celebrated relativistic theory of the electron, one of the great landmarks in the history of science.

Dirac was, from an early age, fascinated by the theory of relativity. He believed, then as later, that a physical theory could be really fundamental only if it lived up to the standards of Lorentz invariance set by Einstein. His first move in quantum mechanics, in October 1925, had been an attempt to make Heisenberg's theory conform to the theory of relativity. Not successful at this, he left the subject for a while. He returned to it half a year later in connection with his theory of the Compton effect (see also Chapter 2). In his paper on this subject, he proposed to treat time as a quantum variable, arguing that "the principle of relativity demands that the time shall be treated on the same footing as the other variables, and so it must therefore be a q-number."[1] Guided by classical Hamiltonian mechanics, he showed that what he called the "quantum time" would be the variable conjugated to $-W$, where W is the energy function:

$$tW - Wt = i\hbar \qquad (3.1)$$

Dirac argued that the relation (3.1) was also suggested by the following formal relativistic argument: The commutation relations $[x_j, p_j] = i\hbar$ ($j = 1,2,3,$) ought also to hold for the fourth components of the corresponding four-vectors, $x_4 = ict$ and $p_4 = iW/c$; in that case one has $[x_4, p_4] = i\hbar$, which is the same as (3.1). The commutation relation (3.1) had earlier been postulated by Pauli, but he had only stated it privately, in a letter to Heisenberg.[2]

In order to prepare for his main task, the calculation of the Compton effect, Dirac wrote down the classical-relativistic Hamiltonian equations of motion. For a free particle, the Hamiltonian is

$$\vec{p}^2 - W^2/c^2 = -m_0^2 c^2 \tag{3.2}$$

For a charged particle in an external electromagnetic field, the corresponding equation is

$$\left(\vec{p} - \frac{e}{c}\vec{A}\right)^2 - \frac{1}{c^2}(W - e\phi)^2 = -m_0^2 c^2 \tag{3.3}$$

where $\vec{A}$ is the vector potential, ϕ the scalar potential, and e the charge of the particle (i.e., $-e$ for an electron). Dirac used equation (3.3) in his theory of Compton scattering, in which he considered the electromagnetic field of the incident radiation to be given by a vector in the direction of the y-axis, that is, $\vec{A} = (0, A_y, 0)$ and $\phi = 0$.

In his first paper on the Compton effect, Dirac did not refer to Schrödinger's new wave mechanics and hence did not attempt to formulate a relativistic wave equation. He did so in his second paper on the Compton effect, but in the meantime the problem had been discussed by several other physicists. To make possible a better appreciation of Dirac's contributions, a brief review of this development may be helpful.[3]

On Schrödinger's original road to wave mechanics, relativistic considerations were of crucial importance.[4] In fact, he first derived a relativistic eigenvalue equation, which he did not publish, mainly because he realized that it did not reproduce the hydrogen spectrum with acceptable accuracy. This first, relativistic Schrödinger equation had the form

$$\nabla^2 \psi + \frac{1}{\hbar^2 c^2}\left[\left(W + \frac{e^2}{r}\right)^2 - m_0^2 c^4\right]\psi = 0 \tag{3.4}$$

for an electron of rest mass m_0 moving in the Coulomb field of a hydrogen nucleus (Schrödinger did not use the symbol $\hbar$). The equation gave a fine structure for the hydrogen spectrum, but not the correct one. The hydrogen fine structure was first given a theoretical explanation in 1915 by Sommerfeld, who worked with a relativistic extension of Bohr's

Dirac: A scientific biography

atomic theory. In his celebrated work, Sommerfeld found that the energy levels of the hydrogen atom were given by the expression

$$W_{n,k} = m_0 c^2 \left[\left(1 + \frac{\alpha^2}{(n - k - \sqrt{k^2 - \alpha^2})^2} \right)^{-\frac{1}{2}} - 1 \right] \qquad (3.5)$$

where α is the fine structure constant (equal to $e/\hbar c$), n the principal quantum number, and k the azimuthal quantum number. Sommerfeld's fine structure formula, or rather its first-order approximation

$$W_{n,k} = - \frac{m_0 e^4}{2\hbar^2} \left[1 + \frac{\alpha^2}{n^2} \left(\frac{n}{k} - \frac{3}{4} \right) \right] \qquad (3.6)$$

was soon verified experimentally by Paschen and other physicists. When the Bohr–Sommerfeld theory was replaced by quantum mechanics, the fine structure formula became a test case for the new theory: if quantum mechanics was to establish its empirical validity, it ought to reproduce equation (3.5) or, at least, equation (3.6). It was this test that Schrödinger's early relativistic equation (3.4) did not pass. Consequently, Schrödinger reported only the non-relativistic approximation of equation (3.4) in his first publication on wave mechanics.

Independently of Schrödinger, the relativistic second-order equation was found in the spring of 1926 by Oskar Klein, who was the first to publish it. During the next half year, it was investigated by several other physicists, including Fock, Gordon, de Broglie, Schrödinger, and Kudar, and eventually became known as the Klein–Gordon equation (KG equation in what follows). In addition to the eigenvalue equation (3.4), the time-dependent KG equation corresponding to equation (3.4) was studied:

$$\hbar^2 c^2 \nabla^2 \psi + \hbar^2 \frac{\partial^2 \psi}{\partial t^2} + 2ie\hbar \frac{\partial \psi}{\partial t} \frac{e^2}{r} + e^2 \left[\frac{e^4}{r^2} - \frac{m_0^2 c^4}{e^2} \right] \psi = 0 \qquad (3.7)$$

The fact that Schrödinger had abandoned his relativistic wave equation to avoid disagreement with experiment was commented on extensively by Dirac. Schrödinger once explained the story to Dirac, probably in 1933 during their stay in Stockholm to receive the Nobel Prize. Later in his life, Dirac considered the story to fit well with his general view of progress in theoretical physics, and he often used it as an illustrative example of how disagreement between theory and experiment should be handled. The disagreement between equation (3.4) and the hydrogen spectrum was, Dirac said,[5]

... most disappointing to Schrödinger. It was an example of a research worker who is hot on the trail and finding all his worst fears realized. A theory which was so beautiful, so promising, just did not work out in practice. What did Schrödinger do? He was most unhappy. He abandoned the thing for some months, as he told me. ... Schrödinger had really been too timid in giving up his first relativistic equation. ... Klein and Gordon published the relativistic equation which was really the same as the equation which Schrödinger had discovered previously. The only contribution of Klein and Gordon in this respect was that they were sufficiently bold not to be perturbed by the lack of agreement of the equation with observations.

According to Dirac, then, Schrödinger should have stuck to his beautiful relativistic theory and not worried too much over its disagreement with experiment; but because of simple psychological reasons – fear that his entire theory could collapse – he was mentally unable to do so. As Dirac, expounding another pet idea of his, further stated in 1971,[6]

It is a general rule that the originator of a new idea is not the most suitable person to develop it because his fears of something going wrong are really too strong and prevent his looking at the method from a purely detached point of view in the way that he ought to.

After this digression, let us return to the situation in 1926.

As mentioned, the relativistic version of wave mechanics was investigated by several researchers in the summer and fall of 1926. Although it was not possible to make the eigenvalue equation fit the fine structure of hydrogen, in other respects the theory looked quite promising. In particular, it proved successful in handling the Compton effect wave mechanically. Also, from a more aesthetic point of view, its four-dimensional form was appealing, especially since it proved possible to define charge and current densities that were parts of a four-vector satisfying the continuity equation. Whereas in usual, non-relativistic wave mechanics the charge and current densities were given respectively by

$$\rho = e|\psi|^2 = e\psi\psi^*$$ (3.8)

and

$$\vec{j} = \frac{e\hbar}{2mi} (\psi^*\nabla\psi - \psi\nabla\psi^*)$$ (3.9)

in the KG theory the corresponding expressions turned out to be

$$\rho = -\frac{e\hbar}{2mc}\left(\psi^* \frac{\partial \psi}{\partial t} - \psi \frac{\partial \psi^*}{\partial t}\right) \tag{3.10}$$

and

$$\vec{j} = \frac{e\hbar}{2mi}(\psi^*\nabla\psi - \psi\nabla\psi^*) \tag{3.11}$$

Some physicists, especially in Germany, adopted another approach in their attempts to reconcile quantum mechanics and relativity. Rather than follow the Klein–Gordon approach, which was based on wave mechanics, they tried to include relativistic effects as perturbations, or corrections, to the non-relativistic theory. This method led to a partial success in the spring of 1926, when Jordan and Heisenberg, developing ideas due to Pauli, were able to derive the fine structure formula in the form (3.6). In doing so, they added to the usual Hamiltonian not only a perturbation term describing the relativistic correction to the kinetic energy but also a term referring to the spin of the electron. However, in spite of its empirical success in accounting for the doublet riddles of spectroscopy, the Jordan–Heisenberg theory was not entirely satisfactory. Since relativity was added as a first-order correction, the theory was not genuinely relativistic, leading to equation (3.6) but not to equation (3.5); also, the spin effect was introduced in an ad hoc manner, being grafted to the theory rather than explained. An entirely satisfactory theory would not only be able to account for the doublet phenomena but would also explain them in the sense of deducing them from the basic principles of relativity and quantum mechanics.

During 1926, Dirac was mainly occupied with developing the formal basis of quantum mechanics. Apparently he did not consider the problem of finding a relativistic wave equation a pressing one at the time. Or perhaps he found it a rather trivial problem after Schrödinger had published his theory. When Dirac first made use of wave mechanics in August 1926, he did consider the Schrödinger equation in its relativistic form.[7] For a free gas molecule (and also for an electron), he wrote it as

$$(\vec{p}^2 - W^2/c^2 + m_0^2c^2)\psi = 0 \tag{3.12}$$

With the substitutions

$$\vec{p} = -i\hbar\nabla \quad \text{and} \quad W = i\hbar\frac{\partial}{\partial t} \tag{3.13}$$

this yielded

$$\left(\nabla^2 - \frac{1}{c^2}\frac{\partial^2}{\partial t^2} - \frac{m_0^2 c^2}{\hbar^2}\right)\psi = 0 \qquad (3.14)$$

which is just the KG equation for a free particle. Although Dirac was one of the first to state this equation, he seems not to have considered it of much importance. In his further treatment of the quantum statistics of ideal gases, he did not use equation (3.12) but used only its non-relativistic approximation.

When Dirac investigated the Compton effect wave mechanically three months later, he also started with the KG equation, but again he used only an approximation.[8] At that time, Klein and Gordon had already treated the Compton effect, but Dirac stressed that his approach was independent and different. "The wave equation," he wrote, referring to the KG equation, "is used merely as a mathematical help for the calculation of the matrix elements, which are then interpreted in accordance with the assumptions of matrix mechanics."[9] While staying at Bohr's institute (where Klein was present) in November 1926, Dirac seems for a short time to have considered the KG theory as a serious candidate for a relativistic quantum mechanics. He even played with Klein's idea of a five-dimensional theory embracing both quantum mechanics and general relativity.[10] But he was soon diverted by other problems, in particular the transformation theory.

When Einstein visited Ehrenfest in Leiden in the autumn of 1926, they used the occasion to discuss Dirac's (first) paper on the Compton effect. On October 1, Ehrenfest wrote to Dirac:[11]

Einstein is currently in Leiden (until Oct. 9th). In the few days we have left, he, Uhlenbeck and I are struggling together for hours at a time studying your work, for Einstein is eager to understand it. But we are hitting at a few difficulties, which – because the presentation is so short – we seem absolutely unable to overcome.

Ehrenfest asked Dirac to explain to him a number of problems, one of which was:

Why do you write the Hamilton equation in the form:

$$W^2/c^2 - p_1^2 - p_2^2 - p_3^2 = m^2c^2$$

and not:

$$mc^2 \sqrt{1 - (p_1^2 + p_2^2 + p_{-3}^2)/m^2c^2} = W.$$

Does it make a difference?

And at the end of his letter:

Please forgive me if some of the questions rest on gross misunderstandings. But despite all efforts we cannot get through them! Of course it would be especially nice if you were to come to Leiden yourself while Einstein is still here. Unfortunately we can only cover the cost of your ship ticket here and back. But you would be our guest here in Holland! It would be wonderful if you yourself arrived here at the same time as your letter!!!!!

Ehrenfest's question concerning the form of the relativistic Hamiltonian equation later proved to be highly relevant with respect to the wave equation of the electron. In 1930, Dirac gave the following argument, which could have been his response to Ehrenfest's question four years earlier:[12]

Equation (3) [the second equation in Ehrenfest's letter], although it takes into account correctly the variation of the mass of the particle with its velocity, is yet unsatisfactory from the point of view of relativity, because it is very unsymmetrical between W and the p's, so much so that one cannot generalize it in a relativistic way to the case when there is a field present. ... Equation (4) [the first of Ehrenfest's equations] is not completely equivalent to equation (3) since, although every solution of (3) is also a solution of (4), the converse is not true.

However, in October 1926, the problem of a relativistic wave equation as an alternative to the equation of Klein and Gordon was scarcely in Dirac's mind.

It was Dirac's preoccupation with the general principles of quantum mechanics, and the transformation theory in particular, that led him to realize that the formal structure of the Schrödinger equation [i.e., the form $(H - i\hbar \partial/\partial t)\psi = 0$] had to be retained even in a future unification of quantum theory and relativity. Since the KG equation is of second order in $\partial/\partial t$, it seemed to Dirac to be in conflict with the general formalism of quantum mechanics. A similar conclusion was reached by Pauli and his Hungarian assistant Johann Kudar at the end of 1926. "Herr Pauli," reported Kudar to Dirac, "regards the relativistic wave equation of *second* order with much suspicion."[13] As an alternative, the Hamburg physicists tended to consider the first-order version

$$(\hbar c \sqrt{m_0^2 c^4 - \nabla^2})\psi = (W - V)\psi \qquad (3.15)$$

as more reasonable. Although Pauli, in a letter to Schrödinger, admitted that this equation was "mathematically rather unpleasant" because of its square root operator, he found it to be "in itself sensible." "On the whole it seems to me that an appropriate formulation of quantum mechanics will only be possible when we succeed in treating space and time as equal to one another," he added.[14] Dirac would have agreed. Heisenberg also recognized what he called "the profound meaning of the linearity of the Schrödinger equations" and for that reason considered equations of the KG type to be without prospect.[15] But it was only Dirac who managed to harvest the rich fruit of this insight. Before we follow Dirac's road to the relativistic equation, a look at the theory of spin may be helpful.

At the end of 1926, it was widely accepted that spin and relativity were intimately related. However, it proved impossible to incorporate spin in the KG theory, and the nature of the relations among spin, relativity, and quantum mechanics remained a problem. In December 1926, Heisenberg and Dirac made a bet in Copenhagen as to when the spin phenomenon would be properly explained. Heisenberg wrote to Pauli, "I made a bet with Dirac that the spin phenomena, like the structure of the nucleus, will be understood in three years at the earliest; while Dirac maintains that we will know for sure about the spin within three months (counting from the beginning of December)."[16] Who won the bet is debatable, but with regard to the spin, at least, Dirac was closer to the mark than Heisenberg.[17] Dirac himself began to think about the spin problem shortly after the bet, probably inspired by discussions with Pauli, who visited Bohr's institute in January 1927. According to his memoirs, Dirac got the idea of representing spin by three spin variables independently of Pauli.[18] But he did not follow up this idea, which was developed by Pauli in his important quantum mechanical theory of spin of May 1927.[19] Pauli proposed a two-component wave function for the electron in order to accommodate spin, which was represented by a new kind of quantum variables, namely, the 2×2 "Pauli matrices"

$$\sigma_x = \begin{pmatrix} 0 & 1 \\ 1 & 0 \end{pmatrix}, \quad \sigma_y = \begin{pmatrix} 0 & -i \\ i & 0 \end{pmatrix}, \quad \text{and } \sigma_z = \begin{pmatrix} 1 & 0 \\ 0 & -1 \end{pmatrix} \quad (3.16)$$

which were related to the spin vector by

$$\vec{s} = \tfrac{1}{2} \hbar \vec{\sigma} \quad (3.17)$$

With this approach the energy eigenvalue equation became two coupled equations of the type

$$i\hbar \frac{\partial \psi}{\partial t} = H\left(q_k, -i\hbar \frac{\partial}{\partial q_k}, \vec{\sigma}\right)\psi \quad (3.18)$$

At about the same time, Darwin independently obtained a spin theory that was equivalent to Pauli's but was expressed in the language of wave mechanics.[20]

The Pauli–Darwin theory was welcomed by physicists because it managed to represent spin quantum mechanically, avoiding the ad hoc feature of the earlier Heisenberg–Jordan theory. On the other hand, it did not solve new empirical problems and did not contribute to the solution of the profound problem of integrating quantum mechanics with relativity. As in the earlier theory, Pauli and Darwin were forced to include relativistic effects only as a first-order correction to the non-relativistic Hamiltonian. This was sufficient from an empirical point of view, but from an aesthetic point of view, as adopted by Dirac, it was a blemish which indicated that the theory was only provisional. Both Pauli and Darwin recognized that "one must require from a final theory that it is formulated relativistically invariant from the outset, and also allows calculation of higher corrections."[21] During 1927, all attempts to improve on the theory of Pauli and Darwin proved fruitless.

Dirac did not try to develop the spin theory beyond the limits set by Pauli and Darwin, but he was very interested in the theory. While wave mechanicians like Schrödinger preferred Darwin's approach, Dirac was much in favor of Pauli's method, which, he argued, was in better agreement with the general theory of quantum mechanics. In lectures given at Cambridge in the fall of 1927, Dirac praised Pauli's theory:[22]

It consists in abandoning from the beginning any attempt to follow the classical theory. One does not try to take over into the quantum theory the classical treatment of some model, which incorporates the empirical facts, but takes over the empirical facts directly into the quantum theory. This method provides a very beautiful example of the general quantum theory, and shows that this quantum theory is no longer completely dependent on analogies with the classical theory, but can stand on its own feet.

At that time Dirac was closely acquainted with the spin theory, including its applications to spectra. In his lecture notes he pointed out its deficiencies too, such as its failure to agree to better than first order with Sommerfeld's fine structure formula. Although preoccupied with spin, Dirac did not attempt to obtain a more satisfactory theory. The problem of getting a relativistic wave equation was in his mind, but he did not yet connect this problem with spin. On the contrary, Dirac thought that his still unborn theory would describe a spinless particle, supposed to be the simplest kind of particle; only after the theory for such a hypothetical particle was established did he expect that spin could be incorporated.

During the Solvay Congress in October 1927, Dirac mentioned to Bohr his concern about a relativistic wave equation:[23]

Then Bohr answered that the problem had already been solved by Klein. I tried to explain to Bohr that I was not satisfied with the solution of Klein, and I wanted to give him reasons, but I was not able to do so because the lecture started just then and our discussion was cut short. But it rather opened my eyes to the fact that so many physicists were quite complacent with a theory which involved a radical departure from the basic laws of quantum mechanics, and they did not feel the necessity of keeping to these basic laws in the way that I felt.

After his return from Brussels, Dirac concentrated on the problem of formulating a first-order relativistic theory of the electron. Within two months he had solved the whole matter.[24]

A few days before Christmas 1927, Darwin went to Cambridge. He was completely surprised to learn of Dirac's new theory, about which he reported to Bohr:[25]

I was at Cambridge a few days ago and saw Dirac. He has now got a completely new system of equations for the electron which does the spin right in all cases and seems to be "the thing." His equations are first order, not second, differential equations! He told me something about them but I have not yet even succeeded in verifying that they are right for the hydrogen atom.

As usual, Dirac had worked alone, almost secretly. He thrived best in this way and very seldom discussed his ideas with other physicists. Mott, who was at the time as close to Dirac as anyone, recalled that "all Dirac's discoveries just sort of fell on me and there they were. I never heard him talk about them, or he hadn't been in the place chatting about them. They just came out of the sky."[26] "The Quantum Theory of the Electron," probably Dirac's greatest contribution to physics, was received by *Proceedings of the Royal Society* on January 2, 1928.

Dirac's theory was a product of his emerging general philosophy of physics. He wanted the theory to be founded on general principles rather than on any particular model of the electron. In contrast to Pauli, Schrödinger, and Darwin, who all imagined that the problem of integrating spin and relativity would probably require some sophisticated model of the electron, Dirac was not at all interested in model-building. "The question remains as to why Nature should have chosen this particular model for the electron instead of being satisfied with a point-charge," he pointed out at the beginning of his paper.[27] Consequently, he considered the electron to be a point charge.

Dirac's point of departure was that "we should expect the interpretation of the relativistic quantum theory to be just as general as that of the non-relativistic theory."[28] In full compliance with his general outlook on physics, he was guided by two invariance requirements: first, the space–time properties of the equation should transform according to the theory of relativity; and second, the quantum properties should transform according to the transformation theory of quantum mechanics. Dirac recognized that the latter requirement excluded the KG theory. Only if the wave equation is linear in $\partial/\partial t$ is the probability interpretation secured. This reasoning suggests the starting procedure

$$i\hbar \frac{\partial\psi}{\partial t} = c \sqrt{m_0^2 c^2 + p_1^2 + p_2^2 + p_3^2}\ \psi \qquad (3.19)$$

for a free electron. This is the same equation Pauli had considered in private communications [see equation (3.15)]. Of course, it faces the same mathematical difficulties because of the square root operator, which seems to yield a differential equation of infinite order. But, Dirac reasoned, it would be a promising start if the square root could be arranged in a linear form in the momenta. Then he faced a purely mathematical problem: How can a square root of four quantities possibly be linearized? At this point he could have consulted the mathematicians; the German algebraists might have supplied him with an answer.[29] But Dirac was not one to ask for assistance. He worked out the problem in his own way, by "playing around with mathematics," as he said.[30] He found a clue in an identity he had noticed when he "played" with the Pauli spin matrices, namely

$$\sqrt{p_1^2 + p_2^2 + p_3^2} = \sigma_1 p_1 + \sigma_2 p_2 + \sigma_3 p_3 \qquad (3.20)$$

At that stage, Dirac may have tried to use the quantity $\vec{\sigma} \cdot \vec{p}$ as the Hamiltonian in a wave equation; that is, he may have considered

$$\vec{\sigma} \cdot \vec{p}\,\psi = W\psi \qquad (3.21)$$

where ψ is a two-component wave function. This equation is Lorentz-invariant, contains a spin of one-half, and is of first order in the time derivative. But since it does not contain a mass term, it does not apply to an electron.[31] Hence Dirac had to reconsider the possible significance of equation (3.19): If it could be generalized to four squares instead of two, it would obviously indicate a solution; for then a linearization of the type wanted

$$\sqrt{p_1^2 + p_2^2 + p_3^2 + (m_0 c)^2} = \alpha_1 p_1 + \alpha_2 p_2 + \alpha_3 p_3 + \alpha_4 m_0 c \qquad (3.22)$$

could be provided. But were there coefficients with this property, and if so, what did they look like? Dirac argued that the linear wave equation, as provisionally given by equations (3.19) and (3.22), has to contain the

KG equation as its square. In this way he was able to deduce the following set of conditions for the coefficients in equation (3.22):

$$\alpha_\mu \alpha_\nu + \alpha_\nu \alpha_\mu = 0 \quad (\mu \neq \nu)$$
$$\alpha_\mu^2 = 1$$

(3.23)

Dirac knew that a set of similar conditions are fulfilled by the spin matrices, of which there are, however, only three. So he naturally tried to take $\alpha_j = \sigma_j$ and sought for another 2×2 candidate for α_4. However, such a candidate does not exist, and Dirac realized that working with 2×2 matrices just would not do. Then he got again one of those invaluable ideas out of the blue: "I suddenly realized that there was no need to stick to quantities, which can be represented by matrices with just two rows and columns. Why not go to four rows and columns?"[32] This idea solved the problem, and he found the explicit form of the α matrices:

$$\alpha_j = \begin{pmatrix} 0 & \sigma_j \\ \sigma_j & 0 \end{pmatrix} \quad \text{and} \quad \alpha_4 = \begin{pmatrix} 1 & 0 & 0 & 0 \\ 0 & 1 & 0 & 0 \\ 0 & 0 & -1 & 0 \\ 0 & 0 & 0 & -1 \end{pmatrix}$$

where $j = x$, y, or z, and σ_j are the Pauli matrices.

With the linearization successfully carried out, the ice was broken. The next stage – to formulate the wave equation for a free electron – was easy. Equations (3.19) and (3.22) immediately yielded

$$(W/c + \vec{\alpha} \cdot \vec{p} + \alpha_4 m_0 c)\psi = 0$$

(3.24)

which is known as the Dirac equation.

Dirac reduced a physical problem to a mathematical one, and mathematics forced him to accept the use of 4×4 matrices as coefficients. This again forced him to accept a four-component wave function $\psi = (\psi_1, \psi_2, \psi_3, \psi_4)$. Though logical enough, this was a bold proposal since there was no physical justification for the two extra components. The conclusion rested upon Dirac's confidence in the power of mathematical reasoning in the realm of physics. Indeed, Dirac's theory of the electron is a beautiful example of what Wigner has called "the unreasonable effectiveness of mathematics in the natural sciences."[33] If Dirac had followed an empiricist logic of science, he would not have introduced such "unphysical" terms as 4×4 matrices. As Darwin acknowledged, in comparing Dirac's work with his own attempt: "Dirac's success in finding the accurate equations shows the great superiority of principle over the previous empirical method."[34]

At this stage the equation was only an inspired guess. Dirac had to prove that it was logically, as well as empirically, satisfactory. It was constructed to conform with the principles of quantum mechanics, and since Dirac could prove its Lorentz invariance, the equation met the formal requirements. But what about its application to experimental reality? To check this, a free electron would not do; it had to be capable of interacting. Dirac placed the electron in an electromagnetic field, using the standard procedure of replacing W with $(W - e\phi)$ and $\vec{p}$ with $(\vec{p} - e/c\,\vec{A})$; that is, he used the Hamiltonian given by equation (3.2). A little manipulation of the α matrices then converted equation (3.24) into

$$\left\{ \left(\frac{W}{c} - \frac{\phi}{c} e \right) + \rho_1 \left(\vec{\sigma} \cdot \left(\vec{p} - \frac{e}{c} \vec{A} \right) \right) + \rho_3 m_0 c \right\} \psi = 0 \quad (3.25)$$

where ρ_1, ρ_3, and $\vec{\sigma} = (\sigma_1, \sigma_2, \sigma_3)$ are new 4×4 matrices derived from the old α matrices. By a further transformation Dirac was then able to show that this differential operator, if squared, included the KG operator and, in addition, the term

$$\frac{e\hbar}{c} \vec{\sigma} \cdot \vec{B}$$

where $\vec{B} = \nabla \times \vec{A}$ is the magnetic field. If divided by $2m$, this term represents an additional energy of the electron corresponding to a magnetic moment $e\hbar\vec{\sigma}/2mc$. "This magnetic moment is just that assumed in the spin electron model" (i.e., in Pauli's theory), Dirac wrote.[35] Without introducing the spin in advance, Dirac was thus able to deduce the correct spin from the first principles upon which his equation was built. This was a great and unexpected triumph:[36]

I was not interested in bringing the spin of the electron into the wave equation, did not consider the question at all and did not make use of Pauli's work. The reason for this is that my dominating interest was to get a relativistic theory agreeing with my general physical interpretation and transformation theory. . . . It was a great surprise for me when I later on discovered that the simplest possible case did involve the spin.

Of course, it was an exaggeration for Dirac to say that he did not make use of Pauli's theory; he did, as we have seen, when playing around with the spin matrices. But the use he made of the spin matrices was heuristic only.

When Dirac's theory appeared, its strength lay at the conceptual and methodological level, not at the empirical level. In fact, at first the theory did not yield even one result or explain even one experimental fact not already covered by earlier theories. Dirac showed in his paper that the new theory in its first approximation led to the same energy levels for the hydrogen atom as those given by the theories of Darwin and Pauli; that is, he deduced the approximate fine structure formula. But he did not attempt to go further, either by including higher corrections or by looking for an exact solution that would, hopefully, yield the exact fine structure formula. To derive this formula, which was still unexplained by quantum mechanics, would have gone far to credit the new theory. One may therefore wonder why Dirac did not attack the problem with more determination. According to his recollections, he did not even attempt to find an exact solution but looked for an approximation from the start:[37]

I was afraid that maybe they [the higher order corrections] would not come out right. Perhaps the whole basis of the idea would have to be abandoned if it should turn out that it was not right to the higher orders, and I just could not face that prospect. So I hastily wrote up a paper giving the first order of approximation and showing it to that accuracy; at any rate, we had agreement between the theory and experiment. In that way I was consolidating a limited amount of success that would be something that one could stand on independently of what the future would hold. One very much fears the need for some consolidated success under circumstances like that, and I was in a great hurry to get this first approximation published before anything could happen which might just knock the whole thing on the head.

However, there are reasons to believe that Dirac's retrospection, based on his hope-and-fear moral, is not quite correct. When he created the theory, he was guided by a strong belief in formal beauty and had every reason to be confident that his theory was true. It seems unlikely that he really would have feared that the theory might break down when applied to the hydrogen atom. After all, the Sommerfeld formula had never been tested beyond its first or second approximation; if relativistic quantum mechanics did not reproduce that formula exactly, it could justifiably be argued that it was not exactly true. Dirac's hurry in publication may have been motivated simply by competition, the fear of not being first to publish. Several other physicists were working hard to construct a relativistic spin theory, a fact of which Dirac must have been aware. Naturally he felt that the credit belonged to him. He did not want to be beaten in the race, a fate he had experienced several times already. If agreement with the fine structure formula had the crucial importance that Dirac later asserted, one would expect that he would have attempted to derive the exact fine structure after he submitted his paper for publication. He did not. I think Dirac was quite satisfied with the approximate agreement and had full confidence that his theory would also provide an exact agreement. He simply did not see any point in engaging in the complicated mathematical analysis required for the exact solution.

Other physicists who at the time tried to construct a relativistic spin theory included Hendrik Kramers in Utrecht; Eugene Wigner and Pascual Jordan in Göttingen; and Yakov Frenkel, Dmitri Iwanenko, and Lev Landau in Leningrad. Kramers obtained an approximate quantum description of a relativistic spinning electron in terms of a second-order wave equation and later proved that his equations were equivalent to Dirac's equation. When he got news of Dirac's theory, he was deeply disappointed, and this feeling evolved into a continuing frustration with regard to Dirac's physics. It is unknown in what direction Jordan and Wigner worked (they never published their work), but it seems to have been toward a relativistic extension of Pauli's spin theory. "We were very near to it," Jordan is supposed to have said, "and I cannot forgive myself that I didn't see that the point was linearization."[38] Although disappointed, Jordan recognized the greatness of Dirac's work. "It would have been better had we found the equation but the derivation is so beautiful, and the equation so concise, that we must be happy to have it."[39] Frenkel, Iwanenko, and Landau engaged in laborious tensor calculations and succeeded in working out theories that in some respects were similar to Dirac's. But apart from being published after Dirac's work, they lacked, like Kramers's theory, the beauty and surprising simplicity that characterized Dirac's theory.[40] Still, there can be little doubt that had Dirac not published his theory in January 1928, an equivalent theory would have been published by other physicists within a few months. Dirac later said that if he had not obtained the wave equation of the electron, Kramers would have.[41]

The news of Dirac's new equation spread rapidly within the small community of quantum theorists. The key physicists knew about it before publication. In Göttingen they learned about the theory from a letter Dirac sent to Born, and Bohr was informed by Fowler (and earlier by Darwin), who, as a Fellow of the Royal Society, had communicated the paper to the *Proceedings*.[42] The reception was enthusiastic. Léon Rosenfeld, who at the time was working with Born in Göttingen, recalled that the deduction of the spin "was regarded as a miracle. The general feeling was that Dirac had had more than he deserved! Doing physics in that way was not done! ... It [the Dirac equation] was immediately seen as *the* solution. It was regarded really as an absolute wonder."[43] From Leipzig, Heisenberg wrote to Dirac about his collaboration with Pauli on quantum electrodynamics, and added: "I admire your last work about the spin in the highest degree. I have especially still for questions: do you get the

Sommerfeld-formula in *all* approximations? Then: what are the currents in your theory of the electron?"[44] A month later, Ehrenfest reported his opinion to the Russian physicist Joffe: "I find Dirac's latest work on electron spin just splendid. Tamm has explained it all to us very well. He is continuing to work on this."[45]

Within two weeks following submission of the paper, Walter Gordon in Hamburg was able to report to Dirac that he had derived the exact fine structure formula from the new equation and that Heisenberg's first question could thus be answered affirmatively. Reporting the main steps in the calculation, Gordon wrote: "I should like very much to learn if you knew these results already and if not, you think I should publish them."[46] A little later, Darwin got the same result. Darwin was impressed by Dirac's genius but found the theory very difficult to understand unless he transcribed it to a more conventional, wave mechanical formalism. As he told Bohr: "I continue to find that though Dirac evidently knows all about everything the only way to get it out of his writings is to think of it all for oneself in one's own way and afterwards to see it was the same thing."[47] The fact that Dirac's equation yielded exactly the same formula for the hydrogen atom that Sommerfeld had found thirteen years earlier was another great triumph. It also raises the puzzling question of how Sommerfeld's theory, based on the old Bohr theory and without any notion of spin, could give exactly the same energy levels as Dirac's theory. But this was a historical curiosity that did not bother the physicists.[48]

A month after the publication of his first paper on the relativistic electron, Dirac completed a sequel in which he investigated various consequences of the theory for the behavior of spectral lines. In a letter to Jordan he reported the main results:[49]

I have worked out a few more things for atoms with single electrons. The spectral series should be classified by a single quantum number j taking positive and negative integral values (not zero) instead of the two, k and j, of the previous theory. The connection between j values and the usual notation is given by the following scheme:

$$j = \quad -1 \quad \underbrace{1 \quad -2} \quad \underbrace{2 \quad -3} \quad \underbrace{3 \quad -4}$$
$$\quad\quad S \quad\quad\quad P \quad\quad\quad\quad D \quad\quad\quad\quad F$$

j and $-(j + 1)$ form a spin doublet. One finds for the selection rule, $j \rightarrow j \pm 1$ or $j \rightarrow -j$ and for the g-value in a weak magnetic field $g = j/j + \frac{1}{2}$. The magnetic quantum number m satisfies $-|j| + \frac{1}{2} \leq m \leq |j| + \frac{1}{2}$.

That is, Dirac showed that all the doublet phenomena were contained in his equation.

The first occasion at which Dirac himself presented his theory to the German physicists was when he delivered a lecture to the *Leipziger Universitätswoche* during June 18–23, 1928.[50] This was the first in a series of annual symposia on current research in physics, and it was arranged by Debye and Heisenberg, the new professors of physics at Leipzig. Dirac gave a survey of his new theory and called attention to a further argument for the linear wave equation. He showed that the charge density associated with the KG theory, equation (3.10), is not positive definite: Since the KG equation is of second order in the time derivative, when $\psi(t_0)$ is unknown, $(\partial\psi/\partial t)_{t=t_0}$ is undetermined and therefore $\psi(t > t_0)$ is also undetermined; and since ρ is a function of ψ and $\partial\psi/\partial t$, knowledge of $\rho(t_0)$ leaves $\rho(t > t_0)$ undetermined, so that the electrical charge $\int\rho dV$ may attain any value. "The principle of charge conservation would thus be violated. The wave equation must consequently be linear in $\partial/\partial t$," asserted Dirac.[51] That Dirac's new equation did not face the same difficulty was explicitly shown by Darwin, who gave expressions for the charge and current densities.[52] In Dirac's theory the probabilities and charge densities took the same form as in the non-relativistic theory, that is, $|\psi|^2$ and $e|\psi|^2$.

Dirac's theory of the electron had a revolutionary effect on quantum physics. It was as though the relativistic equation had a life of its own, full of surprises and subtleties undreamed of by Dirac when he worked it out. During the next couple of years, these aspects were uncovered. The mathematics of the equation was explored by von Neumann, Van der Waerden, Fock, Weyl, and others, and the most important result of this work was the spinor analysis, which built upon a generalization of the properties of the Dirac matrices. Dirac had not worried about the mathematical nature of his four-component quantities; at first, it took the mathematical physicists by surprise to learn that the quantities were neither four-vectors nor tensors. Other theorists attempted to incorporate the Dirac equation into the framework of general relativity or, as in the case of Eddington, to interpret it cosmologically. Producing generalizations of the Dirac equation, most of them without obvious physical relevance, became a pastime for mathematical physicists. Although in the early part of 1928 it seemed that the theory had no particular predictive power or empirical surplus content, it soon turned out to be fruitful for the experimentalists too. In particular, it proved successful in the study of relativistic scattering processes, first investigated by Mott in Cambridge and by Klein, Nishina, and Møller in Copenhagen.

By the early thirties, the Dirac equation had become one of the cornerstones of physics, marking a new era of quantum theory. Its undisputed status was more a result of its theoretical power and range than of its empirical confirmation. In fact, several of the predictions that followed from Dirac's theory appeared to disagree with experiment. For example, in 1930 Mott predicted, on the basis of Dirac's theory, that free electrons could be polarized, a result that not only was contradicted by experiment but also ran counter to Bohr's intuition.[53] The negative outcome of the experiments threatened to discredit not only Mott's theory but also, by implication, Dirac's. In the mid-1930s, the feeling was widespread in some quarters that "the Dirac equation needs modification in order to account successfully for the absence of polarization."[54] But although the Dirac equation was confronted with this and other apparent failures, its authority remained intact.[55] Even in the spectrum of hydrogen – a show-piece for the Dirac equation – anomalies turned up. During the thirties, improved experiments showed a small but significant discrepancy between the experimentally determined fine structure of the hydrogen lines and that predicted by the Dirac theory. In 1934, William V. Houston and Y. M. Hsieh at Caltech made careful calculations of the Balmer lines which forced them to conclude that "the theory, as we have used it, is inadequate to explain the observations."[56] However, although the discrepancies between theoretical predictions and observed values were confirmed by several other studies, the exact validity of Dirac's theory was not seriously questioned until 1947.

Other difficulties faced Dirac's theory in connection with cosmic radiation and the new field of nuclear physics. For example, the theory was believed also to apply to protons, for which it predicted a magnetic moment of one nuclear magneton. When Otto Stern and Otto Frisch succeeded in measuring that quantity in 1933, their result was almost three times as large as predicted.[57] But this anomaly also was unable to seriously discredit Dirac's theory, which was, after all, a theory of electrons. Most physicists concluded that the theory just did not apply to nuclear particles. Bohr was of the opinion that "we cannot . . . expect that the characteristic consequences of Dirac's electron theory will hold unmodified for the positive and negative protons."[58] As a last example of the difficulties that confronted Dirac's theory, we may mention the so-called Meitner–Hupfeld effect, an anomalously large scattering found when high-energy gamma rays were scattered in heavy elements. Although the Meitner–Hupfeld effect, first reported in the spring of 1930, disagreed with the predictions of Dirac's electron theory, the theory was not found guilty; instead, the anomaly was thought to arise from intranuclear electrons.[59]

The real difficulties for the theory were connected with the physical interpretation of its mathematical structure, in particular, with the negative-energy solutions. This problem, which in its turn led to new and amazing discoveries, will be considered in more detail in Chapter 5. Let it suffice to mention here that Dirac had already noticed the difficulty in his first paper, where it caused him to label the theory "still only an approximation." In his Leipzig address he commented briefly on a related difficulty, namely, that the theory allows transitions from charge $+e$ to $-e$. Dirac had no answer to this problem, which was already much discussed in the physics community, except the following vague remark: "It seems that this difficulty can only be removed through a fundamental change in our previous ideas, and may be connected with the difference between past and future."[60] Heisenberg seems to have been a bit disappointed that Dirac had not yet come up with a solution. Deeply worried over the situation, Heisenberg wrote to Jordan after Dirac's lecture: "Dirac has lectured here only on his current theory, giving a pretty foundation for it that the differential equations must be linear in $\partial/\partial x_\mu$. He has not been able to solve the well-known difficulties. . . ."[61] And a month later he wrote to Bohr: "I am much more unhappy about the question of the relativistic formulation and about the inconsistency of the Dirac theory. Dirac was here and gave a very fine lecture about his ingenious theory. But he has no more of an idea than we do about how to get rid of the difficulty $e \rightarrow -e$ . . ."[62]

CHAPTER 4

TRAVELS AND THINKING

THE relativistic theory of the electron made Dirac in great demand at physics conferences and centers around the world, and in general increased his status as a scientist. In early 1928, Klein visited Cambridge and took back to Copenhagen the latest news of Dirac's ideas. Two months later, Schrödinger went to Cambridge and gave a talk to the Kapitza Club on the "Physical Meaning of Quantum Mechanics." In April, Dirac spent a few weeks at Bohr's institute and then traveled on to Leiden, where he discussed physics with Ehrenfest and other Dutch physicists and also gave a few lectures.[1] Dirac had begun working on a book on quantum mechanics, to be published two years later as *The Principles of Quantum Mechanics,* and he took the opportunity to test the first chapters on his audience in Leiden. Ehrenfest was much impressed by Dirac but found it difficult to understand the papers of this British wizard of quantum mechanics. He was therefore delighted when on one occasion Dirac was asked a question to which he, just for once, could not give an immediate and precise answer. "Writing very small he [Dirac] made some rapid calculations on the blackboard, shielding his formulae with his body. Ehrenfest got quite excited: 'Children,' he said, 'now we can see how he really does his work.' But no one saw much; Dirac rapidly erased his tentative calculations and proceeded with an elegant exposition in his usual style."[2] After spending about a month with Ehrenfest, Dirac continued to Leipzig to attend the *Universitätswoche* and to discuss the latest developments in physics with Heisenberg. After Leipzig the tour went on to Göttingen, where Dirac stayed until the beginning of August. In Göttingen he met, among others, the twenty-four–year–old Russian physicist George Gamow, who had just succeeded in explaining alpha-radioactivity on the basis of quantum mechanics. In all these places – Copenhagen, Leiden, Leipzig, and Göttingen – Dirac lectured on his electron theory.

Among the physicists in Leiden was Igor Tamm, a thirty-two–year–old visitor from Moscow State University. In a letter to a relative, he reported being eager to meet Dirac, of whom he had been told the strangest things: "It is now definite that on April 23 Dirac will come here for three months. So I shall be able to learn something from that new physics' great genius. Though they say that Dirac is not a great one for words and you have to try very hard to start up a conversation with him. He seems to talk only with children, and they have to be under ten. . . ."[3] Dirac lived up to Tamm's great expectations, and the two became closely acquainted. "The criteria to go by now are Dirac's," wrote Tamm. "And compared to him I am just a babe in arms. Of course, it is still more stupid to measure oneself by a man of genius. . . . With great patience Dirac is teaching me the way I should go about things; I am very proud that we have come to be friends."[4] During their stay in Holland, Tamm taught Dirac to ride a bicycle, and together they went for long tours. "I have not forgotten the cycling that I learnt in Leiden," Dirac later wrote from Cambridge. "I have already cycled about 2000 km in the neighbourhood of Cambridge."[5] When Dirac went to Leipzig, Tamm accompanied him. On his return to Moscow, Tamm told about his experiences abroad:[6]

While abroad, I lived for five months in Holland and for two in Germany. . . . What pleased me most was coming together with Dirac. He and I kept company for three months and came to be very close. Dirac is a true man of genius. Do not smile that it sounds high-flown; I really mean it. I know that when I grow old I'll be telling my grandchildren with pride about that acquaintance of mine.

In the summer of 1928, Heisenberg was professor at Leipzig, a position he had obtained at only twenty-six years of age. Dirac received his first offer of a chair at the same age, when Milne, who had occupied the professorship in applied mathematics at Manchester University since 1924, was elected a professor of mathematics at Oxford. Dirac was asked in July if he was interested in the vacant chair, but he declined the offer.[7] He wanted to continue his own style of life and be free to cultivate the specialized research in which he was an expert. Neither the prestige of a professorship nor the prospect of much-improved material and economic conditions tempted him:[8]

I feel greatly honoured by being considered a possible successor to Prof. Milne, but I am afraid I cannot accept the appointment. My work is of too specialised a nature to be satisfactorily carried on outside a great centre such as Cambridge where there are others interested in the same subject, and my knowledge of and interest in mathematics outside my own special branch are too small for me to be competent to undertake the duties of a Professor of applied mathematics.

A few months later, he was asked by A. H. Compton to accept a new chair of theoretical physics to be created at the University of Chicago.[9] The offer was rewarding and so was the salary of $6,000 a year. Dirac declined.

Bohr wanted Dirac to come to Copenhagen again in September to participate in a conference, but Dirac decided to go to the Soviet Union instead. To Klein he wrote:[10]

I am afraid I shall not be able to come to Copenhagen in September as I intended. I am going to the Physical Congress in Russia (on the Volga) and I expect to return to England via Constantinople. I very much regret that I shall not be meeting you and Prof. Bohr, and I hope to be able to visit Copenhagen next year. I have not met with any success in my attempts to solve the $\pm e$ difficulty. Heisenberg (whom I met in Leipzig) thinks the problem will not be solved until one has a theory of the proton and electron together. I shall be leaving for Leningrad on about August 2nd. Best wishes to you and Prof. Bohr, and Nishina if he is still with you.

In the decade 1925–35, there was relatively close contact between Western and Soviet science. Soviet authorities wanted to develop the national science rapidly and stimulated contact with the West, which included the employment of foreign scientists in Russia, the arrangement of international conferences, publishing in foreign journals, and sending Soviet scientists abroad to Western institutions. Around 1930, theoretical physics in the USSR, and particularly in Leningrad, could compete with that in any Western nation. Several of the best Russian theorists, including Tamm, Frenkel, Landau, Iwanenko, and Fock, worked in the same areas as Dirac and looked forward to meeting him.[11]

The sixth All-Union Conference on physics took place in August and September 1928, arranged by Abraham Joffe, president of the Russian Association of Physicists. Apart from the Russian hosts, several Western physicists participated: Brillouin from France, G. N. Lewis from the United States, Dirac and Darwin from Great Britain, and Born, Pringsheim, Scheel, Pohl, Ladenburg, and Debye from Germany. In the company of Born and Pohl, Dirac went from Göttingen to Leningrad, which he found to be the most beautiful city he had ever seen, and only joined the congress in Moscow some days after it opened on August 4. A week later, the congress went to Nisjni-Nowgorod, from where it continued on board a steamer along the Volga river, visiting Gorki, Kazan, Saratow, and Tiflis.[12] During the Volga trip, Dirac bathed in the Volga and learned to eat caviar and watermelons. He lectured on his theory of the electron and after the congress traveled alone through the Caucasus to Batoum and the Black Sea coast. While in the Caucasus, Dirac participated in an excursion that took him to a height of about 3,000 meters, which was, he reported to Tamm,[13]

On the Volga. Dirac in company with Yakov Frenkel (next to him) and Alfred
Landé at the 1928 physics congress in Russia. Reproduced with permission of
AIP Niels Bohr Library.

a good deal higher than my previous record. . . . I spent three days in Tiflis, mostly
resting and making up for lost sleep, and then went to Batoum to try to get a boat
for Constantinople. . . . From Constantinople I took a ship to Marseilles, visiting
Athens and Naples on the way, and then came home across France and ended a
most pleasant holiday.

He arrived back in Cambridge after half a year's traveling around Europe.
This was not a comfortable life, but lack of comfort never bothered Dirac.

After his return to Cambridge, Dirac resumed his solitary style of life,
spending most of his time working in his college room or in one of the
libraries. He completed a work on statistical quantum mechanics and
started preparing his lecture notes for a book, which would become his
great textbook on quantum mechanics. In January 1929, he met again
with Gamow, who, after a stay in Copenhagen, spent about a month in

Cambridge. Later in the year, Dirac was appointed to the more secure (insured for three years) position of University Lecturer at St. John's College, a post that carried a basic annual stipend of two hundred pounds. Knowing that other universities had offered Dirac a position, the college was anxious to give him the best conditions in order to retain him. Dirac was therefore given very little teaching and administrative work and could continue to spend most of his time on research.

After half a year in Cambridge, Dirac next went to the United States in order to spend two months as a visiting professor at the University of Wisconsin in Madison, a job for which he was paid $1,800. American theoretical physics was in a state of transformation, which in the course of a few years would make the United States the leading nation in the field. But in the late twenties, American universities were still relatively weak in theoretical physics and made considerable use of visiting physicists from Europe. Dirac did not know much about physics in the United States, and what he knew did not impress him. When asked by Edward Condon in 1927 if he would like to visit America, he had replied, "There are no physicists in America."[14] Earlier invitations to visit America, issued by A. H. Compton in Chicago and his brother, Karl T. Compton, at Princeton, had also failed to tempt Dirac.[15]

Dirac arrived in New York on March 20, 1929, and proceeded the following day to Princeton where he met with the mathematician Oswald Veblen.[16] From there he went to Wisconsin to be welcomed by Van Vleck, who had recently been appointed professor of mathematical physics at the university. As part of the terms for his position, Van Vleck was to invite a foreign theorist to visit the university each year; Dirac was the first, to be followed by Wentzel and Fowler. He lectured on quantum mechanics, mostly his transformation theory, and alone or together with Van Vleck went for walks in nearby Minnesota. During this first stay in America, Dirac was associated with the University of Wisconsin and later with the summer school of the University of Michigan; during the spring vacation, he gave lectures at the University of Iowa. The content of the lectures Dirac planned to give in Madison was the subject of a letter he wrote to Van Vleck in December 1928; through this letter we also learn that he had by then begun to write his great quantum mechanics text:[17]

I think that in my lectures at Madison the best place to begin is with the transformation theory. A good knowledge of this is necessary for all the later developments. I could deal with this subject assuming my audience have only an elementary knowledge of Heisenberg's matrices and Schrödinger's wave equation, or alternatively I could present it in a way which makes no reference to previous forms of the quantum theory, following a new method which I am now incorporating in a book. This alternative would perhaps be a little more difficult for non-mathematical students and would need more time. After dealing with the transformation theory I propose to apply it to problems of emission and absorption, the quantization of continuing media, and the relativity theory of the electron.

Dirac's introversive style and his interest in abstract theory were rather foreign to the scientists at the University of Wisconsin. They recognized his genius but had difficulties in comprehending his symbolic version of quantum theory.[18] The Americans also found him a bit of a strange character. A local newspaper, the *Wisconsin State Journal,* wanted to interview the visiting physicist from Europe and assigned this task to a humorous columnist known as "Roundy." His encounter with Dirac is quoted here in extenso because it not only reveals some characteristic features of Dirac's personality but also is an amusing piece of journalism:[19]

I been hearing about a fellow they have up at the U. this spring – a mathematical physicist, or something, they call him – who is pushing Sir Isaac Newton, Einstein and all the others off the front page. So I thought I better go up and interview him for the benefit of the State Journal readers, same as I do all the other top notchers. His name is Dirac and he is an Englishman. He has been giving lectures for the intelligentsia of the math and physics department – and a few other guys who got in by mistake.

So the other afternoon I knocks at the door of Dr. Dirac's office in Sterling Hall and a pleasant voice says "Come in." And I want to say here and now that this sentence "come in" was about the longest one emitted by the doctor during our interview. He sure is all for efficiency in conversation. It suits me. I hate a talkative guy.

I found the doctor a tall youngish-looking man, and the minute I see the twinkle in his eye I knew I was going to like him. His friends at the U. say he is a real fellow too and good company on a hike – if you can keep him in sight, that is.

The thing that hit me in the eye about him was that he did not seem to be at all busy. Why if I went to interview an American scientist of his class – supposing I could find one – I would have to stick around an hour first. Then he would blow in carrying a big briefcase, and while he talked he would be pulling lecture notes, proof, reprints, books, manuscripts, or what have you, out of his bag. But Dirac is different. He seems to have all the time there is in the world and his heaviest work is looking out of the window. If he is a typical Englishman it's me for England on my next vacation!

Then we sat down and the interview began. "Professor," says I, "I notice you have quite a few letters in front of your last name. Do they stand for anything in particular?"

"No," says he.

"You mean I can write my own ticket?"

"Yes," says he.

"Will it be all right if I say that P. A. M. stands for Poincare Aloysius Mussolini?"

"Yes," says he.

"Fine," says I, "We are getting along great! Now doctor will you give me in a few words the low-down on all your investigations?"

"No," says he.

"Good," says I. "Will it be all right if I put it this way – 'Professor Dirac solves all the problems of mathematical physics, but is unable to find a better way of figuring out Babe Ruth's batting average'?"

"Yes," says he.

"What do you like best in America?" says I.

"Potatoes," says he.

"Same here," says I. "What is your favorite sport?"

"Chinese chess," says he.

That knocked me cold! It sure was a new one to me! Then I went on: "Do you go to the movies?"

"Yes," says he.

"When?" says I.

"In 1920 – perhaps also 1930," says he.

"Do you like to read the Sunday comics?"

"Yes," says he, warming up a bit more than usual.

"This is the most important thing yet Doctor," says I. "It shows that me and you are more alike than I thought. And now I want to ask you something more: They tell me that you and Einstein are the only two real sure-enough high-brows and the only ones who can really understand each other. I wont ask you if this is straight stuff for I know you are too modest to admit it. But I want to know this – Do you ever run across a fellow that even you cant understand?"

"Yes," says he.

"This will make great reading for the boys down to the office," says I. "Do you mind releasing to me who he is?"

"Weyl," says he.

The interview came to a sudden end just then, for the doctor pulled out his watch and I dodged and jumped for the door. But he let loose a smile as we parted and I knew that all the time he had been talking to me he was solving some problem no one else could touch.

But if that fellow Professor Weyl ever lectures in this town again I sure am going to take a try at understanding him! A fellow ought to test his intelligence once in a while.

While Dirac stayed in Madison, Heisenberg, invited by Compton, lectured at the University of Chicago. Just after Dirac had arrived in Madison, Heisenberg had written from Cambridge, Massachusetts: "Next Friday I will arrive in Chicago; since we then are not more than 200 miles apart from another, I would like to establish the connection between us."[20]

At the end of May, Dirac completed his lectures in Wisconsin and then traveled west alone as a tourist, visiting the Grand Canyon, Yosemite National Park, Pasadena, and Los Angeles.[21] Then he went east again, to

Ann Arbor, where he lectured at the University of Michigan's summer school in physics. In early August, he dined with Heisenberg in Chicago and then left by train for the West Coast. The two physicists met again in Berkeley, where Dirac lectured on the quantum mechanics of many-electron systems and Heisenberg talked on ferromagnetism and quantum electrodynamics.

Since both of the scientists had been invited to lecture in Japan, Heisenberg had suggested as early as February 1928 that they return to Europe together, going westward via Japan,[22] and he made the necessary arrangements for their travel.[23] On August 16, 1929, they left San Francisco on a Japanese steamer that brought them to Hawaii four days later. According to one account, they decided to visit the University of Hawaii in Honolulu to pass some of the time until the boat resumed its voyage. They introduced themselves to the chairman of the physics department, but apparently quantum mechanics had not yet reached Hawaii. The chairman told the two youthful visitors that if they would like to attend some of the physics lectures at the university they would be welcome to do so![24] Another story from this journey concerns Dirac's dislike of reporters. As the boat approached Yokohama in Japan, a reporter wanted to interview the two famous physicists, but Dirac evaded him and with Heisenberg's help avoided being interviewed. When the reporter, who did not know which of the passengers was Dirac, ran into Heisenberg and Dirac, he said to Heisenberg "I have searched all over the ship for Dirac, but I cannot find him." Instead of identifying Dirac, who was standing right beside him, Heisenberg offered to answer the reporter's questions about Dirac. "So I just stood there, looking in another direction, pretending to be a stranger and listening to Heisenberg describing me to the reporter," Dirac later recounted.[25] One would expect that Heisenberg and Dirac used much of their time onboard together to discuss questions of physics, but the tour was primarily intended to be a holiday, so they hardly talked physics at all.[26] In fact, Heisenberg spent most of his time practicing table tennis.

Heisenberg and Dirac arrived in Japan on August 30 and spent about a month there, partly as tourists and partly giving lectures to the Japanese physicists. Of these Dirac knew Yoshio Nishina, their main host, well. Nishina had recently returned to Japan after several years' stay with Bohr in Copenhagen. He had visited Dirac in Cambridge in November 1928 and had then mentioned the possibility of having him visit Japan. Nishina was delighted that his efforts succeeded and only regretted that Heisenberg and Dirac's stay was so short.[27] Another of the Japanese hosts was Hantaro Nagaoka, an older physicist, who as far back as 1904 had proposed a Saturnian atomic model and had since then contributed much to the growth of Japanese physics. Dirac and Heisenberg lectured at Tokyo's Institute for Physical and Chemical Research and at Kyoto's Imperial University. The lectures were successful and did much to stimulate the young Japanese physicists, some of whom (e.g., Yukawa, Tomonaga, Sakata, and Inui) would later make important contributions to theoretical physics. The content of the Tokyo lectures was later published as a book.[28] From the Miyako Hotel in Kyoto, Dirac informed Tamm about his further travel plans.[29]

I shall leave Japan for Moscow on Sept. 21st. I shall take the Northern route from Vladivostok, which does not go through China at all, as this is the only way now open. I shall leave Vladivostok in the early morning of Oct. 3rd at about 8 o'clock. I cannot remember the exact time of my arrival in Moscow and have left my timetable in Tokyo, but as there is now only one train a week from Vladivostok I expect you will be able to find it out without difficulty. I am afraid I will not be able to stay in Moscow for more than about two days (perhaps till the evening of Oct. 5th) as the term in Cambridge begins soon after.

After their Japanese intermezzo, Dirac and Heisenberg separated. Heisenberg returned to Germany via Shanghai, Hong Kong, India, and the Red Sea while Dirac followed the route oulined in his letter to Tamm. Originally he had planned to go through Manchuria, but on Nishina's advice he decided to avoid the area. At the time, Nishina reported, there were political troubles at the Chinese–Soviet border and danger that the Russians might close it.[30] From Vladivostok Dirac went on the Trans-Siberian Railroad to Moscow via Chabarowsk and Tchita. From the Russian capital he went to Leningrad by train, and from there to Berlin by airplane, which was an unusual way of traveling at that time.

Right after his return to Cambridge, Dirac began to prepare for two lectures he had agreed to present in Paris in December, and it was also in November 1929 that he worked out his revolutionary idea of negative-energy electrons (see Chapter 5). In the following years, he was primarily occupied with developing this idea and improving the theory of quantum electrodynamics. As far as his many travels allowed him, he continued to give lectures on quantum mechanics in Cambridge and also found time to deal with other subjects that did not belong to his own specialized field of research. In seminars at Cambridge he dealt with recent developments in quantum molecular theory, the theory of magnetism, and other areas in which he never published.[31] In February 1930, Dirac was elected a Fellow of the Royal Society (FRS), a title of the highest prestige. To a British scientist, being able to use the initials FRS ranks almost with winning a Nobel Prize. Dirac was now, among other things, allowed to communicate papers to the *Proceedings,* his favorite vehicle of publication.

In the period 1928–30, he also published important works on statistical quantum theory and atomic theory. Following earlier work by von Neumann, Dirac in October 1928 examined statistical quantum mechanics when applied to a Gibbsean ensemble.[32] For such an ensemble he showed that there exists a close analogy between the classical and quantum equations. Just before leaving for Wisconsin, he published an important work on the calculation of atomic properties of many-electron systems, and he followed it up the next year with a further developed theory.[33] These works were in atomic theory proper, a branch that attracted many of the best physicists but one in which Dirac had not previously shown much interest. In his work of 1929, Dirac studied the exchange interaction of identical particles, which he related to the permutations of the coordinates. He introduced permutations as dynamical variables (operators) and built up a vector model of spin that he applied to the interaction of two or more electrons in an atom. Dirac's theory was extended by Van Vleck, who applied it to complex spectra and ferromagnetism.[34] The calculation of atomic properties of atoms with many electrons had previously been worked out with various approximation methods, in particular by Thomas (1926), Hartree (1927), and Fermi (1928). Douglas Hartree's method of the so-called self-consistent field was given a better theoretical basis by Vladimir Fock in early 1930, but the Hartree–Fock method proved too complicated to be of much practical use for systems with very many electrons. In his work from 1930, Dirac supplied Thomas's model with a theoretical justification and proposed a calculational improvement that yielded a better approximation. Dirac's theory of 1929 included a general method for calculating atomic and molecular energy levels, which was, however, generally overlooked. Just after Dirac's paper appeared, Slater published his important work on the wave function determinant method, which was at once adopted by the atomic physicists.[35] Most physicists and chemists preferred Slater's wave mechanical theory to Dirac's more abstract version. Dirac's works of 1929–30 were major contributions to atomic and, it turned out, solid-state physics, as eponymized in such terms as "the Fock–Dirac atom" and "the Thomas–Fermi–Dirac method." However, in Dirac's career they were mere parentheses.

Since the fall of 1927, Dirac had given a course of lectures on quantum mechanics at Cambridge.[36] The content of these lectures formed the basis of his celebrated book *The Principles of Quantum Mechanics,* the first edition of which was published in the summer of 1930. The book was written at the request of Oxford University Press – and not, remarkably, Cambridge University Press – which was preparing a series of monographs in physics. The general editors of the series were two of Dirac's friends, the Cambridge physicists Fowler and Kapitza. The author and science journalist James Crowther arranged the publication of Dirac's book for Oxford University Press. "When I first called on Dirac," he recalled, "he was living in a simply furnished attic in St. John's College. He had a wooden desk of the kind which is used in schools. He was seated at this, apparently writing the great work straight off."[37] Dirac started writing *Principles* in 1928, but because of his travels progress was slow. In January 1929, he wrote to Tamm:[38]

The book is progressing with a velocity of about 10^{-8} Frenkel. I have started writing it again in what I hope is the final form and have written about 90 pages. I shall try hard to finish it before going to America. It is to be translated into German. Have you seen Weyl's book on "Gruppentheorie und Quantenmechanik"? It is very clearly written and is far the most connected account of quantum mechanics that has yet appeared, although it is rather mathematical and therefore not very easy.

Van Vleck, at the time on sabbatical leave from the University of Wisconsin, visited Cambridge in March 1930 and was allowed to read the proof-sheets of Dirac's text. "What I have read so far I like very much," he wrote Dirac.[39]

Principles became a success. It went through several editions and translations and is still widely used.[40] Its Russian translation, in particular, became very popular; while the first English edition sold two thousand copies, *Printsipy Kvantovoi Mekhaniki* sold three thousand copies in a few months. In the thirties, *Principles* was the standard work on quantum mechanics, almost achieving a position like that which Sommerfeld's *Atombau und Spektrallinien* had before quantum mechanics. The nuclear physicist Philip A. Morrison recalled, with some exaggeration, that "everybody who had ever looked at books had a copy of Dirac."[41] Unlike most other textbooks, *Principles* was not only of use to students in courses on quantum mechanics but was probably studied as much by experienced physicists, who could find in it a concise presentation of the mathematical principles of quantum mechanics, principles that were likely to be of eternal validity. When Heisenberg received the fourth edition of *Principles* in 1958, he gave Dirac the following fine compliment: "I have in the past years repeatedly had the experience that when one has any sort of doubt about difficult fundamental mathematical problems and their formal representation, it is best to consult your book, because these questions are treated most carefully in your book."[42]

The book expressed Dirac's personal taste in physics and possessed a style unique to its author. Regarded as a textbook, it was and is remarkably abstract and not very helpful to the reader wanting to obtain physical insight into quantum mechanics. Unlike most other modern textbooks, *Principles* is strictly ahistorical and contains very few references and no illustrations at all. Neither is there any bibliography or suggestions for further reading. The first edition did not even include an index.

Principles was based on what Dirac called "the symbolic method," which "deals directly in an abstract way with the quantities of fundamental importance (the invariants, &c., of the transformations)." This method, Dirac said, "seems to go more deeply into the nature of things."[43] In accordance with the symbolic method, he wanted to present the general theory of quantum mechanics in a way that was free from physical interpretation. "One does not anywhere specify the exact nature of the symbols employed, nor is such specification at all necessary. They are used all the time in an abstract way, the algebraic axioms that they satisfy and the connexion between equations involving them and physical conditions being all that is required."[44] Thanks to the wide distribution of the book, Dirac's interpretation of quantum mechanics was disseminated to a whole generation of physicists, who through it learned about the formal aspects of the Copenhagen School's views of the measurement process and the nature of quantum mechanical uncertainty.

The Copenhagen spirit and the vague idealism associated with *Principles* did not go unnoticed by Soviet commissars, who supplied the Russian translation with a word of warning; although Dirac's book was valuable to Russian physics and thus to the Soviet Union, ideologically it was all wrong:[45]

The publishers are well aware that there is contained in this work a whole series of opinions, both explicit and implicit, which are totally incompatible with Dialectical Materialism. But it is precisely the necessity for a smashing attack on the theoretical front against idealism, against mechanism, and against a whole series of eclectic doctrines, that makes it the duty of the publisher to provide Soviet scientists with the concrete material that plays a crucial part in the foundation of these theories in order that, critically assimilated, their material may be employed on the front for the fight for Dialectical Materialism.

Still, one may safely assume that Russian physicists studied *Principles* to absorb the abstract theory of quantum mechanics, not to launch a smashing blow against bourgeois idealism.

Most physicists welcomed Dirac's exposition and praised it for its elegance, directness, and generality. To Einstein it was the most logically perfect presentation of quantum mechanics in existence.[46] Some years before *Principles* appeared, Eddington had praised Dirac's method for its symbolism and its emancipation from visualizable models, a characteristic that was also valid with respect to the book:[47]

If we are to discern controlling laws of Nature not dictated by the mind it would seem necessary to escape as far as possible from the cut-and-dried framework into which the mind is so ready to force everything that it experiences. I think that in principle Dirac's method asserts this kind of emancipation.

Although Dirac preferred an abstract or symbolic approach to physics, a kind of pictorial model appeared frequently in his works. But these models had very little in common with the traditional models of classical physics. Dirac used models, metaphors, and pictures to think about premature physical concepts and to transform vague ideas into a precise mathematical formalism. "One may," he stated in *Principles,* "extend the meaning of the word, 'picture' to include any *way of looking at the fundamental laws which makes their self-consistency obvious.* With this extension, one may gradually acquire a picture of atomic phenomena by becoming familiar with the laws of the quantum theory."[48]

When Pauli reviewed *Principles* in 1931, he recommended it strongly. But he also pointed out that Dirac's symbolic method might lead to "a certain danger that the theory will escape from reality." Pauli complained that the book did not reveal the crucial fact that quantum mechanical measurement requires real, solid measuring devices that follow the laws of classical physics and is not a process that merely involves mathematical formulae.[49] This was an important point in Bohr's conception of the measurement process in quantum mechanics, a conception that Pauli shared. According to Bohr and his disciples, the classical nature of the measuring apparatus is crucial, but this point was not appreciated by Dirac.

For all its qualities, *Principles* was not a book easily read or one that suited the taste of all physicists. It reflected Dirac's aristocratic sense of physics and his neglect of usual textbook pedagogy. Ehrenfest studied it very hard, only to find it "ein greuliches Buch" that was difficult to understand. "A terrible book – you can't tear it apart!" he is said to have exclaimed.[50] In their reviews of *Principles,* both Oppenheimer and Felix Bloch emphasized its generality and completeness. Oppenheimer compared the book with Gibb's *Elementary Principles in Statistical Mechanics* (1902) and warned that it was too difficult and abstract to be a suitable text for beginners in quantum theory.[51]

In the second edition of *Principles,* published in 1935, Dirac revised the book considerably. Apart from correcting some mistakes, he added a chapter on field theory and in general presented his subject in a slightly less abstract form. The major change was in his use of the concept of a "state," a notion central to Dirac's exposition: whereas in 1930 he had used the word in its relativistic sense, referring to the conditions of a dynamical system throughout space–time, he now argued that "state" should refer to conditions in a three-dimensional space at one instant of time. This means that Dirac's theory was built on a non-relativistic concept, a fact he saw as indicating serious problems in quantum mechanics rather than in his exposition of it. The second edition was reviewed by Darwin, who was happy to notice that there had been no change in the author's refusal to waste time over rigorous discussions of unimportant difficulties. He paraphrased Dirac's relaxed attitude to mathematical rigor with the words, "Though something seems a bit wrong, it can't be really serious, and with reasonable precautions there is no danger."[52] I shall have more to say about Dirac's view on mathematics and rigor in Chapter 14.

At the time of the publication of *Principles,* Dirac's ideas about the more philosophical aspects of quantum mechanics were fully developed. He did not change them much during his later career. We shall now deal in a more systematic way with Dirac's position around 1930 in the apparently everlasting debate over the interpretation of quantum mechanics (see also Chapter 13).

By and large, Dirac shared the positivist and instrumentalist attitude of the Copenhagen–Göttingen camp, including its belief that quantum mechanics is devoid of ontological content. He thought that the value of quantum mechanics lay solely in supplying a consistent mathematical scheme that would allow physicists to calculate measurable quantities. This, he claimed, is what physics is about; apart from this, the discipline has no meaning. Instrumentalism was part of Dirac's lectures on quantum mechanics from 1927 onward. In the introductory remarks to the lecture notes he wrote:[53]

The main feature of the new theory is that it deals essentially only with observable quantities, a very satisfactory feature. One may introduce auxiliary quantities not directly observable for the purpose of mathematical calculation; but variables not observable should not be introduced merely because they are required for the description of the phenomena according to ordinary classical notions . . . The theory enables one to calculate only observable quantities . . . and any theories which try to give a more detailed description of the phenomena are useless.

This does not mean that Dirac was ever a positivist in any reflective sense; he just did not care about ontological problems or problems of wider philosophical significance. From 1926 on through the thirties, he was rightly regarded as belonging to, or at least being an ally of, the Copenhagen school. In agreement with the views of Bohr, Jordan, Heisenberg, and other members of the Copenhagen circle, Dirac taught that the indeterminacy relations were not the result of an incompleteness of quantum mechanics but instead expressed a fundamental feature of nature. He wrote, for example:[54]

There is . . . an essential indeterminacy in the quantum theory, of a kind that has no analogue in the classical theory, where causality reigns supreme. The quantum theory does not enable us in general to calculate the result of an observation, but only the probability of our obtaining a particular result when we make the observation. [This] lack of determinacy in the quantum theory should not be considered as a thing to be regretted.

Still, there were differences between the views of Dirac and those of the more orthodox Copenhageners. As to quantum mechanical indeterminacy, Dirac put emphasis in a different place than did his colleagues in the Copenhagen school:[55]

One of the most satisfactory features of the present quantum theory is that the differential equations that express the causality of classical mechanics do not get lost, but are retained in symbolic form, and indeterminacy appears only in the application of these equations to the results of observations.

Neither did Dirac join Bohr, Heisenberg, and Jordan in their strict rejection of microscopic causality. Instead of completely abandoning causality, he wanted to revise the concept so that it still applied to undisturbed atomic states. "Causality will still be assumed to apply to undisturbed systems and the equations which will be set up to describe an undisturbed system will be differential equations expressing a causal connexion between conditions at one time and conditions at a later time."[56] However, in spite of his slightly unorthodox views, Dirac never showed any interest in the opposition waged against Bohr's views by Einstein, Schrödinger, or de Broglie, nor was he aroused by later theories involving hidden variables. Basically, he was not very interested in the interpretation debate and did not feel committed to argue either for the Copenhagen philosophy or for opposite views. When questions about the objectivity and completeness of quantum mechanics became much discussed in the thirties (in, for example, the Bohr–Einstein–Podolsky–Rosen debate), Dirac was silent.

In 1936, Born portrayed what he called Dirac's *l'art pour l'art* attitude to physics:[57]

Some theoretical physicists, among them Dirac, give a short and simple answer to this question [concerning the existence of an objective nature]. They say: the existence of a mathematically consistent theory is all we want. It represents everything that can be said about the empirical world; we can predict with its help unobserved phenomena, and that is all we wish. What you mean by an objective world we don't know and don't care.

No doubt, this is a fair characterization of Dirac's position. This instrumentalist and aristocratic attitude is also recognizable in Dirac's early
works. At the 1927 Solvay Congress, all the key contributors to the new
atomic theory (with the exception of Jordan) met. For the first time,
Dirac met Einstein, a man he admired more than any other physicist. But
Einstein's views on physics differed much from Dirac's, and consequently
Einstein reacted negatively to Dirac's early contributions to quantum
mechanics. "I have trouble with Dirac. This balancing on the dizzying
path between genius and madness is awful," Einstein told Ehrenfest in
August 1926. In another letter he wrote, "I don't understand Dirac at all
(Compton effect)."[58]

The 1927 Solvay Congress is famous for the debate that occurred there
between Einstein and Bohr over the interpretation of quantum mechanics. Of this Dirac later recalled: "In this discussion at the Solvay Conference between Einstein and Bohr, I did not take much part. I listened to
their arguments, but I did not join in them, essentially because I was not
very much interested. I was more interested in getting the correct equations."[59] However, he participated in the discussion following the report
of Born and Heisenberg, and after Bohr's address he gave a more elaborate account of his own view. He agreed with Bohr that classical determinism had to be abandoned. Quantum physics, he said, consists essentially in relating two sets of numbers: one referring to an isolated system,
and the other to the system when perturbed. In order to measure the system, the observer forces it into a certain state by means of a perturbation
and, "*It is only the numbers describing these acts of free will which can be
taken as initial numbers for a calculation in the quantum theory.* Other
numbers describing the initial state of the system are essentially unobservable and are not revealed in the quantum theoretical treatment."[60] He
further analyzed the nature of the measurement process as follows:[61]

This theory [quantum mechanics] describes the state of the world at any given
moment by a wave function ψ which normally varies according to a causal law in
such a way that its initial value determines its value at any later moment. It may
happen, however, that at a given moment τ_1, ψ may be expanded into a series of
the type $\psi = \sum_r c_n \psi_n$ in which the ψ_n's are wave functions of such a kind that they
are unable to interfere mutually at a moment later than t_1. Should that be the case,
the state of the world in moments further removed from t_1 will be described not
by ψ, but by one of the ψ_n's. One could say that it is nature that chooses the particular ψ_n that is suitable, since the only information given by the theory is that
the probability that any one of the ψ_n's will get selected is $|c_n|^2$. Once made, the
choice is irrevocable and will affect the entire future state of the world. The value
of n chosen by nature can be determined by experiment and the results of any
experiment are numbers that describe similar choices of nature.

Dirac's interpretation was supported by Born, who mentioned that it was in perfect agreement with the still unpublished work of von Neumann. Heisenberg, although he agreed in general with Dirac's exposition, objected to his statement that nature makes a choice when something is observed. It is not nature, but the observer, who chooses one of the possible eigenfunctions, Heisenberg maintained.[62] That is, Heisenberg tended to conceive of nature as a product of the free will of the human observer. Against this subjectivistic notion, Dirac held a more moderate position: While the observer decides what type of measurement to make, and thus fixes which set of eigenfunctions is relevant, nature chooses the particular eigenfunction that is to signify the result of the measurement.[63]

However, one should not exaggerate the difference of opinion between Dirac and Heisenberg. At least after 1927, Dirac seemed in most respects to be in line with the Copenhagen school, including its tendency toward subjectivism. This was part of the message of the preface to *Principles,* which contained a rather full exposition of how he conceived of the philosophy of the new physics. Borrowing phrases from Eddington's philosophy, Dirac stated his position as follows:[64]

[Nature's] fundamental laws do not govern the world as it appears in our mental picture in any very direct way, but instead they control a substratum of which we cannot form a mental picture without introducing irrelevancies. The formulation of these laws requires the use of the mathematics of transformations. The important things in the world appear as the invariants (or more generally the nearly invariants, or quantities with simple transformation properties) of these transformations. . . . The growth of the use of transformation theory, as applied first to relativity and later to the quantum theory, is the essence of the new method in theoretical physics. Further progress lies in the direction of making our equations invariant under wider and still wider transformations. This state of affair is very satisfactory from a philosophical point of view, as implying an increasing recognition of the part played by the observer in himself introducing the regularities that appear in his observations, and a lack of arbitrariness in the ways of nature, but it makes things less easy for the learner of physics.

While Dirac had much in common with Heisenberg, the situation was somewhat different with respect to Bohr, the other leader of the Copenhagen school. Dirac was a mathematically minded physicist who did not really understand Bohr's insistence on the primary of physical – not to mention philosophical – considerations over mathematical formalism. According to Heisenberg, Bohr "feared . . . that the formal mathematical structure would obscure the physical core of the problem, and in any case, he was convinced that a complete physical explanation should absolutely precede the mathematical formulation."[65] Bohr's entire *"nur die Fülle führt zur Klarheit"* philosophy, as it most cogently manifested itself in the complementarity principle, was foreign to Dirac's mind.[66] "I didn't altogether like it," Dirac said in 1963, referring to the complementarity principle. He argued that "it doesn't provide you with any equations which you didn't have before."[67] For Dirac, this was reason enough to dislike the idea. Although it did not appeal to Dirac, the principle of complementarity may have influenced his way of thinking. It may be argued that Dirac's emphasis on invariance transformations, as stated in the preface to *Principles* and elsewhere, emerged as a response to Bohr's notion of complementarity.[68]

The complementarity principle was first introduced in Bohr's Como address of September 1927, although the principle was only published in April 1928.[69] Dirac helped Bohr with the proofs of the article although, as Bohr realized, Dirac did not fully agree with its content. It seems that Dirac, in accordance with his statement at the Solvay Congress, criticized Bohr for making too much room for subjectivism in quantum mechanics. This difference of opinion can be glimpsed from a letter Bohr wrote to Dirac shortly before his Como address appeared in print:[70]

I do not know, however, whether you are quite in sympathy with the point of view, from which I have tried to represent the paradoxes of the quantum theory.... Of course I quite appreciate your remarks that in dealing with observations we always witness through some permanent effects a choice of nature between the different possibilities. However, it appears to me that the permanency of results of measurements is inherent in the very idea of observation; whether we have to do with marks on a photographic plate or with direct sensations the possibility of some kind of remembrance is of course the necessary condition for making any use of observational results. It appears to me that the permanency of such results is the very essence of the ordinary causal space–time description. This seems to me so clear that I have not made a special point of it in my article. What has been in my mind above all was the endeavour to represent the statistical quantum theoretical description as a natural generalisation of the ordinary causal description and to analyze the reasons why such phrases like a choice of nature present themselves in the description of the actual situation. In this respect it appears to me that the emphasis on the subjective character of the idea of observation is essential.

As we shall soon see (in Chapters 5 and 6), the intuitions of Bohr and Dirac about the future development of quantum theory differed much during the thirties. However, even though he disagreed with Bohr in many respects, Dirac continued to admire him greatly. In Dirac's opinion, Bohr occupied the same position with regard to atomic theory that Newton did with respect to macroscopic mechanics.[71]

In October 1930, Dirac participated in his second Solvay Congress, which was, as usual, held in Brussels. The meeting's main theme was solid-state physics and magnetism, but the conference eventually became better known as the site of the second round of the Bohr–Einstein debate. Dirac gave no lecture at the conference but discussed problems in the interpretation of quantum mechanics with Bohr. This was evidently a subject that at the time interested Dirac, although he kept a low profile in public with regard to this interest. After returning to Cambridge, he wrote to Bohr and continued the oral discussion begun in Brussels; the letter is reproduced in extenso:[72]

Dear Professor Bohr, 30-11-30.

I would like to thank you for your very interesting talks to me in Brussels about uncertainty relations. I have been thinking over the last problem about coherence for two light quanta that are emitted in quick succession from an atom A.

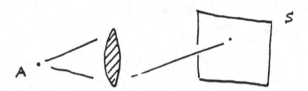

I think it is quite certain that the two light quanta will both fall on the same point of the screen S, even when one measures accurately the initial and final momentum of the atom A. One must look at the question from the point of view of the many-dimensional wave function to get a definite answer. If X, x_1, x_2 denote the positions of the atom and the two light-quanta, then just after the first emission we shall have a 6-dimensional wave function that vanishes everywhere except where $X - x_1$ is very small. Just after the second emission we shall have a nine-dimensional wave function that vanishes everywhere except where both $X - x_1$ and $X - x_2$ are very small. We can now express the total wave function ψ as the sum of a number of terms

$$\psi = \psi_1 + \psi_2 + \psi_3 + \cdots$$

such that each term ψ_r vanishes except when x_1 and x_2 have definite numerical values just after the second emission, these two values being nearly equal to each other, but different for different r. Each ψ_r will now give both light-quanta falling on the same point of the screen, this point being different for the different ψ_r's.

In looking over the question of the limit to the accuracy of determination of position, due to the limit c to the velocity of the shutter, I find I cannot get your result, and I am afraid I have missed an essential point somewhere. I should be very glad if you would kindly repeat the argument to me briefly. There are several people in Cambridge who are interested in the question and would like to know exactly how it goes.

Yours sincerely

P. A. M. Dirac

After this lengthy digression on Dirac's general views on quantum theory, we now proceed to consider his most important contribution to physics during the period, the theory of anti-particles.

CHAPTER 5

THE DREAM OF PHILOSOPHERS

WHEN Dirac published his wave equation of the electron in January 1928, he was acutely aware of the fact that in a formal sense the equation had a surplus value: Its solutions referred not only to ordinary electrons but also to electrons with negative energy – that is, to nonexisting entities. This peculiarity is not related to the linear wave equation in particular but is, as Dirac emphasized, common to all relativistic wave equations, a fact that is evident from the classical-relativistic energy expression for a free particle

$$E^2 = \vec{p}^2 c^2 + m_0^2 c^4 \qquad (5.1)$$

which has the two (positive *and* negative) energy solutions

$$E = c\sqrt{m_0^2 c^2 + \vec{p}^2} \quad \text{and} \quad E = -c\sqrt{m_0^2 c^2 + \vec{p}^2}$$

Since the quantum mechanical wave equation was essentially a translation of the classical equation by means of $H\psi = E\psi$, the double value was carried over into quantum mechanics. The Klein–Gordon equation contained negative-energy solutions, too, but these were dismissed as trivially unphysical until Dirac called attention to the problem of doing this in quantum theory.[1]

One gets over the difficulty of the classical theory by arbitrarily excluding those solutions that have negative W. One cannot do this in the quantum theory, since in general a perturbation will cause transitions from states with W positive to states with W negative. Such a transition would appear experimentally as the electron suddenly changing its charge from $-e$ to e, a phenomenon which has not been observed.

The Dirac equation for the free electron, equation (3.24), has four linearly independent solutions of the plane wave form

87

$$\psi_\mu = a_\mu(\vec{p}) \exp\{-i\hbar(\vec{p} \cdot \vec{r} - Et)\}, \qquad \mu = 1,2,3,4$$

The corresponding eigenvalue equation shows that there are nonvanishing solutions only for

$$E = \pm c\sqrt{m_0^2 c^2 + \vec{p}^2}$$

as would be expected from the classical analogy, equation (5.1). For $\mu = 1$ and $\mu = 2$ the positive sign holds, while $\mu = 3$ and $\mu = 4$ are associated with the negative energies. The wave system that represents the electron is complete only if the negative-energy solutions are included. Hence they cannot be easily dismissed. Another way of stating the difficulty is by referring to the electrical charge: Only half of the solutions to the Dirac equation for an interacting electron refer to the particle's negative charge; the other half can be shown to refer to a particle of positive charge.

Although in early 1928 Dirac realized that the problem of the negative energies could not be dismissed as "unphysical," he was not much disturbed and decided to bypass the problem for the time being. As he later recalled: "I was well aware of this negative energy difficulty right at the beginning, but I thought it was a less serious difficulty than the others, less serious than our not being able to apply the transformations of the general transformation theory."[2]

However, the "± difficulty" was immediately taken up by other physicists who regarded it as a most serious problem. When Dirac's theory appeared, Jordan became depressed *(Trübsinnig)* because he realized that his own work toward constructing a relativistic wave equation had been made obsolete.[3] In a review of the light-quantum hypothesis in 1928, Jordan referred with some reserve to Dirac's theory, which "certainly is not final." His main objection was "the murky problem of *the asymmetry of the forms of electricity,* that is, the inequality of mass for positive and negative electrons in relation to which this theory is entangled in temporarily insoluble difficulties."[4] This uneasiness was shared by Jordan's colleagues. Having discussed the matter with Dirac at the Leipzig meeting (see Chapter 3), Heisenberg reported to Pauli about what he called "the saddest chapter of modern physics":[5]

The matrix elements which correspond to the transition $+mc^2 \rightarrow -mc^2$ are of the order $a_0 \cdot \alpha$ in the position coordinates (a_0 = hydrogen radius, α = fine structure constant), of the order c^2 in the velocity coordinates, and of order mc^3/h in the acceleration. The spontaneous radiation transitions $+mc^2 \rightarrow -mc^2$ are thus much *more frequent* than other spontaneous transitions.

Although it did not cause him sleepless nights, Dirac too was concerned with the ± difficulty. "I have not met with any success in my attempts to solve the ± difficulty," he admitted to Klein in July. "Heisenberg (whom I met in Leipzig) thinks the problem will not be solved until one has a theory of the proton and electron together."[6] That, eventually, was the direction in which Dirac developed his theory more than a year later.

The problem of the negative energies was highlighted at the end of 1928 when Klein showed that in some simple cases the Dirac electron would behave in a patently absurd way.[7] "Klein's paradox," as it soon became known in the literature, took as an example an electron moving against a potential barrier, a simple and well-known example from non-relativistic quantum mechanics. Klein proved that the Dirac equation predicted a result very different from that of the Schrödinger equation, a plainly unreasonable result: If the potential is sufficiently large (larger than $E + mc^2$, where E is the kinetic energy of the electron), the transmitted wave will not be exponentially damped but will retain its oscillatory form with a negative kinetic energy. Despite its absurdity, this result followed logically from Dirac's theory. Other absurd consequences were also recognized in 1928. For example, it is obvious that the energy expression $E = -\sqrt{p^2c^2 + m_0^2c^4}$ implies that the energy of a negative-energy electron will decrease the faster it moves. Also, it can be proved that its acceleration will be in the opposite direction to the force impressed on it.

In spite of the fact that Dirac's theory from its very birth was troubled by conceptual difficulties with regard to the negative energies, this aspect of the theory seemed unavoidable. A few months before Nishina returned to Japan from Copenhagen, he worked out, in collaboration with Klein, a theory for scattering of photons by free electrons according to the Dirac theory.[8] The result, known as the Klein–Nishina scattering formula, called further attention to the intrinsic oddity of Dirac's theory since it relied crucially on positive as well as negative energies. In 1930, the Swedish physicist Ivar Waller proved that the Klein–Nishina formula would give correct results in the limit of low energies (i.e., would reduce to the experimentally confirmed Thomson scattering formula) only if the negative-energy states were included.[9] A couple of years later the formula was also vindicated for higher energies, and it then supplied Dirac's theory with considerable empirical power. In retrospect, the Klein–Nishina theory helped to convince physicists that the negative energies had to be taken seriously in one way or another. In 1934, after the positron theory had largely cleared up the paradoxes, Bohr wrote to Nishina, "The striking confirmation which this [Klein–Nishina] formula has obtained became soon the main support for the essential correctness of Dirac's theory when it was apparently confronted with so many grave difficulties."[10] But in 1929 the situation seemed grave indeed.

When Dirac returned to Cambridge from his tour around the world, he came upon a startling new idea that, he believed, offered an answer to the ± difficulty. His idea was probably indebted to a suggestion made by

Dirac: A scientific biography

Weyl in the spring of 1929. Weyl presented a unitary theory of gravitation, electromagnetism, and quantum theory in which the electromagnetic field was brought in as an accompaniment of the quantum wave field and not, as in other unitary theories, of the gravitational field. In this context he argued that the electron must be described by a two-component Dirac wave function; the other two ("superfluous") components should be ascribed to the proton: "It is natural to expect that the electron belongs to one of the two pairs of components of Dirac quantities, while the proton belongs to the other."[11] Perhaps Dirac was made aware of Weyl's proposal through Heisenberg. In March 1929, after arriving in America, Heisenberg wrote to Dirac:[12]

I would be especially interested to hear your opinion about Weyl's work. Weyl thinks to have the solution to the ± difficulty. I have not heard any detail of Weyls work; could you be kind – if it is not too much work for you – to write to me the main point in Weyls work.

Weyl's proposal was taken up by Dirac in his paper "A Theory of Electrons and Protons," which was completed in early December. A couple of weeks earlier, Bohr, who had heard about Dirac's new idea through Gamow, wrote Dirac a letter in which he expressed his worries over the consequences of the new theory of the electron:[13]

The difficulties in relativistic quantum mechanics might perhaps be connected with the apparently fundamental difficulties as regards conservation of energy in β-ray distintegration and the interior of stars. My view is that the difficulties in your theory [of 1928] might be said to reveal a contrast between the claims of conservation of energy and momentum on one side and the conservation of the individual particles on the other side.

Dirac replied promptly, outlining his new approach; the letter is reproduced in extenso:[14]

Dear Professor Bohr,

Many thanks for your letter. The question of the origin of the continuous β-ray spectrum is a very interesting one and may prove to be a serious difficulty in the theory of the atom. I had previously heard Gamow give an account of your views at Kapitza's club. My own opinion of this question is that I should prefer to keep rigorous conservation of energy at all costs and would rather abandon even the concept of matter consisting of separate atoms and electrons than the conservation of energy.

There is a simple way of avoiding the difficulty of electrons having negative kinetic energy. Let us suppose the wave equation

$$\left[\frac{W}{c} + \frac{e}{c}A_0 + \rho_1(\vec{\sigma}\cdot\vec{\gamma} + \frac{e}{c}\vec{A}) + \rho_0 mc\right]\psi = 0$$

does accurately describe the motion of a single electron. This means that if the electron is started off with a +ve energy, there will be a finite probability of its suddenly changing into a state of negative energy and emitting the surplus energy in the form of high-frequency radiation. It cannot then very well change back into a state of +ve energy, since to do so it would have to absorb high-energy radiation and there is not very much of this radiation actually existing in nature. It would still be possible, however, for the electron to increase its velocity (provided it can get the momentum from somewhere) as by so doing its energy would be still further reduced and it would emit more radiation. Thus the most stable states for the electron are those of negative energy with very high velocity.

Let us now suppose there are so many electrons in the world that all these most stable states are occupied. The Pauli principle will then compel some electrons to remain in less stable states. For example if all the states of −ve energy are occupied and also few of +ve energy, those electrons with +ve energy will be unable to make transitions to states of −ve energy and will therefore have to behave quite properly. The distribution of −ve electrons will, of course, be of infinite density, but it will be quite uniform so that it will not produce any electromagnetic field and one would not expect to be able to observe it.

It seems reasonable to assume that not all the states of negative energy are occupied, but that there are a few vacancies or "holes." Such a hole which can be described by a wave function like an X-ray orbit would appear experimentally as a thing with +ve energy, since to make the hole disappear (i.e. to fill it up,) one would have to put −ve energy into it. Further, one can easily see that such a hole would move in an electromagnetic field as though it had a +ve charge. These holes I believe to be the protons. When an electron of +ve energy drops into a hole and fills it up, we have an electron and proton disappearing simultaneously and emitting their energy in the form of radiation.

I think one can understand in this way why all the things one actually observes in nature have a positive energy. One might also hope to be able to account for the dissymmetry between electrons and protons; one could regard the protons as the real particles and the electrons as the holes in the distribution of protons of −ve energy. However, when the interaction between the electrons is taken into account this symmetry is spoilt. I have not yet worked out mathematically the consequences of the interaction. It is the "Austausch" effect that is important and I have not yet been able to get a relativistic formulation of this. One can hope, however, that a proper theory of this will enable one to calculate the ratio of the masses of proton and electron.

I was very glad to hear that you will visit Cambridge in the spring and I am looking forward to your visit. With kind regards from

Yours sincerely

P. A. M. Dirac

The main feature of Dirac's new theory was outlined in this letter. Bohr found Dirac's proposal as interesting as he found it objectionable. After discussing the letter with Klein, he answered:[15]

We do not understand, how you avoid the effect of the infinite electric density in space. According to the principles of electrostatics it would seem that even a finite uniform electrification should give rise to a considerable, if not infinite, field of force. In the difficulties of your old theory I still feel inclined to see a limit of the fundamental concepts on which atomic theory hitherto rests rather than a problem of interpreting the experimental evidence in a proper way by means of these concepts. Indeed according to my view the fatal transition from positive to negative energy should not be regarded as an indication of what may happen under certain conditions but rather as a limitation in the applicability of the energy concept.

As another problem Bohr pointed to the astrophysical consequences of Dirac's electron–proton annihilation hypothesis:[16]

Thus Eddington's theory of the equilibrium of stars seems to indicate that the rate of energy production per unit mass ascribed to such annihilation is larger in the earlier stages of stellar evolution where the density in the interior is smaller than in the latter stages where the interior density is larger. As far as I can see any views like yours would claim a variation with the density in the opposite direction. On the whole it seems to me that an understanding of the laws of stellar evolution claims some new radical departure from our present view regarding energy balance.

Dirac answered at once, using the opportunity to stress the difference between his and Bohr's conception of the then current difficulties of quantum theory. Opposed to Bohr's revolutionary strategy, including his willingness to give up energy conservation for some atomic processes, Dirac favored a piecemeal strategy. Again, the entire letter is reproduced.[17]

Dear Professor Bohr,

Many thanks for your letter. I am afraid I do not completely agree with your views. Although I believe that quantum mechanics has its limitations and will ultimately be replaced by something better, (and this applies to all physical theories,) I cannot see any reason for thinking that quantum mechanics has already reached the limit of its development. I think it will undergo a number of small changes, mainly with regard to its method of application, and by these means most of the difficulties now confronting the theory will be removed. If any of the concepts now used (e.g. potentials at a point) are found to be incapable of having an exact meaning, one will have to replace them by something a little more general, rather than make some drastic alteration in the whole theory.

Have you seen the paper by Pokrowski, Z.f.Physik 58, 706, where he claims to show that radioactive processes are not independent of one another. I think it is on some such basis as this, even if Pokrowski is not right, that one must look for an explanation of the continuous β-ray spectrum. Perhaps theoretically it depends on some hitherto overlooked resonance or "austausch" interaction between the electrons in different nuclei.

There is one case where transitions of electrons from positive to negative energy levels does give rise to serious practical difficulties, as has been pointed out to me by Waller. This is the case of the scattering of radiation by an electron, free or bound. A scattering process is really a double transition, consisting of first an absorption of a photon with the electron jumping to any state and then an emission with the electron jumping to its final state (as in the Raman effect) (or also of first the emission and then the absorption). The initial and final states of the whole system have the same energy, but not the intermediate state, which lasts only a very short time. One now finds, for radiation whose frequency is small compared with mc^2/h, that practically the whole of the scattering comes from double transitions in which the intermediate state is of negative energy for the electron. Detailed calculations of this have been made by Waller. If one says the states of negative energy have no physical meaning, then one cannot see how the scattering can occur.

On my new theory the state of negative energy has a physical meaning, but the electron cannot jump down into it because it is already occupied. There is, however, a new kind of double transition now taking place, in which first one of the negative-energy electrons jumps up to the proper final state with emission (or absorption) of a photon, and secondly the original positive-energy electron jumps down and fills up the hole, with absorption (or emission) of a photon. This new kind of process just makes up for those excluded and restores the validity of the scattering formulas derived on the assumption of the possibility of intermediate states of negative energy.

I do not think the infinite distribution of negative-energy electrons need cause any difficulty. One can assume that in Maxwell's equation

$$\mathrm{div}E = -4\pi\rho,$$

the ρ means the difference in the electric density from its value when the world is in its normal state, (i.e. when every state of negative energy and none of positive energy is occupied.) Thus ρ consists of a contribution $-e$ from each occupied state of positive energy and a contribution $+e$ from each unoccupied state of negative energy.

I have not made any actual calculation of the transition probabilities from $+$ve to $-$ve energy, but I think they are fairly small.

With kind regard to you and Klein,

Yours sincerely

P. A. M. Dirac

Bohr read Dirac's reply with equal amounts of interest and skepticism. He did not find Dirac's answer to the difficulty of the electric field satisfactory and explained to Hendrik Casimir, a young Dutch physicist studying in Copenhagen, that it was "nonsense" and that even with Dirac's suggestion the divergence of the electric field would have to be infinite.[18]

Dirac publicly presented the main features of his new theory at a conference held in Paris on December 13–20, 1929.[19] But it was through the informal channels of the physics community that the news first spread, its primary source being Dirac's letter to Bohr of November 26. Gamow, who had received a firsthand account of the theory from Dirac in Cambridge, reported about Dirac's ideas to his colleagues in Germany and Russia.[20] Heisenberg, Jordan, and Pauli knew about the theory in early December, and so did Iwanenko, Frenkel, Tamm, and Fock. Ehrenfest arrived in Leningrad on December 10 and brought with him a copy of Dirac's letter to Bohr.[21] Within a few months, a Russian translation of Dirac's paper appeared.

The earliest reception of Dirac's ideas varied between enthusiastic and skeptical. Tamm and Iwanenko were enthusiastic, while Fock expressed reserve and, like Bohr, raised several objections against Dirac's audacious hypothesis.[22] Even before Heisenberg had seen Dirac's paper, he made a rough calculation of what the electron–proton interaction would look like according to the new theory. He concluded that the electron and the Dirac proton had to have the same mass. On December 7, he wrote to Dirac, "I think I understand the idea of your new paper; it is certainly a great progress." But then he objected: "I cannot see yet, how the ratio of the masses etc. will come out. It seems to me already very doubtful, whether the terms of the electron (i.e. Sommerfeld formula) will not be completely changed by the interaction with the negative cells."[23] And two weeks later, he wrote to Bohr: "Dirac has already written a new work on the $\pm e$ business; you probably also know that I am, however, really skeptical, because the proton mass comes out equal to the electron mass, so far as I can see."[24]

After Heisenberg had studied a reprint of Dirac's paper, he returned to the mass question, now more confident in his criticism: "One can prove, that electron and proton get the *same* mass." The only way to change this conclusion, he continued, was to admit radical changes in the Coulomb interaction, in which case Sommerfeld's fine structure formula would also be changed. "So I feel," Heisenberg wrote to Dirac, "that your theory goes very far away from any correspondence to classical laws, and also from experimental facts." And he expressed more objections:[25]

For the normal scattering (coherent sc.) the initial and the final state of the electron is the same. (Take esp. an electron in a discret state). How can the negative-energy electron go up to the final level, which is already occupied? Of course, the neg-en. electron can go up to any other level and then drop down again, but I cannot see, how this effect can compensate for the scattering of the upper electron. But this difficulty might be negligible compared to the other difficulties. – I hope you won't get angry about this criticism, which is certainly unjustified in so far, as I don't know anything better than your theory.

We shall now take a closer look at Dirac's reasoning. As mentioned, his point of departure was the recognition, then generally accepted, that "we cannot ignore the negative-energy states without giving rise to ambiguity in the interpretation of the theory."[26] But with which physical entities (or "things," as Dirac preferred it) should these states be associated? Formal reasons seemed to indicate that the negative-energy states could be represented by positive-energy electrons with positive charge. Dirac had no difficulty in showing that, according to the relativistic wave equation, "an electron with negative energy moves in an external field as though it carries a positive charge."[27] However, he did not simply identify the negative-energy solutions with either protons or positively charged electrons. Particles with negative energy have no reality in physics, and Dirac emphasized that such ghost-entities as positively charged, negative-energy particles would produce physical absurdities. But he also added the straight empirical argument that "no particles of this nature have ever been observed."[28] Consequently, he rejected Weyl's suggestion which implied that protons should be described by negative-energy wave functions.

His way out of the dilemma, as mentioned in the letter to Bohr, was to introduce a world of negative-energy states uniformly occupied by an infinite number of electrons. As a crucial point, he supposed the distribution of negative-energy electrons to be governed by Pauli's exclusion principle. If the distribution of negative-energy electrons in the "Dirac sea" is exactly uniform, they will be unobservable, merely serving to define a state of normal electrification. But if a few of the negative-energy states are unoccupied, these vacant states, or "holes," will appear as observable physical entities: "*Only the small departures from exact uniformity, brought about by some of the negative-energy states being unoccupied, can we hope to observe.* . . . These holes will be things of positive energy and will therefore be in this respect like ordinary particles."[29]

According to an interesting – but no doubt apocryphal – story, Dirac came upon his idea of holes unconsciously, while dreaming of a problem posed in a competition arranged by the Cambridge Students' Mathematical Union.[30] More seriously, the main source of Dirac's daring theory seems to have been an argument by analogy, taken from the fields of X-ray theory and chemical atomic theory.[31] For example, in the inert gases the electrons fill up closed shells, resulting in chemical inactivity (which means unobservability in the chemical sense); a hole in the outer shell yields a halogen atom with some distinct chemical properties. More than forty years later, Dirac recalled: "It was not really so hard to get this idea [of holes], once one had the proper understanding of what one needed, because there was a very close analogy provided by the chemical theory of valency."[32] In his Paris address Dirac stressed the analogy to X-rays rather than to chemical atomic theory. He demonstrated how the empty state of an inner energy level, created by the absence of an electron, could be described by a usual wave function and hence be interpreted as an ordinary physical entity.

Another source of inspiration may have been Dirac's theory of radiation from 1927.[33] In this theory the absorption and emission of photons was pictured as taking place relative to an unobservable state of zero-energy photons; the emission of a photon was considered to be a jump from this zero-energy state to a state of positive energy.

With the idea of holes, Dirac managed to account for the negative-energy solutions without introducing observable negative-energy particles. He then became confronted with the problem of supplying the holes with a physical identity. Two possibilities were worth considering, the proton and the positive electron. Dirac's first formal proposal was for the wrong candidate, the proton; he wrote, "We are . . . led to the assumption that *the holes in the distribution of negative-energy electrons are the protons.*"[34] However, in his later recollections he claimed that his first inclination was actually toward what turned out to be the correct candidate, the positive electron, because of the appealing symmetry this choice would fulfill between the masses of the electron and the hole. In one interview he said, "I really felt that it [the mass of the hole] should be the same [as the mass of the electron] but I didn't like to admit it to myself," and at another occasion he remarked, "as soon as I got this idea [of holes] it seemed to me that the negative-energy states would have to correspond to particles having a positive charge instead of the negative charge of the electron, and also having the same mass as the electron."[35] But regardless of this first leaning, he initially introduced protons as the holes, although he was quite aware of the inherent difficulties of this idea and, indeed, retrospectively thought of it as "rather sick."[36]

Why did Dirac consider the identification of the hole with the proton to be sick? First of all, the Dirac wave equation is symmetric with respect to negative and positive charges (electrons and anti-electrons), while nature shows no symmetry between the electron and the much heavier proton. Dirac, lacking a better solution, expressed the hope that it might be possible to account for the difference in mass by means of a future theory of the interactions of protons and electrons; but he admitted that he was at the time unable to work out such a theory.[37]

If Dirac so clearly recognized the difficulties of the proton theory, then why did he propose it in place of his original idea of positive electrons? At least two motives were operative, one being empirical and the other formal. In 1930, physicists almost universally believed that matter consisted of only two material particles, electrons and protons. Neither theory nor experiment indicated that there were other fundamental particles in nature (the neutron still existed only as a name for an electron–proton composite, and the neutrino, although recently proposed, was not taken seriously). Naturally, Dirac preferred to base his hole theory on known entities rather than to postulate a new elementary particle for which there was at the time no empirical evidence whatsoever. In 1930, the climate in the infant branch of particle physics was very conservative with respect to new elementary particles, and Dirac's fear of introducing an unobserved particle may have been the result of a sociological constraint, determined by the paradigms of the period. (Of course, it may also be viewed from a psychological perspective.[38]) This conservatism of the physics community may have inhibited Dirac from making a proposal that he would have made under different sociological conditions. However, another motive also played an important role in Dirac's considerations, significantly catalyzing his suggestion of the proton theory. He considered the identification of protons with vacant negative-energy states to be a highly attractive idea because it promised a reduction of the known elementary particles to just one fundamental entity, the electron.

Dirac gave an account of his theory at the ninety-ninth meeting of the British Association for the Advancement of Science, held in Bristol on September 3–10, 1930. In his address he emphasized the attractiveness of the unitary view:[39]

It has always been the dream of philosophers to have all matter built up from one fundamental kind of particle, so that it is not altogether satisfactory to have two in our theory, the electron and the proton. There are, however, reasons for believing that the electron and proton are really not independent, but are just two manifestations of one elementary kind of particle. This connexion between the electron and proton is, in fact, rather forced upon us by general considerations about the symmetry between positive and negative electric charge, which symmetry prevents us from building up a theory of the negatively charged electrons without bringing in also the positively charged protons.

After the lecture, Dirac was asked to discuss his theory further, but he "shook his head, saying he could not express his meaning in simpler language without being inaccurate." When a reporter from the *New York Times* wanted a comment, it came not from Dirac but from Oliver Lodge; the old physicist was fascinated by Dirac's theory, which appealed greatly to his imagination, he said.[40]

Dirac thought that the dream of philosophers was on its way to being realized, that the quantum theory had now intimated a positive answer to the age-old question of the unity of matter. His grand idea was not

Dirac at the height of his career. Photograph from about 1930. Reproduced with permission of AIP Niels Bohr Library.

without similarity to one fostered years earlier by another great Cambridge physicist, J. J. Thomson. In the original verson of his atomic theory, developed during the period 1897–1907, Thomson tried to build up a theory of matter based solely on electrons. Although the proton was not yet recognized as a particle at the time, Thomson still, of course, faced the problem of how to account for positive electricity within a unitary theory of electrons. He never solved the problem and could only express the hope that someday the positive electricity would somehow be explained as a manifestation of electrons.[41] It seems that Dirac's devotion to the principle of unity deceived him into believing that the proton model was "rather forced upon us by general consideration." In fact, those general considerations, that is, the consequences implied by the mathematical structure of the wave equation, showed that the mass of the hole was the same as the mass of the electron. This point was already made by Heisenberg in his letter to Dirac of January 16 (see above), but at that time Dirac was too fascinated by the unitary view to be convinced of the point.

Having concluded that the proton was the most likely candidate for the hole, Dirac was faced with several difficulties. For example, one would expect that a positive-energy electron might occasionally make a quantum transition to fill a hole, under which circumstance the two particles would annihilate, that is

$$p^+ + e^- \to 2\gamma$$

Dirac considered this hypothetical process in detail in a paper of March 1930, and there he also showed that the hole theory can be brought into agreement with the Klein–Nishina formula.[42] At the British Association for the Advancement of Science meeting in Bristol, he said:[43]

There appears to be no reason why such processes should not actually occur somewhere in the world. They would be consistent with all the general laws of Nature, in particular with the law of conservation of electric charge. But they would have to occur only very seldom under ordinary conditions, as they have never been observed in the laboratory.

In 1930, the idea of annihilation processes was not new. In connection with astrophysical speculations on energy production in stars, proton-electron annihilation had been proposed as a possible candidate since World War I. Eddington and Jeans, in particular, in the twenties advocated the view that such processes should have a prominent place in astrophysics, although they admitted that the mechanism of annihilation was unknown. Annihilation and hypothetical fusion processes were also discussed by Walther Nernst and, in connection with the origin of cosmic rays, by Robert A. Millikan and his co-workers.[44] The idea even figured in cosmology, when Richard Tolman argued that the recently discovered expansion of the universe was caused by annihilation processes.[45] In July 1931, James Jeans reviewed the annihilation hypothesis and stated that "the majority of astronomers think it probable that annihilation of matter constitutes one of the fundamental processes of the universe, while many, and perhaps most, physicists look upon the possibility with caution and even mistrust."[46] Jeans did not have in mind Dirac's theory, to which he only referred casually. This theory soon supplied the astonomers' view with physical credibility, even though it did not introduce the notion of annihilation into astronomy or physics.

The lack of experimental evidence for proton–electron annihilation did not worry Dirac too much. He was content to observe that the predicted process was "consistent with all the general laws of Nature." The other main problem that faced the theory was a more serious one, namely, the difference in mass between the electron and its assumed anti-particle, the proton.[47] If the difference in mass could not be explained, then "the dream of philosophers" would remain a dream. Although Dirac did not succeed in accounting for the mass difference in terms of interaction forces, he did not consider the flaw as fatal to the theory. Lacking a better idea, he referred to the unorthodox views of Eddington as possibly supplying a solution to the problem.[48]

If the astrophysical evidence for proton–electron annihilation was uncertain, surely the annihilation frequency had to be very small. Terrestrial matter, which according to the current view consisted solely of protons and electrons, was stable, which placed a very low limit on the transition probability. Accordingly, Dirac tried to calculate the transition probability of the process. Or, more accurately, he calculated the annihilation cross section between an electron and a "proton" with the same mass as an electron. Because of the lack of explanation of the proton's mass, Dirac had to neglect interaction effects altogether, which "prevents one from attaching much physical importance to the results."[49] Dirac found an effective collison area (cross section) of the order of magnitude of the size of the classical electron, a result much too large to agree with the stability of matter. In spite of the result, he remained optimistic, hoping that interaction effects would reduce the cross section considerably. At about the same time, a similar result was obtained by Igor Tamm in Moscow. Tamm worked on the annihilation probability when he was informed by his friend Dirac about his still unpublished calculations.[50] Like Dirac, Tamm got "an impossibly low value" for the mean lifetime of atoms. But although he admitted that this constituted "a fundamental difficulty" for the theory, he assumed with Dirac that inclusion of interaction effects would probably improve the result.

The Dirac–Tamm result for the mean lifetime of a proton can be expressed as

$$T = \frac{m^2 c^3}{\pi e^4 n_e}$$

where n_e is the electron density of the surrounding matter and m is the mass of the anti-electron (the proton in Dirac's theory). With $m = m_p$, the formula gives about 10^{-3} seconds; with $m = m_e$, the result is about 10^{-9} seconds.

Very few physicists shared Dirac's faith in his unitary theory. When the theory was first proposed, it was received with great interest but also with skepticism, to say the least. Informally, it was regarded as nonsense in most quarters. According to the reminiscences of von Weiszäcker, the Fermi group in Rome "held some sort of 'judicial meeting' about Dirac who, in absentia, was doomed. I think they had decided to 'bastonare' him because he had written such a nonsense."[51] Pauli formulated what in inner circles became known as the Second Pauli Principle: Whenever a physicist proposes a theory, it should immediately become applicable to himself; therefore Dirac should annihilate! In September 1930, Landau and Max Delbrück attended the British Association for the Advancement of Science meeting in Bristol in order to learn whether Dirac had come up with some new ideas. Landau was disappointed to learn that this was not the case, and after the talk he went to the nearest post office, where he sent to Bohr's institute a telegram containing just the word "nonsense." It was the code agreed on with Gamow to summarize Dirac's talk.[52]

Oppenheimer, who had recently returned to California after a stay with Pauli in Zürich, was quick to respond to Dirac's theory. His interest derived from his conviction that the new quantum electrodynamics of Pauli and Heisenberg (see Chapter 6) was unable to account for the interaction between particles and the electromagnetic field. When he read Dirac's paper in January 1930, he hoped it would supply an answer to the problem. But he soon concluded that the Dirac theory was not tenable in the form proposed by its author. In a short paper from February – that is, slightly before the corresponding papers of Dirac and Tamm – Oppenheimer calculated the transition probability for proton-electron annihilation. For the mean lifetime of a free electron in matter, he obtained the result

$$T = \frac{(m + M)^2 c^3}{64 \pi^5 e^4 n_p}$$

where n_p is the density of protons in the area surrounding the electron, and M and m are the protonic and electronic masses.[53] In usual matter

n_p is of the order of magnitude 10^{25} protons/cm^3, which leads to the absurdly low value $T \simeq 10^{-9}$ seconds. Instead of abandoning the Dirac theory totally, Oppenheimer concluded that the theory needed to be modified into a theory of two independent "Dirac seas":[54]

If we return to the assumption of two independent elementary particles, of opposite charge and dissimilar mass, we can resolve all the difficulties raised in this note, and retain the hypothesis that the reason why no transitions to states of negative energy occur, either for electrons or protons, is that all such states are filled. In this way we may accept Dirac's reconciliation of the absence of these transitions with the validity of the scattering formulae.

Oppenheimer's proposal thus ruled out proton–electron annihilation on the grounds that there were no holes in the negative-energy sea to which the particles could pass. Dirac did not like Oppenheimer's theory. It broke with the unitary view Dirac found so attractive: "One would like, if possible, to preserve the connexion between the proton and electron, in spite of the difficulties it leads to, as it accounts in a very satisfactory way for the fact that the electron and proton have charges equal in magnitude and opposite in sign."[55]

Another early response to Dirac's short-lived unitary theory was supplied by the Leningrad physicists Iwanenko and Victor Ambarzumian.[56] They proposed to apply the theory in order to account for beta radioactivity, one of the most perplexing problems of the period. What happened in a beta process, according to the Russian physicists, was a transition of a negative-energy electron into a positive energy state; in this way an observable beta electron would emerge together with a proton, born in the nucleus. The mechanism suggested by Iwanenko and Ambarzumian thus did not rely on the standard electron–proton model of the nucleus.

While, as mentioned, Dirac was not much impressed by the empirically based objections raised by Oppenheimer and others, he was deeply concerned with objections deduced from the general theory of quantum mechanics. In the fall of 1930, Pauli reached the same conclusion as Heisenberg, that Dirac's unitary theory was inconsistent. Tamm, who had met Pauli during a conference in Odessa, reported to Dirac: "Pauli told us that he rigorously proved . . . that on your theory of protons the interaction of electrons can't destroy the equality of the masses of an electron & a proton."[57] Another argument was advanced by Weyl in the second edition of his *Gruppentheorie und Quantenmechanik*. Weyl showed that according to Dirac's own theory of the electron the hole must necessarily have the same mass as an ordinary electron.[58] When, in the early part of 1931, Dirac gave up his unitary theory and proposed the anti-electron as a separate particle, he quoted both Oppenheimer's and Weyl's objections as decisive arguments.[59] Probably Weyl's argument, which was in spirit close to Dirac's own methodological preferences, was of more importance.

In his various recollections, Dirac repeatedly pointed out that his disinclination to postulate positively charged electrons was rooted in a lack of boldness both to rely on the mathematics of his wave equation and to disregard the restrictions set by current empirical knowledge. Had he only been faithful to the power of pure mathematical reasoning and not been led astray by what was known empirically, he immediately would have postulated the positive electron. Indeed, this was what Weyl did, although only implicitly. On several occasions, Dirac attributed this success of Weyl's to his mathematical approach to physics. This was an approach that Dirac strongly recommended:[60]

Weyl was a mathematician. He was not a physicist at all. He was just concerned with the mathematical consequences of an idea, working out what can be deduced from the various symmetries. And this mathematical approach led directly to the conclusion that the holes would have to have the same mass as the electrons. Weyl . . . did not make any comments on the physical implications of his assertion. Perhaps he did not really care what the physical implications were. He was just concerned with achieving consistent mathematics.

Dirac's new version of the hole theory appeared rather casually in a remarkable paper of May 1931. The main content of the paper will be dealt with in Chapter 10. The anti-electron was now introduced for the first time as "a new kind of particle, unknown to experimental physics, having the same mass and opposite charge to an electron."[61] Dirac was not principally concerned with the question of the empirical existence of the anti-electron, although he did, of course, believe in the particle's existence. Perhaps, he mentioned, anti-electrons might be produced in a high-vacuum region by means of energetic gamma rays. He predicted the annihilation process

$$e^- + e^+ \rightarrow 2\gamma$$

as well as the pair-production process

$$\gamma + \gamma \rightarrow e^- + e^+$$

In 1931, Dirac considered only pair production resulting from a collision between two gamma quanta (photons), not the interaction between a gamma quantum and the Coulomb field of a heavy nucleus. The probability of the former process is extremely small compared to that of the latter. Experimentalists probably saw no reason to examine pair production as predicted by Dirac, because of the small probability of observing electron pairs.

According to the new view, protons were, as suggested by Oppenheimer, unrelated to electrons. To take this stand was not easy for Dirac, because it implied a farewell to the unitary theory. He regretted that "theory at present is quite unable to suggest a reason why there should be any differences between electrons and protons,"[62] and only reluctantly accepted this as a contingent fact. Because the proton was a separate species of particle, it would, Dirac argued, probably have its own species of unfilled hole, that is, the anti-proton. In a few lines he doubled the number of elementary particles (and this disregards the magnetic monopole, which he also introduced in the paper). A few years later, Dirac went a step further, speculating about matter made up entirely of anti-particles. Because of the complete symmetry between positive and negative charges, he held it probable that anti-protons, and anti-atoms too, existed. In 1933, he ended his Nobel lecture with the following speculation:[63]

We must regard it rather as an accident that the Earth (and, presumably, the whole solar system), contains a preponderance of negative electrons and positive protons. It is quite probable that for some of the stars it is the other way about, these stars being built up mainly of positrons and negative protons. In fact, there may be half the stars of each kind. The two kinds of stars would both show exactly the same spectra, and there would be no way of distinguishing them by present astronomical methods.

As far as I know, the anti-proton was first introduced by Dirac in 1931. In 1933, it was considered in more detail by the Rumanian-French physicist Jean Placintéanu, who attempted to build up a scheme of elementary particles based on Dirac's theory. The idea of stars made up of anti-particles was also put forth by Reinhold Fürth at the German University of Prague. In 1934, Gamow suggested that negative protons might be constituents of the atomic nucleus.[64] And in the same year, the Yugoslavian physicist Stjepan Mohorovicic suggested the existence of what later became known as positronium, atoms made up of electrons and positrons.[65]

The Dirac holes of 1930–1 were not the only "holes" that appeared in the physics of the period. In 1929, Rudolf Peierls introduced the idea of the "hole" in solid-state theory, a notion more fully elaborated by Heisenberg in 1931. The solid-state hole of Peierls and Heisenberg was an unfilled state in the energy band that behaved exactly like an electron with positive charge. There was therefore a rather close analogy between Dirac's holes and the holes of solid-state physics. Since the ideas emerged during the same period, it would be reasonable to expect a genetic connection between them. However, the two ideas seem to have been invented independently. Heisenberg, in his paper of 1931, did not refer to Dirac's theory, and Dirac never referred to the solid-state hole of Peierls and Heisenberg.[66]

During this period of hectic scientific activity, Dirac continued his quiet life as a bachelor and scientist at St. John's College, a life interrupted only by several trips abroad. In the summer of 1930, he went to Russia for the third time; in Kharkov he met with the German physicist Walter Elsasser, who worked at the University of Kharkov during 1930–1. Dirac left Russia by way of Odessa at the end of July. He sailed to Venice and then went by train back to England.[67]

As a result of his scientific reputation, Dirac came in closer contact with other scientists, on a professional level and also on a personal and social level. In Cambridge he often went to the house of the Mott family and the Peierls family. Dirac helped them in various ways, sometimes driving them in a car he had bought. In March 1931, Nevill Mott wrote to his mother: "I went down to London yesterday in Dirac's car – very cold. Dirac ran – very gently – into a back of a lorry and smashed a headlamp."[68] In 1933, Mott went to the University of Bristol, where Dirac had begun his career, and founded an important school of solid-state physics. Much later, when Mott returned to Cambridge as a Cavendish professor, he became a neighbor of Dirac. At Dirac's invitation, Tamm spent two months in Cambridge in the spring of 1931. They toured England and Scotland in Dirac's car, and Dirac used the occasion to teach Tamm how to drive. Later, when they returned to Cambridge, he talked Tamm into taking a driving test.[69]

In February 1931, Dirac was elected a corresponding member of the Academy of Science of the USSR. The following month he stayed for some days in Copenhagen, where his hole theory became the subject of heated discussions with Bohr and his colleagues. To Van Vleck he reported:[70]

Things are to be fairly quiet now in theoretical physics. I went to Copenhagen at the end of March and met Bohr, Pauli and others. Bohr is at present trying to convince everyone that the places where relativistic quantum theory fails are just those where we would expect it to fail from general philosophical considerations. Pauli is now studying English, as he has to lecture in Ann Arbor this summer, and is rather alarmed at the slow rate of progress he is making.

In the same letter, Dirac made arrangements for a second tour of the United States. In early 1930, he had been approached by the mathematician Oswald Veblen, who wanted him to come to Princeton University, but at that time Dirac had had to decline the invitation. It was instead arranged that he would go to Princeton for the fall term of 1931, to lecture on a subject in mathematical physics to be chosen by himself.[71] The salary would be $5,000. Veblen was happy to get Dirac and wrote to him, "There will be rather a concentration of quanta in this neigborhood at that time for von Newmann [sic], Wigner, Condon and Robertson will be here as well as yourself."[72] Dirac arrived in Montreal in mid-August, visited Niagara Falls, and then went west to Glacier National Park, where he walked and camped in company with Van Vleck. Dirac was an ideal tent-mate, preferring to retire early and arise late. Van Vleck recalled that Dirac, when dining in American hotels, found the ice water very cold and the soup very hot, "a situation he solved with his characteristic directness by transferring an ice cube from the water to the soup."[73] On his way back east he revisited Yellowstone National Park and came to Princeton via Madison and Washington. To Van Vleck he wrote:[74]

Many thanks for your hosp'.ability in Madison. I went to Washington and stopped there till 10 o'clock, which gave me plenty of time to look over the Capitol as I forgot about breakfast. I got to Princeton about 2:30 that afternoon and spent a couple of days looking for rooms, but am settled down now. Wigner has obtained the following general result: If an atom with an odd number of electrons is put in *any* electric field, (not necessarily a uniform one) there will be a two-fold degeneracy of its states. He got this result by considering the reflection operator which changes t to $-t$. He would like to know whether you have already found this result and are putting it into your book or whether he should write a paper about it. Pauli came to Princeton last night and he and I gave a colloquium on neutrons and magnetic poles. He sails for Italy this evening. How is the oil burner?

The curious reference to the oil burner relates to an incident that took place while Dirac was visiting Van Vleck in his new house: Because of a malfunctioning furnace, some spilled oil caught fire and threatened to set the entire house in flames. Dirac, calm and logical, suggested that they move outside and close all the doors in order to suffocate the fire. His plan succeeded, and only minor smoke damage was done.[75]

In Princeton Dirac lectured on quantum mechanics, largely following the content and structure of his recently published *Principles of Quantum Mechanics.* He briefly outlined his new theory of anti-electrons, stating:[76]

An anti-electron ought to have the same mass as an electron and this appears to be unavoidable; we should prefer to get a much larger mass so as to identify the anti-electrons with the protons, but this does not seem to work even if we take a Coulomb interaction into account. . . . This idea of the anti-electrons doesn't seem to be capable of experimental test at the moment; it could be settled by experiment if we could obtain two beams of high frequency radiation of a sufficiently great intensity and let them interact.

After his return to Cambridge in the early days of 1932, Dirac wrote Tamm about his impressions of his second trip to the United States:[77]

I had an interesting time in Princeton and learnt a good deal of pure mathematics there (about group theory and differential geometry) which I must now try to apply to physics. I enjoyed myself very much in the Rockies, although I did not do any difficult mountaineering but mostly kept to the trails. The scenery is very fine, particularly in the canyon country, which I visited after Glacier Park.

By that time, Dirac had twice been offered a professorship. In February, he was approached again, this time with an offer to succeed John C. McLennan as a professor at the University of Toronto for a salary of $4,000 a year.[78] But the offer did not tempt Dirac, who knew that the prestigious Lucasian Chair at Cambridge was soon to be vacant. Since 1903, it had been occupied by Joseph Larmor, who had been inactive for some time and was scheduled to retire at the end of September 1932. The prestige of the chair was related to the fact that it was once held by Isaac Newton for more than 30 years; Newton, in turn, had taken it over from its first holder, his teacher Isaac Barrow. Since then, it had been occupied by Charles Babbage, Gabriel Stokes, and Joseph Larmor, among others. When Dirac was appointed Lucasian Professor of Mathematics in the fall of 1932, the chair was finally occupied by a scientist comparable to the great Newton. Dirac remained Lucasian Professor for thirty-seven years. Several times he was offered other, more economically rewarding positions, but he always declined. For example, in 1935 he was invited to accept the Jones Professorship in mathematical physics at Princeton University, which included a salary of $12,000 a year.

When Dirac proposed the anti-electron in March 1931, it was a purely hypothetical particle. Within a year or two, the situation changed drastically, with the result that Dirac's brainchild turned into a real particle, the positron. The discovery of the positron has been examined by historians and philosophers of science and need not be detailed again here.[79] But it may be useful to make a digression into the prehistory of the positive electron, especially since this aspect is missing in most historical analyses.

Dirac pictured the vacuum state as the state of lowest possible energy, in which all the negative-energy states (and none of the positive states) were occupied. This picture may bear some resemblance to the once popular picture of the ether. Indeed, as we shall see in Chapter 9, this resemblance is more than superficial. Some of the classical ether models – long ago discarded and in 1930 serving only as examples of methodologically unsound science – were in fact strikingly similar to Dirac's negative-energy world. For example, Osborne Reynolds had proposed in 1902 that the ether might consist of minuscule particles that were unobservable in their normal configuration; only deviations from the normal configuration would be observable and then appear as material particles.[80] Another Victorian analogy to anti-matter was, with hindsight and good will, the "aether squirt" hypothesis of Karl Pearson. According to Pearson's speculations, atoms of "aether sinks" had to exist in order to keep up with ether (or matter) conservation. These atoms were believed to correspond to "negative matter," resulting in a mutual repulsion between two lumps of such matter. Pearson's strange idea won little support, but it was independently entertained by Arthur Schuster in what he admitted was a holiday dream. Schuster's vision included not only "anti-atoms" but also worlds made up of "anti-matter."[81]

As to the positive electron, by 1932 it had quite a long prehistory. It was part of nineteenth-century electrodynamics as developed by Weber, among others, and it figured frequently in the classical electron theories of Lorentz, Wien, Wiechert, and others. In several of the pre-Rutherford atomic models, positive electrons were assumed to be the stuff of positive electricity instead of Thomson's positive fluid.[82] Nernst was convinced of the reality of positive electrons, and so were Jean Becquerel and Robert Wood, who even claimed to have detected the particles.[83] As a result of Rutherford's conception of the atom, it became the general view in the second decade of the twentieth century that the "positive electron" was in fact the proton. There seemed to be no need for a positive particle with the same mass as the electron, and speculations concerning true positive electrons largely vanished.[84] However, in quite another context the positive electron turned up occasionally, namely, in the early attempts to develop a unified field theory. The first to consider such an entity within the framework of general relativity was probably Pauli, but it also figured in Einstein's work.[85] However, none of the theories referred to here can fairly be considered as predecessors of Dirac's theory. Dirac's prediction of the positive electron was genuinely new and quite unconnected with earlier speculations.

Around 1930, the cosmic radiation emerged as a new and interesting field in physics, studied by Millikan and his group in California, among others. The experimental discovery of the positive electron was an unsuspected, and at first rather unwelcome, result of Millikan's research program of investigating the nature of the cosmic radiation. Carl D. Anderson, a former student of Millikan, investigated the cosmic rays of the upper atmosphere by means of cloud chambers; he noticed some tracks that he interpreted as the result of positive light-weight particles.[86] At first, Anderson did not interpret the tracks as those of positive electrons; in accordance with Millikan's research program, he believed that they were protons, or perhaps electrons passing the opposite way through the chambers. It was only in a second paper of March 1933 that Anderson abandoned what he later called "the spirit of scientific conservatism" and suggested that he had discovered a positively charged electron.[87] In the same paper he proposed the name "positron," which was soon generally accepted. It is notable that Anderson's discovery of the positron was experimental and not indebted at all to Dirac's theory. As Anderson himself later remarked: "Despite the fact that Dirac's relativistic theory of the electron was an adequate theory of the positron, and despite the fact that the existence of this theory was well known to nearly all physicists, it played no part whatsoever in the discovery of the positron."[88] Instead of interpreting the positive electron to be a result of pair production, Anderson believed that it was ejected from an atomic nucleus split by an incoming cosmic ray photon. This interpretation was in accordance with Millikan's theory but not, of course, with Dirac's. Anderson probably knew about Dirac's theory of anti-electrons, which was discussed at Oppenheimer's seminars in which Anderson participated; but the fact remains that he did not identify the positron with Dirac's anti-electron.

It was only after Patrick Blackett and the Italian physicist Guiseppe Occhialini had reported new cosmic ray experiments, also carried out with cloud chambers, that it was realized that Anderson had in fact discovered the Dirac particle.[89] The work of Blackett and Occhialini was important because it explicitly linked Dirac's hole theory with the cosmic ray experiments. Blackett was acquainted with Dirac and his theory and interpreted the cloud chamber tracks as the result of pair production in Dirac's sense.[90] In their paper, Blackett and Occhialini acknowledged valuable discussions with Dirac on the subject. They cited a calculation, made by Dirac, of the mean free path in water for annihilation; according to this calculation, the mean lifetime of a positron in water would be around 3.6×10^{-10} seconds. Referring to Dirac's theory, they concluded: "There appears to be no evidence as yet against its validity, and in its favour is the fact that it predicts a time of life for the positive electron that is long enough for it to be observed in the cloud chamber but short enough to explain why it had not been discovered by other methods."[91]

When Dirac learned about the discovery of the positron, probably through Blackett, he was, of course, happy to see his hypothesis confirmed. Yet to Dirac the main thing was not so much that the positron actually existed in nature; he thought it more important that Anderson's discovery vindicated his theoretical ideas. Witness the following conversation with Thomas Kuhn from 1963:[92]

K: Was it [the discovery of the positron] a great vindication?
D: Yes.
K: Does that sort of event generate great immediate excitement and satisfaction?
D: I don't think it generated so much satisfaction as getting the equations to fit.

In 1932–3, the American mathematician and mathematical physicist Garrett Birkhoff stayed at Cambridge University. Before Blackett and Occhialini's paper appeared, he reported his impressions of Dirac in an illuminating letter to Edwin Kemble:[93]

As Prof. Dirac has just concluded his lectures for the year, and as they are the only physical lectures which I am attending, I thought that you might be interested in having a confidential report on my impressions of him, and of his work. . . . I think that Dirac's principal present concern, is with the attempt to reconcile Schrödinger's equations with spin and relativity. He believes that this can be accomplished by suitable modifications, alterations, and suppressions of various terms – such as of spin interaction. He opposes extensions of the principles of quantum mechanics which introduce new experimentally meaningless concepts. . . . He also has an important secondary concern in the question of the existence of a positively charged particle with the mass of an electron. The recent researches of Blackett at the Cavendish on ionization caused by cosmic rays (using Wilson's cloud-chamber method), clearly indicate the existence of a positively charged particle with mass much smaller than that of a proton. A report of the reasons for this will be in the forthcoming issue of the Cambridge Philosophical Magazine. However, following the communication by Blackett of his researches to the Cambridge Philosophical Society, Lord Rutherford observed that we could not yet feel certain of even this much, and that he would prefer to be able to produce the particle by laboratory methods. Dirac's interest proceeds from the fact that he predicted the existence of such a particle – at first with a vague notion of identifying it with the proton – three or four years ago. Now he hopes in addition for a negatively charged particle with the mass of an electron [sic (proton?)] . . .

Kemble answered that he was willing to accept the discovery of the positron as a confirmation of Dirac's theory:[94]

We have all been very much stirred up over Blackett's confirmation of the discovery of the positive electron as well as by the original announcement of the part of Anderson. It seems to be less embarrassing to theoretical physics than the discovery of the neutron and I judge that, by confirming Dirac's theory of the positive and negative energy states, it bids fair to clear up to a very large extent the difficulties in the relativistic formulation of quantum mechanics.

However, at first Kemble's view was not shared by most physicists. Although Dirac's theory of the anti-electron was far from unanimously accepted (see below), the positron was quickly acknowledged as a new elementary particle. Very soon after Anderson's discovery, the positron was reported in other experiments, in the cosmic radiation as well as in the recently discovered artificial radioactivity. The theorists, too, studied the positron eagerly. They worked out formulae for the probability of pair production and annihilation and checked them with a growing amount of empirical data, mostly based on the cosmic radiation. By the end of 1934, a fairly reliable knowledge of the behavior of positrons had been established.[95]

In spite of the favorable response to the positron, there was much skepticism toward Dirac's theory. Not surprisingly, this negative attitude was more marked before the positron was discovered, that is, during the period 1930–2. The negative energies that appeared as physically significant states in Dirac's theory continued to cause worry. To mention just one response, Schrödinger worked out an alternative theory in which the negative energies did not appear. His equation coincided with Dirac's in the field-free case but not if fields were present.[96] Although Schrödinger was able to reproduce hydrogen's fine structure, as well as other results of Dirac's theory, his alternative won no acceptance. The wave equation was not Lorentz-invariant, which was regarded as reason enough to discard it. After studying Schrödinger's alternative, Dirac concluded that "the new wave equation, though, is not accurately relativistic, so this is not a satisfactory solution of the problem [of ± energies]."[97] Some physicists apparently misunderstood Dirac's theory; for example, some interpreted it as supplying the negative-energy electrons with observable properties.[98]

Some of the opposition to Dirac's theory rested on philosophical grounds. "One often hears people to say, that the unobservable negative electrification of the world, produced by the negative energy electrons, is a metaphysical notion," Tamm wrote to Dirac, comparing the negative-energy world with Lorentz's unobservable ether.[99] Positivist ideals, which at that time enjoyed support among many physicists associated with the Copenhagen program, were difficult to reconcile with Dirac's metaphysics. Thus Landau and Peierls argued in 1931 that the "senseless results" of Dirac's theory were to be expected from a version of the uncertainty principle extended in the spirit of Bohr: "One cannot be surprised if the formalism leads to all kinds of infinities. On the contrary, it would be quite remarkable if it had any sort of resemblance to reality," they wrote.[100] Landau and Peierls's dismissal of Dirac's ideas was strongly influenced by Bohr and relied on an extreme empiricist attitude. They stated as a matter of fact that "the meaning of every physical theory lies in linking the results of one experiment to the results of later experiments."[101] This was a doctrine to which Dirac too would have claimed to subscribe, but in his scientific practice he did not follow it (see also Chapter 13). That there is no one-to-one correspondence between philosophical doctrines and scientific thoughts is shown by a paper written by the American physicists Edward Condon and Julian Mack shortly after the proposal of Dirac's first hole theory. They argued for a subjectivist interpretation of quantum mechanics in accordance with Bridgman's "operationalism" and supplied on this basis a defense of Dirac's idea of a negative-energy electron sea. Arguing that Pauli's exclusion principle was to be understood as an instance of the subjectivity of human knowledge, Condon and Mack deduced that a world governed by the Pauli principle was the only one that could possibly be observed. Since Dirac's hole theory rested crucially on the exclusion principle, they concluded, "We see why it is that the Pauli Principle World is the only one which can be built out of Dirac's relativistic electrons."[102] This was rather the opposite conclusion to that drawn by Landau and Peierls.

In Copenhagen the negative-energy Dirac electrons were jocularly referred to as "donkey electrons," a name suggested by Gamow, because negative-energy electrons, like donkeys, would move slower the harder they were pushed! In April 1932, the institute in Copenhagen celebrated the tenth anniversary with a conference in which Pauli, Delbrück, Klein, Heisenberg, Ehrenfest, Dirac, and Bohr, among others, participated. On this occasion the participants staged a collectively authored parody of Goethe's *Faust,* adapted to the current situation in theoretical physics. Part of this memorable play dealt with Dirac's hole theory and his theory of the magnetic monopole (see Chapter 10).[103]

Bohr did not like Dirac's hole theory at all. He believed, as shown in his letters to Dirac, that the difficulties in relativistic quantum mechanics would require some drastic alteration of the concepts of space and time, probably including the abandonment of energy conservation in atomic processes. Bohr, and with him many of his disciples, hesitated to interpret the discovery of the positron as a vindication of Dirac's ideas. Faced with the experimental evidence in favor of the positron, Bohr reportedly said that "even if all this turns out to be true, of one thing I am certain: that it has nothing to do with Dirac's theory of holes!"[104] Pauli also took care to distinguish between the Anderson–Blackett particle and the hole–particle of Dirac's theory. In May 1933, he told Dirac, "I do not believe on your perception of 'holes,' even if the existence of the 'antielectron' is proved."[105] Pauli believed at that time that the discovery of the positron supported the existence of a neutrino. "If the positive and negative electron both exist," he wrote to Blackett, "it is not so phantastic to assume a neutral particle consisting of both together."[106]

Pauli's intense dislike of the hole theory is evident from his correspondence as well as from his published works. Together with the French physicist Jacques Solomon, he attempted to construct a unification of Dirac's theory and the general theory of relativity in which the negative-energy electrons would not appear.[107] In his authoritative *Handbuch* survey of quantum mechanics in 1933, which was completed shortly before Anderson announced his discovery, Pauli was very critical of the hole theory. The difficulty related to the negative energies, wrote Pauli, could "neither be denied away, nor resolved in any simple way." He rejected Dirac's and Oppenheimer's ideas of anti-particles with the following argument:[108]

The actual lack of such particles would then be traced back to a special initial state in which there happens to be present only one kind of particle. This appears already unsatisfactory because the laws of nature in this theory are strictly symmetrical with respect to electrons and anti-electrons. Thus γ-ray photons – at least two, in order to satisfy the conservation laws of energy and momentum – must be able to transform spontaneously into an electron and an anti-electron. Accordingly we do not believe that this way out should be seriously considered.

Within the framework of quantum field theory, Pauli was able to show that the arguments which had originally led Dirac to his wave equation were no longer valid. He dismissed Dirac's original objection to the Klein–Gordon theory – that it contradicted the transformation theory – as "pure nonsense."[109] In 1935, when lecturing at the Institute for Advanced Study in Princeton, Pauli had the following to say about Dirac's theory:[110]

The success seems to have been on the side of Dirac rather than on logic. His theory consisted in an number of logical jumps. . . . There is no longer a conservation of the total number of particles, when one considers positrons, and so Dirac's argument for the form of the wave equation is no longer cogent because there no longer exists any *a priori* reason that the wave equation shall be of the first order and the charge density shall be a sum of squares.

Pauli's criticism, though valid, does not diminish the greatness of Dirac's theory. After all, the criticism was based on the knowledge of the mid-1930s, not the knowledge of early 1928. It may be true that Dirac's theory of the electron and positron consists of "a number of logical jumps" (though "logical" is hardly the right term) and this, with hindsight, would be methodologically objectionable. But is that not a hallmark of all truly great discoveries? Perhaps it was no accident that Heisenberg, appraising Dirac's theory of anti-electrons, described it as a "jump." Heisenberg once called it "perhaps the biggest jump of all big jumps in physics of our century."[111]

Pauli's skepticism remained long after the positron was discovered.

When in 1933–4 he and Victor Weisskopf developed a quantum field version of the Klein–Gordon theory for scalar particles, it was consciously developed as an alternative to "my old adversary, the Dirac theory of the spinning electron."[112] Pauli hoped that the new theory would lead to "a more complete liberation from Diracian approaches and ways of thinking."[113] He and Weisskopf succeeded in proving that such concepts as "pair creation," "annihilation," and "anti-particles" could be established without accepting the idea of a vacuum filled with negative-energy particles. However, since the Pauli–Weisskopf theory did not refer to electrons and positrons, but to hypothetical particles of spin zero, it was not a proper alternative to Dirac's theory. At about the same time, Oppenheimer and Wendell Furry also showed that anti-particles could be accounted for by quantum field theory without introducing the Dirac sea of unobservable negative-energy particles.[114]

Fermi was, like Bohr and Pauli, skeptical of Dirac's theory. In his important theory of β-decay in 1934, he followed Pauli in picturing the process as the transmutation of a neutron into a proton, an electron, and a neutrino. But he stressed that "this possibility [of creation and annihilation of particles] has, however, no analogy with the creation or disappearance of an electron–positron pair."[115] Other physicists, including Guido Beck and Kurt Sitte, did, however, make use of Dirac's hole theory in alternative theories of β-decay.[116] So did Gian Wick, who in 1934 gave a theoretical explanation of the recently discovered artificial radioactivity. Wick extended Dirac's hole theory to cover neutrinos also and introduced the anti-neutrino as a hole in a neutrino-sea in Dirac's sense.[117] Opposition to the Dirac positron was not limited to the theorists. For example, Anderson, the discoverer of the positron, followed Millikan in denying the validity of the Dirac theory as far as the positron was concerned. Even in 1934, the Caltech experimentalists stuck to the view that the cosmic ray positrons preexisted in atomic nuclei and therefore were not to be identified with Dirac positrons.

In spite of the reservations, the identification of the Dirac anti-electron with the Anderson–Blackett positive electron became vindicated during 1933. In July, Peierls wrote to Hans Bethe, who at that time was trying to obtain a position outside Nazi Germany: "The positive electrons certainly have their origin in a collision between a light quantum and a nucleus, that is, a process which in the Dirac theory can be described as a photoelectric effect from the negative to the positive part of the energy spectrum."[118] Three weeks later, Peierls reported that he had stopped working with the theory of metals and had begun to study positron theory: "Recently . . . I have roamed about the Dirac holes = Blackett electrons. Perhaps this Dirac theory makes reasonable sense in some approximation."[119]

The most desired reward in science, the Nobel Prize, was awarded to Dirac in 1933 for his "discovery of new fertile forms of the theory of atoms and for its applications," as the Nobel Committee put it. Dirac was first suggested for the 1929 Nobel Prize by the Viennese physicist Hans Thirring, who also suggested de Broglie and, as his first choices, Heisenberg and Schrödinger.[120] However, the time was not yet ripe for honoring the founders of quantum mechanics with a prize. According to a statement from the committee in 1929, the theories of Heisenberg and Schrödinger "have not as yet given rise to any new discovery of a more fundamental nature"![121] De Broglie alone received the prize for 1929. In 1933, two prizes in physics were to be awarded, one for 1932 and one for 1933. William Bragg again suggested Schrödinger, Heisenberg, and Dirac. "It is so difficult to distinguish between these men that I wondered whether it would be possible to establish a new precedent by dividing the prize between the three, particularly as a prize was not awarded last year. I feel that such an award would be generally considered as just and would give universal pleasure."[122] The committee in Stockholm decided to follow Bragg's proposal, awarding Heisenberg the full prize for 1932 and dividing the prize for 1933 between Schrödinger and Dirac (Pauli, who was also a candidate in 1933, had to wait another twelve years). The decision was announced on November 9, 1933.

Carl Wilhelm Oseen, professor at the University of Uppsala and a close friend of Bohr, wrote the memorandum on Dirac for the Nobel Committee. In his careful, twenty-eight–page evaluation, Oseen stressed that Dirac, though an original and productive scientist, was not really a pioneer with respect to the foundations of physics:[123]

However highly one must value Dirac's work, still it remains that this work is not fundamental in the same sense as Heisenberg's. Dirac is in the front rank of the group of researchers who have set themselves the task to realize Heisenberg's bold thought. Compared to Born and Jordan he is independent. The data, just mentioned, shows this and a study of the papers confirms it. But Dirac is a successor in relation to Heisenberg.

It was no doubt the 1928 theory of the electron – which was, according to Oseen, "the work which so far has most contributed to his fame" – that contributed most heavily to Dirac's Nobel Prize. But he was awarded the prize for his entire production since 1925, not for any particular work. In the conclusion of his memorandum, Oseen had this to say to the Nobel Committee:[124]

If one asks if Dirac is a scientific pioneer of the same dimensions as a Planck, an Einstein, or a Bohr, the answer to this question must for the present be, I think, a definite no. But then it should be recognized that it does not only depend on the scientist himself if he becomes a great pioneer. It also depends on the time in which he lives. When Dirac opened his eyes to the world of science he undoubtedly saw the development of Heisenberg's thought as the most important and immediate task. Dirac committed himself to this task with life and soul. So far it has not left him the time for really great innovative work. But it is not at all impossible that such work may still come. It is noteworthy that Dirac's most original papers stem from the last years.

One may question Oseen's rather critical evaluation of Dirac's contributions to physics. It indeed seems that Oseen underestimated the revolutionary character of Dirac's work. This character is more visible today than it was in 1933, but even then most specialists in quantum theory recognized Dirac as an innovative genius comparable to a Bohr or a Planck.

In 1933, the Nobel physics committee consisted of five Swedish physicists, of whom only two – Oseen and Erik Hulthén – had a strong command of modern quantum theory. The other three members were H. Pleijel from the Royal Institute of Technology; Manne Siegbahn, an X-ray specialist and himself a Nobel laureate; and Vilhelm Carlheim-Gyllensköld, a 74-year-old physicist with no contact to modern physics. The Nobel Committee recommends candidates on the basis of suggestions from a number of nominators, but it has no obligation to follow the majority of nominators. Moreover, the final decision does not lie with the committee but with the plenary session of the Swedish Royal Academy of Science. Had it not been for these rules, Dirac probably would not have won his prize in 1933. He was nominated by only two physicists (Bragg and Bialobrzeski) and by neither of them as first choice. By contrast, Schrödinger received eleven nominations and was suggested by, among others, Einstein, Bohr, Oseen, Franck, Maurice de Broglie, and Louis de Broglie. Other nominees for 1933 included Sommerfeld, Percy Bridgmann, Clinton Davisson, and Friedrich Paschen, all of whom received more nominations than Dirac.

One would imagine that any thirty-one–year–old scientist would enthusiastically welcome a Nobel Prize. But Dirac was never much for rewards and honors. At first he did not want to accept the distinguished Swedish prize; he feared the publicity it would inevitably bring with it. Only after Rutherford made clear to him that a refusal would bring even more publicity did Dirac accept it.[125] But, of course, the Nobel Prize did make Dirac a public person, much to his dismay. A London newspaper portrayed him under the headline "The genius who fears all women" and described him to be "as shy as a gazelle and modest as a Victorian maid."[126]

Physicists welcomed the fact that the Nobel Prize had been awarded for the creation and development of quantum mechanics, which was by far the most important physical theory of the preceding decade. Dirac received many letters of congratulation, one of them from Niels Bohr:[127]

I need not repeat my heartiest congratulations to this well merited appreciation of your extraordinary contributions to atomic physics, but I should like to take this external occasion as an opportunity of expressing the deepfelt happiness I have felt by witnessing through the work of the younger generation the realization beyond any expectation of dreams and hopes as regards the development of the atomic theory which I have cherished through so many years. Not least it has been an extreme pleasure to me to come in so close contact and friendship with you, which I hope will grow still more in the coming years.

Dirac was moved by Bohr's letter and answered in words that, from his pen, were unusually emotional:[128]

Many thanks for your very nice letter – so nice that I find it a little difficult to answer. I feel that all my deepest ideas have been very greatly and favourably influenced by the talks I have had with you, more than with anyone else. Even if this influence does not show itself very clearly in my writings, it governs the plan of all my attempts at research.

To accept the prize, Dirac went to Stockholm in company with his mother. He was permitted by the Nobel Committee to invite both of his parents, but he did want his father to go with him. He could not forget the traumatic experiences of his childhood, for which he blamed his father, with whom he came to want to have as little contact as possible. In the Swedish capital Paul Dirac received the Nobel Prize from the Swedish king on December 11 and delivered the traditional Nobel lecture, which dealt with the theory of electrons and positrons. On his way back to England, Dirac and his mother spent a few days in Copenhagen after being invited by Bohr to stay at his new residence, the Carlsberg mansion.[129] Heisenberg and his mother also stayed with the Bohr family on the way back from Stockholm.

Nobel Prize winners have the right to nominate scientists for future Nobel Prizes. Most Nobel Prize winners take advantage of this right, realizing that the award is an important instrument of science policy. As far as I know, Dirac never nominated anyone.

CHAPTER 6

QUANTA AND FIELDS

IN 1925, the problem of supplying the electromagnetic field and, in particular, blackbody radiation with a proper explanation in terms of quantum theory was an old one. Earlier theories, such as those due to Ehrenfest (1906), Debye (1910), and Einstein (1909, 1916–17), were only partly satisfactory and, when quantum mechanics came into being, they were quickly superseded by new theories that pioneered what is now known as quantum field theory or quantum electrodynamics (I shall use the two terms without discriminating between them). Before looking at Dirac's important contributions to this area, we shall briefly review earlier contributions, starting with Einstein's work, which was of particular importance in the period.[1]

In 1909, Einstein showed that the energy fluctuations of a blackbody radiation enclosed in a cavity of volume V could be written as a sum of two terms, one referring to the quantum properties of the radiation, and the other to its wave properties.[2] For the mean square energy fluctuation, Einstein found the expression

$$\langle (E - \langle E \rangle)^2 \rangle = \langle E \rangle h\nu + \frac{c^3 \langle E \rangle^2}{8\pi\nu^2 d\nu V} \tag{6.1}$$

In an important paper of 1917, he went a step further.[3] He introduced the probability coefficients of induced emission and absorption (B_{mn}, B_{nm}) and of spontaneous emission (A_{mn}). For the rate of absorption processes ($n \rightarrow m$), he assumed the formula

$$\frac{d}{dt} W_{nm} = N_n \rho B_{nm} \tag{6.2}$$

where N_n is the number of oscillators in the energy level E_n and ρ is the spectral density, that is, the radiation energy per unit volume in the frequency interval between ν and $\nu + d\nu$. The corresponding result for emission processes $(m \rightarrow n)$ was assumed to be

$$\frac{d}{dt} W_{mn} = N_m(\rho B_{mn} + A_{mn}) \qquad (6.3)$$

By means of these two expressions, Einstein was able to give a simple derivation of Planck's radiation law. He also showed that the A and B coefficients were related by

$$A_{mn} = B_{mn} \frac{8\pi h\nu^3}{c^3} \qquad (6.4)$$

but he was unable to calculate the quantities in terms of quantum theory. This would require, he remarked, "an exact theory of electrodynamics and mechanics," and such a theory still did not exist.[4] In the old quantum theory Einstein's radiation probabilities played a part in particular in connection with the correspondence principle. By means of correspondence arguments, Bohr was able to show in 1918 that the Einstein probabilities were associated with the electric dipole moment of the radiators by

$$A_{mn} = \frac{(2\pi)^4 \nu^3}{3hc^3} |P_{mn}|^2 \qquad (6.5)$$

where P_{mn} denotes the Fourier coefficients of the dipole moment. However, Bohr's result, based on the correspondnece principle, could only claim validity in the region of large quantum numbers. It turned out to be impossible to extend it to small quantum numbers within the quantum theory prior to 1926.

Shortly after Heisenberg's invention of quantum mechanics in the summer of 1925, the new kinematics was applied to the radiation problem. The leading force in these preliminary attempts to establish a new quantum theory of the electromagnetic field was Jordan, already a specialist in the quantum theory of radiation.[5] In two of the pioneering papers of quantum mechanics, written in the fall of 1925, one with Born and one with Born and Heisenberg, Jordan applied the idea of noncommutative quantum theory to the electromagnetic field.[6] In the *Dreimännerarbeit* he showed that matrix mechanics was able to deal with electromagnetic radiation in a cavity, when the radiation was considered as an ensemble of Planck oscillators with energy $E = (n + \frac{1}{2})h\nu$. Among other things, Jordan derived Einstein's formula for the energy fluctuations of a radiation field, equation (6.1). Although Planck's formula had been derived several times earlier from ideas of field quantization (first by Debye in 1910), it had not been possible to obtain the fluctuation formula from first principles with the old quantum theory. Jordan's result was therefore highly valued as "particularly encouraging with respect to the further development of the theory."[7] It was a promising start, but the radiation theory turned out to be premature and difficult to develop further. Jordan's theory dealt with quantization of free waves only and was unable to throw light on the Einstein coefficients. Neither did Jordan succeed in providing an explanation of the new statistics of Bose and Einstein in terms of quantum mechanics. What was needed in order to establish a proper quantum field theory was a theory of the interaction of radiation and atoms. Such a theory was created by Dirac in February 1927 in what has been generally recognized (though this is rather unfair to Jordan) to be the founding paper of modern quantum electrodynamics.[8]

As demonstrated by his first contributions to physics, Dirac was from an early date interested in radiation theory. As soon as the formal basis of quantum mechanics had been reasonably well established, it was natural for him to turn to the connection – hitherto cultivated only by Jordan – between radiation and the new mechanics. Dirac's pioneering work, entitled "The Quantum Theory of the Emission and Absorption of Radiation," was written at the end of his stay at Bohr's institute, that is, in January–February 1927. In the introduction to this paper, Dirac wrote:

The new quantum theory, based on the assumption that the dynamical variables do not obey the commutative law of multiplication, has by now been developed sufficiently to form a fairly complete theory of dynamics. One can treat mathematically the problem of any dynamical system composed of a number of particles with instantaneous forces acting between them, provided it is describable by a Hamiltonian function, and one can interpret the mathematics physically by a quite definite general method. On the other hand, hardly anything has been done up to the present on quantum electrodynamics. The questions of the correct treatment of a system in which the forces are propagated with the velocity of light instead of instantaneously, of the production of an electromagnetic field by a moving electron, and of the reaction of this field on the electron have not yet been touched.

However, Dirac's theory of February 1927 was not his debut in quantum electrodynamics. The theory should rather be seen as the development of a more preliminary theory that went back to the summer of 1926. In what follows we shall briefly look at the main content, as far as radiation theory is concerned, of the papers that constituted Dirac's trilogy on quantum electrodynamics.

Radiation theory was the subject of the last section of the important paper "On the Theory of Quantum Mechanics" (see also Chapter 2). There Dirac considered a system of atoms subjected to an external perturbation that could vary arbitrarily with the time. Of course, the particular perturbation he had in mind was an incident electromagnetic field, but, characteristically, he stated the problem in the most general way possible. For the undisturbed system governed by the Schrödinger equation $(H - W)\psi = 0$, the wave function for a particle was expanded in terms of eigenfunctions, $\psi = \Sigma c_i \psi_i$. In the same way, the wave function of the perturbed system, governed by the wave equation $(H - W + A)\psi = 0$, where A is the (electromagnetic) perturbation energy, was written as $\Sigma a_i \psi_i$. The absolute squares of the expansion coefficients, $|c_n(t)|^2$ and $|a_n(t)|^2$, were interpreted as the number of atoms in state n at time t. Specifying the perturbation to be an electromagnetic wave, Dirac proved that the number of transitions from state n to m caused by the incident radiation could be written as

$$\frac{8\pi^3}{3h^2c} |P_{nm}|^2 |c_n|^2 I_\nu \qquad (6.6)$$

where P_{nm} denotes the appropriate matrix component of the total polarization and I_ν is the intensity of frequency ν per unit frequency. For induced emission from state m to n, the same expression was shown to hold, with $|c_m|^2$ instead of $|c_n|^2$. Since $|c_i|^2$ was interpreted as the number of atoms in state i, one could write the probability for either process as

$$B_{nm} = B_{mn} = \frac{2\pi}{3h^2c} |P_{nm}|^2 \qquad (6.7)$$

a result that was, as Dirac remarked, "in agreement with the ordinary Einstein theory," that is, with the quantum mechanical derivation of the B coefficients that occurred in Einstein's theory of 1917. Since he made use of a classical description of the electromagnetic field, Dirac was not at the time able to proceed further, and he noted, "One cannot take spontaneous emission [i.e. the A coefficients] into account without a more elaborate theory. . . ."[9] This more elaborate theory was ready less than half a year later. In the meantime, a wave mechanical calculation of the Einstein B coefficients, similar to that made by Dirac, was independently published by Slater, who also treated a special case of spontaneous emission.[10] Slater referred to Dirac's paper in a note added in proof, and he there acknowledged Dirac's priority as far as the absorption problem was concerned.

According to Dirac's recollections, the fundamental idea of his 1927 radiation theory was obtained unsystematically and unexpectedly. It was "one of those ideas out of the blue" from which so much of his science originated.[11] The work was the result of "playing about with [the Schrödinger equation] . . . [and] seeing what happens when you make the wave functions into a set of noncommuting variables."[12] This idea out of the blue led to what was later called "second quantization."[13] Instead of treating the energies and phases of radiation as c-numbers, as he had done in his 1926 paper, Dirac treated them as q-numbers. In general, second quantization involves considering the wave function as an operator instead of just a number. In working out his radiation theory of 1927, Dirac consciously designed it to accord with his recently completed transformation theory. The radiation theory can be regarded as the first major fruit of the program that had begun with the transformation theory, the second fruit becoming the relativistic theory of the electron a year later.

Dirac worked with a representation (later known as "number representation") based on the q-numbers N_r and θ_r, where N_r is the probable number of particles in state r and θ_r is a phase quantity canonically conjugate to N_r. He then defined a more convenient set of variables

$$b_r = \sqrt{N_r + 1}\; e^{-(i/\hbar)\theta_r} = e^{-(i/\hbar)\theta_r} \sqrt{N_r}$$

and

$$b_r^\dagger = \sqrt{N_r}\, e^{(i/\hbar)\theta_r} = e^{(i/\hbar)\theta_r} \sqrt{N_r + 1}$$

These new variables are canonical too:

$$[b_r, b_s^\dagger] = \delta_{rs} \quad \text{and} \quad [b_r, b_s] = 0$$

The b and N variables are furthermore related by the formula

$$N_r = b_s^\dagger b_r$$

Dirac interpreted the new variables, defined above, as absorption (b) and emission ($b^\dagger$) operators, although at the time he did not use this terminology; and N_r was taken to be the occupation number operator, specifying the number of systems in the state r. The eigenvalues of N_r are the whole numbers $n = 0, 1, 2, \ldots$. The effect of the Dirac operators is as follows:

$$b_r^\dagger \psi(n_1, \ldots, n_r, \ldots) = \sqrt{n_r + 1}\, \psi(n_1, \ldots, n_r + 1, \ldots)$$
$$b_r \psi(n_1, \ldots, n_r, \ldots) = \sqrt{n_r}\, \psi(n_1, \ldots, n_r - 1, \ldots)$$
$$N_r \psi(n_1, \ldots, n_r, \ldots) = n_r \psi(n_1, \ldots, n_r, \ldots)$$

Dirac applied this formalism to a system of photons; in this case, $b_r^{\dagger}$ has the effect of creating an additional photon in state r while none of the other photon states are affected, b_r works similarly by destroying a photon in state r, and N_r leaves the occupation number unchanged. In order to calculate the probabilities of absorption and emission of a photon, he constructed a Hamiltonian that described the interaction between a photon and an atom. The Hamiltonian had to incorporate the fact that photons are not conserved but may be spontaneously created or destroyed. Dirac included this feature in his Hamiltonian by introducing unobservable, or spurious, photons, that is, photons with zero energy and momentum:[14]

When a light-quantum is absorbed it can be considered to jump into this zero state, and when one is emitted it can be considered to jump from the zero state to one in which it is physically in evidence, so that it appears to have been created. Since there is no limit to the number of light-quanta that may be created in this way, we must suppose that there are an infinite number of light-quanta in the zero-state.

Dirac introduced the idea of zero-state photons because he believed such entities were necessitated by formal reasons. From the point of view of methodology, it should be noted that unobservable, zero-state photons were dubious entities according to the positivist, Heisenberg-inspired view of physical theory that Dirac by and large accepted. This only illustrates that Dirac's positivism was not strict. Also worth noticing is that in 1930, when introducing the hole theory of electrons, Dirac proceeded in a way that was conceptually similar to the way he followed in formulating his radiation theory (see Chapter 5).

Armed with his Hamiltonian, Dirac was able to calculate by means of perturbation theory the probabilities of absorption, emission, and scattering of a photon originally in a state r. He proved that these probabilities were proportional to n_r, $n_r + 1$ and $n_r(n_s + 1)$, respectively.[15] The latter result agreed with results obtained by Pauli on the basis of the old quantum theory.[16] The other results led to Einstein's law of radiation. The ratio between the emission and absorption probabilities is $(n_r + 1)/n_r$. Dirac further showed that the intensity was related to the number of photons by $I_r - (h^3/c^2)n_r$, so that the ratio is then

$$I_r^{-1}\left(I_r + \frac{h\nu^3}{c^2}\right)$$

Turning to Einstein's theory of 1917, equations (6.2) and (6.3), the same ratio can be written

$$\rho^{-1} \left(\rho + \frac{A_{mn}}{B_{mn}} \right)$$

where use has been made of $B_{mn} = B_{nm}$. Since the intensity and density functions are related by $I = \rho c/4\pi$, Dirac's theory gave the result

$$A_{mn} = B_{mn} \frac{4\pi h \nu^3}{c}$$

which, apart from a factor of 2, was Einstein's old formula. But the difference in the factor of 2 was not important; as Dirac pointed out, it was merely a result of two different ways of accounting for the polarization components.

In the last section of his paper, Dirac turned to consider the interaction between an atom and radiation from the wave point of view. He again calculated the Einstein coefficients, getting the same result he had obtained when using the photon point of view. He concluded that[18]

> ... the Hamiltonian which describes the interaction of the atom and the electromagnetic waves can be made identical with the Hamiltonian for the problem of the interaction of the atom with an assembly of particles moving with the velocity of light and satisfying the Einstein–Bose statistics, by a suitable choice of the interaction energy for the particles. ... There is thus a complete harmony between the wave and light-quantum descriptions of the interaction.

Also in his radiation theory, Dirac competed unknowingly with John Slater, who, in fact, submitted his own theory four days before Dirac. But, as Slater later admitted, "It was obvious that I would never catch up with Dirac to the point of being clearly ahead of him."[19] Rather than compete with Dirac, Slater decided to switch his focus to other problems of quantum physics. He did so with success.

Shortly after leaving Copenhagen for Göttingen, Dirac further developed his theory of radiation. In Göttingen he wrote up what may be considered the third part of his trilogy on quantum electrodynamics. The subject, dispersion, was one of the classic ones of quantum theory. In the creation of matrix mechanics, dispersion had played an important role, particularly as it was analyzed by Kramers and Heisenberg in the winter of 1924–5 (that is, before the advent of quantum mechanics). They managed to derive a satisfactory dispersion formula based on a sophisticated use of Bohr's correspondence principle, and in 1925 Heisenberg derived the same formula on the basis of matrix mechanics. As soon as wave mechanics appeared, this theory was also applied to the dispersion problem, first by Schrödinger, who derived a formula similar to that of Kramers and Heisenberg. Schrödinger's theory was further developed by other physicists who made use of the Klein–Gordon version of relativistic wave mechanics.[20]

Dirac was aware that yet another derivation of the dispersion formula would not be of much interest. But he was not satisfied with the theoretical basis of the earlier dispersion theories. "These methods," he wrote, referring to the wave mechanical theories of Schrödinger, Gordon, and Klein, "give satisfactory results in many cases, but cannot even be applied to problems where the classical analogies are obscure or non-existent, such as resonance radiation and the breadths of spectral lines."[21] With regard to the latter problem, Dirac was already able to supply an answer in February. He reported to Bohr that he had calculated from his radiation theory the probable distribution in frequency of an emitted radiation: "I have been able to integrate the equations of motion for the interaction of an atom and a field of radiation in a certain simple case, and thus obtain an expression for the breadth of a spectral line on the quantum theory."[22] In this letter to Bohr, Dirac went over the main steps in his calculation and wrote down his result for the frequency distribution as

$$\frac{|\alpha|^2}{h^2} \frac{1}{\gamma^2 + 4\pi^2(\nu - \nu_0)^2}$$

where γ is one half of the inverse lifetime and $|\alpha|^2$ is a quantity proportional to γ. Remarkably, Dirac did not publish this important result, which was only rediscovered three years later by Weisskopf and Wigner.[23]

Two months after his spectral line width calculation, Dirac presented a complete theory of dispersion, including derivations of the Kramers–Heisenberg formula and the Thomson formula for scattering of radiation by atoms. He was also able to treat the case of resonance, which theretofore had eluded quantum radiation theory. In Copenhagen, Dirac's work was followed with interest by Klein, who had just published an ambitious theory in which Compton scattering and dispersion were treated by means of five-dimensional relativistic wave mechanics.[24] Klein had derived a dispersion formula similar to the Kramers–Heisenberg formula, but Dirac did not find Klein's result or approach satisfactory. He wrote Klein "a very curious letter" in which he gave his own dispersion formula and argued that Klein's result, because it was based on a wrong theory, was not identical with the empirically verified Kramers–Heisenberg result. In a letter responding to Dirac's objections, Klein showed that his dispersion formula differed from that of Kramers and Heisenberg only in form, and that Dirac's objection was therefore unfounded. "I felt a little proud because it was very rare that one found an error in Dirac's work," Klein later recalled. Referring, in his reply, to Dirac's work on radiation theory, he wrote: "Thank you for the reprint of your last paper, which I read with great interest. I think it has quite convinced me that the quantum field theory ought to come on the lines you trace there."[25]

In his dispersion paper Dirac introduced a concept of great importance, later known as the concept of "virtual states." These states appeared in his discussion of "double scattering processes" – as opposed to "true scattering processes" – which he introduced as follows:[26]

Radiation that has apparently been scattered can appear by a double process in which a third state, n say, with different proper energy from m and m', plays a part. . . . The scattered radiation thus appears as the result of the two processes $m' \rightarrow n$ and $n \rightarrow m$, one of which must be an absorption and the other an emission, in neither of which is the total proper energy even approximately conserved.

In his paper, Dirac began, as usual, by looking for a Hamiltonian for the system under consideration, an atom perturbed by radiation. The atom was considered to be a single electron moving in an electrostatic field, and the perturbing radiation was given by a magnetic vector potential $\vec{A}$. In this case the classical-relativistic Hamiltonian is

$$ H = c \sqrt{ m_0^2 c^2 + (\vec{p} + \frac{e}{c} \vec{A})^2 } - e\phi $$

where ϕ is the scalar potential of the electrostatic field. However, Dirac did not use the relativistic Hamiltonian but only the approximation

$$ H = H_0 + \frac{e}{c} \vec{v} \cdot \vec{A} + \frac{e^2}{2m_0 c^2} \vec{A}^2 $$

where H_0 is the unperturbed Hamiltonian. Following a procedure unlike that used in earlier papers, he now quantized the vector potential by writing its components essentially as $N_r \cos(\theta_r/\hbar)$ and substituting into the classical Hamiltonian. This step amounted to considering the vector potential itself as a q-number, or an operator – a completely new idea. As Gregor Wentzel described it many years later, "Today the novelty and boldness of Dirac's approach to the radiation problem may be hard to appreciate. . . . Dirac's explanation in terms of the quantized vector potential came as a revelation."[27]

In Dirac's calculations mild forms of divergent integrals turned up. He handled these infinities with a characteristic lack of worry. In fact, he was quite ready to pass over the inconvenient integrals without examining them in detail, assuming that in a less approximate version of the theory the divergences would disappear. Statements like "we again obtain a divergent integral . . . which we may assume becomes convergent in the most exact theory" occur repeatedly in the paper.[28] As we shall soon see, Dirac's relaxed attitude toward infinities did not last long.

Dirac's publications on quantum electrodynamics in 1927 completed the scheme of quantum mechanics. At the same time, they initiated a new field of research that soon was to move to the forefront of theoretical physics. The strength of Dirac's theory lay in its conceptual innovations rather than in its ability to yield new physical results. The results obtained by Dirac, such as Einstein's radiation laws and the Kramers–Heisenberg formula of dispersion, were well known in 1927. Dirac supplied them with new and more satisfactory derivations, but he could not fail to recognize that to some extent his theory was like a new bottle containing old wine. His ambition reached higher, and consequently he was not entirely satisfied: "It was a bit of disappointment to find that nothing really new came out of the idea [of second quantization]. I thought at first it was a wonderful idea and was very much looking forward to getting something really new out of it, but it turned out to be just a new way of going back to the idea of an assembly satisfying Bose statistics."[29] In addition, he was fully aware of the limitations of the theory. For one thing, it was not relativistically invariant, and for another, it was restricted to particles satisfying Bose–Einstein statistics, that is, photons. In his three papers on quantum electrodynamics from 1926–7, Dirac did not attempt to extend his method also to cover Fermi–Dirac particles such as electrons.

Dirac's radiation theory served as the foundation in the early phase of quantum electrodynamics, from 1927 to 1932. Attempts to build up a relativistic quantum theory of matter and fields, based on the pioneering contributions of Jordan and Dirac, were made by a small but growing number of physicists. The majority showed little interest in the field, which seemed quite frightening because of its conceptual difficulties and mathematical complexity. Quantum electrodynamics was for some years a rather mysterious theory that was only understood by a few specialists. This situation remained until Fermi published an extensive review of the theory in 1932, presenting it in a clear and pedagogical way that made it accessible to a larger part of the physics community.[30] Slater recalled the advent of Fermi's article as follows: "I was very much heartened to hear some time later that Fermi had written his paper, which was a set of lecture notes for some lectures he had given at the University of Rome, because he had also not been able to get anything out of Dirac's paper [of

1927], and felt he would never understand it until he had worked it out for himself."[31]

Attempts to build a comprehensive relativistic theory that included quantization of radiation as well as of matter waves had been initiated even before Dirac's quantum electrodynamics appeared in print. In February 1927, Pauli proposed an ambitious program that was developed during the following years; he outlined it in a letter to Heisenberg, who answered:[32]

I strongly agree with your program regarding electrodynamics, but not completely with the analogy: Qu[antum]-W[ave]-Mechanics : classical mech[anics] = Qu[antum]-electrodyn[amics] : class[ical] Maxw[ell]-Th[eory]. I certainly believe that one should quantize the Maxwell equations in order to obtain light quanta, etc. a la Dirac, but perhaps one should then quantize de Broglie waves, too, in order to get charge and mass and statistics (!!) of the electrons and nuclei.

Pauli joined forces with Jordan and, a little later, with Heisenberg. A fourth member of the informal team was Oppenheimer, who at the time was in Göttingen and later worked with Pauli in Zürich. The first fruit of the program appeared in 1927 with a paper written by Jordan and Klein.[33] Considering a system of momentarily interacting particles, they succeeded in quantizing the matter waves and setting up commutation relations between the ψ-waves. Their results were found largely by copying Dirac's method. Jordan also generalized Dirac's method of second quantization to include a gas of Fermi–Dirac particles (such as electrons) and developed the theory further in collaboration with Wigner.[34] Jordan and Wigner took over the creation and destruction operators from Dirac's theory but modified them to be applicable to fermions. They introduced 2×2 matrix operators given by

$$
\begin{aligned}
b_r &= e^{-(i/\hbar)\theta_r}\sqrt{N_r} = \sqrt{1 - N_r}\, e^{-(i/\hbar)\theta_r} \\
b_r^\dagger &= \sqrt{N_r}\, e^{-(i/\hbar)\theta_r} = e^{-(i/\hbar)\theta_r}\sqrt{1 - N_r} \\
N_r &= b_r^\dagger b_r
\end{aligned}
$$

where N_r was again interpreted as an occupation number operator. The b_r operators were shown to satisfy the anti-commutation relations

$$
\{b_r, b_s^\dagger\} = \delta_{rs}, \qquad \{b_r, b_s\} = \{b_r^\dagger, b_s^\dagger\} = 0
$$

where the symbol $\{a, b\}$ stands for $ab + ba$. The above relations should be compared with the corresponding relations for bosons in Dirac's theory. For the Jordan–Wigner operators it follows that

$$N_r N_r = b_r^\dagger b_r b_r^\dagger b_r = b_r^\dagger (1 - b_r^\dagger b_r) b_r = N_r$$

or

$$N_r(N_r - 1) = 0$$

so that N_r must be either one or zero. The natural conclusion is that state r can accommodate either one electron or none. In this way, Jordan and Wigner proved that Pauli's exclusion principle, valid only for fermions, was incorporated in their theory.

The Jordan–Wigner theory did not appear until the spring of 1928, but Dirac knew it from a copy Pauli sent him before publication.[35] The essence of Jordan's theory was discussed at the 1927 Solvay Congress, where Dirac first learned about it. He did not find it appealing because it did not live up to his standards of clarity and formal beauty. At discussions at the Solvay Congress, Dirac objected that Jordan's theory was artificial and ad hoc: "In order to obtain the Fermi statistics, Jordan had to use a peculiar method of wave quantization especially chosen to yield the desired result."[36] Much later, he gave the following account of how he felt about the Jordan–Wigner theory:[37]

When I first heard about this work of Jordan and Wigner, I did not like it. The reason was that in the case of the bosons we had our operators that were closely connected with the dynamical variables that describe oscillators. We had operators that had classical analogues. In the case of the Jordan–Wigner operators, they had no classical analogues at all and were very strange from the classical point of view. The square of each of them was zero. I did not like that situation. But it was wrong of me not to like it, because, actually, the formalism for fermions was just as good as the formalism I had worked out for bosons. I had to adapt myself to a rather different way of thinking. It was not so important always to have classical analogues for everything. . . . If there is a classical analogue, so much the better. One can picture the relationships more easily. But if there is no classical analogue, one can still proceed quite definitely with the mathematics. There were several times when I went seriously wrong in my ideas in the development of quantum mechanics, and I had to adjust them.

Another major contribution to quantum electrodynamics from the same period was made by Pauli and the indefatigable Jordan. At the beginning of 1928, they published a relativistically invariant theory of the free radiation field.[38] Among other results, they obtained Lorentz-invariant commutation relations for the electromagnetic field quantities.

Dirac and Jordan held markedly different views with regard to the wave-versus-particle interpretation of quantum electrodynamics. Jordan's research program was based on his aim to derive the corpuscular properties of radiation – and in a wider sense to explain the material particles themselves – in terms of the quantization conditions imposed on wave-propagating quantum fields. In 1927, Jordan stated his program as follows:[39]

The results obtained allow us to see that one may construct a quantum wave theory in which electrons are represented by means of quantum waves in three-dimensional space. . . . The basic fact of electronic theory, the existence of discrete particles of electricity, is explained as a characteristic quantum effect, or, in other words, it means that matter waves appear only in discrete quantum states.

Jordan continued for some time to develop this program. He wanted to break radically with classical physics, to explain material particles as a quantum effect, and hence to give priority to (quantum) waves over particles. Dirac's ideas were very different. He had no trouble in accepting the corpuscular nature of radiation and saw no reason that one should explain why material particles exist. To Dirac the discontinuity of matter was simply a primary property, a contingent matter of fact not deducible by means of second quantization. Dirac's insistence on the corpuscular nature of radiation (and matter) was a natural consequence of his general view of quantum theory.[40] According to this view, quantum theory was a generalization of classical Hamiltonian theory; since this theory was a corpuscular, mechanical theory, the quantum theory of radiation could not dispense with particles and merely picture them, as Jordan would have it, as epiphenomena of waves. Dirac's commitment to this view also helps explain why he refused to accept quantization procedures, such as Jordan's, which had no classical analogy. Dirac wanted to restrict the procedure of second quantization to electromagnetic fields, while Jordan, Wigner, and others applied it to particles too.

In February 1928, Pauli wrote to Dirac about the progress and problems of quantum electrodynamics: "In addition, Heisenberg and I have occupied ourselves . . . with the question of the relativistic invariant formulation of the electrodynamical interaction of particles."[41] Pauli and Heisenberg wanted to establish a very general, relativistic quantum field theory that would comprise all of the results theretofore obtained. They wanted to avoid earlier assumptions such as instantaneous interaction and restriction to the pure radiation field. However, Pauli had to admit that their ambition had not yet been fulfilled and that they had run into difficulties. He asked for Dirac's advice:

[I would like] to ask your opinion about a substantial physical difficulty, which appears in the system of Heisenberg and myself, and which we have not been able to eliminate. Our theory will only be complete and ready to be compared with experiment if we establish conservation laws of energy and momentum for the whole system (light and matter waves).

Pauli's difficulty was that the energy-momentum tensor, even for a single particle, turned out to be infinite, a situation about which he asked Dirac:[42]

What do you think about this? At the moment I know of no satisfactory way out. And my own impression is even that one would have to undertake deep changes in the foundation of our views in order to avoid these difficulties. Finally, in reality energy can hardly be pieced together additively from two parts (the contribution of matter waves and that of electromagnetic fields) which are logically independent of one another. On the whole, a conception which would allow light and matter waves to appear essentially identical (the former as a particular case of the latter) would be much to prefer.

When the Heisenberg–Pauli theory finally appeared, most of the difficulties had been cleared away, but by means of mathematical tricks with seemingly dubious physical significance.[43] The theory was indeed general and impressive, but it was also tedious and complicated. Seventy-six pages filled with complex mathematical formulae made it indigestible for the majority of physicists. Pauli and Heisenberg's tour de force led to canonical equations for the combined matter and radiation fields. They proved that all of the electromagnetic field quantities commuted with the matter field operators and that the general commutation rules embraced the more special commutation rules of earlier theories. But Pauli and Heisenberg did not succeed in presenting a theory that was entirely free of infinities. In particular, the self-energy of the electron turned out to be infinite in the theory. This was pointed out by Jordan at the Kharkov conference in May 1929, just after the Heisenberg–Pauli paper had appeared in print.[44] Jordan did not share the optimism of Pauli and Heisenberg, and shortly afterwards he largely stopped publishing on quantum electrodynamics.

The problem of quantization of matter waves was supplied with a new dimension in 1928 with Dirac's relativistic equation of the electron. To the extent that relativistic field equations of fermions had been previously examined, the Klein–Gordon equation was the only model used in theoretical speculations prior to 1928. Pauli and Heisenberg, for example, had originally based their theory on the similar relativistic equation

$$\left(\frac{\partial}{\partial x_k} + \frac{ie}{c}\phi_k\right)^2 \psi + m_0^2 c^2 \psi = 0$$

where ψ is a matter field operator.[45] As soon as the Dirac equation became known, the German physicists attempted to incorporate it into quantum field theory. In the above-quoted letter from Pauli to Dirac, which was written a few days after the appearance of Dirac's paper, Pauli remarked:[46]

In this connection I have thought about the question, *what is the relationship between your new quantum theory of the electron and your earlier quantum electrodynamics* (Vol. 114 of the Proceedings of the Royal Society)? If the principles underlying your quantum theory of the electron have general validity, one could perhaps expect that even with the introduction of the numbers N_r and phases θ_r of light quanta as q-numbers (in the sense of your earlier work) a Hamiltonian function should exist, which (1) is relativistically invariant, and (2) only includes the operators p_0, p_1, p_2, p_3 linearly. But I have not succeeded in finding such a Hamiltonian. *Do you think this would work?*

The Dirac equation and its corresponding Lagrangian were then incorporated into the part of the Pauli–Heisenberg theory that dealt with electrons and protons.

Dirac did not publish again on quantum electrodynamics for five years, partly because he was dissatisfied with the way the subject was developed by his German colleagues. As mentioned, he did not like Jordan's approach, and neither did the theory of Pauli and Heisenberg gain his approval. He examined carefully and critically the latter theory in private notes, in which he stated the following as among his reasons for disapproval: "These [practical applications] are rather disappointing since they make so many approximations that all the special relativistic features of the present paper disappear and they get results which could have been obtained from a much simpler non-relativistic theory."[47]

In March 1932, when Dirac returned publicly to quantum field theory, he was explicit in his rejection of the Heisenberg–Pauli approach, which he criticized from a methodological point of view.[48] By that time, Dirac had developed a characteristic style and structure in his papers that became a kind of personal stamp: He would begin with a lengthy introduction, discussing methodological problems in a broad and general way, without the use of mathematics; would then continue with the main part, presented in his usual concise mathematical style; and would typically conclude by applying his proposed theory to a simple example, usually too simplified to have any empirical relevance. In the 1932 paper, Dirac claimed that a proper conception of the nature of observation would provide a clue for a new and better foundation for quantum field theory:[49]

If we wish to make an observation on a system of interacting particles, the only effective method of procedure is to subject them to a field of electromagnetic radiation and see how they react. Thus the role of the field is to provide a means for making observations. *The very nature of an observation requires an interplay between the field and the particles.* We cannot therefore suppose the field to be a dynamical system on the same footing as the particles and thus something to be observed in the same way as the particles. The field should appear as something more elementary and fundamental.

Dirac hoped that this idea would supply the "simplification and unification which are entirely lacking in the Heisenberg–Pauli theory," in which the field itself was regarded as a dynamic system. He believed that Pauli and Heisenberg had betrayed the fundamental methodological principle that in their hands had proved so successful in 1925, namely, the operationalist principle that only observable quantities should appear in physical theories: "The Heisenberg–Pauli theory . . . involves many quantities which are unconnected with results of observation and which must be removed from consideration if one is to obtain insight into the underlying physical relations."[50]

Dirac thus hoped to establish a new basis for quantum electrodynamics by adopting an operationalist viewpoint that would be even more thoroughgoing than that of the German physicists. It is somewhat remarkable that "aesthetic" considerations – and perhaps even more surprising that "philosophical" motivations – provided the sole grounds for Dirac's dismissal of the Heisenberg–Pauli theory. Although Dirac was only concerned with methodology, implicitly his idea also involved an ontological commitment: If, in the context of observation, the field *appears* as "something more elementary and fundamental" than the particle, then it is but a small step to assume that it *is* more elementary. To explicitly draw such a realist conclusion, however, would not be consistent with the positivist spirit of Dirac's work.[51] Dirac was not a new Schrödinger. He did not believe that waves were ontologically primary to particles, although some physicists probably (and not without reason) believed he did.[52]

In the case of a single electron interacting with a field of radiation, Dirac considered the radiation to be resolved into an ingoing and an outgoing wave. While in the Heisenberg–Pauli theory problems that referred only to ingoing waves (or only to outgoing waves) were allowed, this was a meaningless notion according to the philosophy adopted by Dirac. He further assumed that the field was resolvable into plane waves, which implied that no Coulomb force was introduced. The equation of motion for a spinless electron was expressed as

$$F\psi = \left[\left(\frac{\hbar}{i}\nabla + e\bar{A} \right)^2 - m_0^2 c^4 \right]\psi = 0$$

The field is determined by the potentials, which were taken to be operators. In the case of two interacting electrons, Dirac had to do without introducing a Coulomb interaction energy in the wave equation. Instead he expressed the interaction by coupling the motion of both electrons to the same field, which was accomplished by requiring the wave function to satisfy the two equations

$$F_1\psi = 0 \quad \text{and} \quad F_2\psi = 0$$

Here $\psi = \psi(t_1, x_1, \ldots; t_2, x_2, \ldots)$, where each set of subscripts refers to one of the two electrons; and F_1 and F_2 are operators – corresponding to F in the equation for the single spinless electron – that can be obtained by substituting $\partial/\partial t_1, \partial/\partial x_1, \ldots$, and $\partial/\partial t_2, \partial/\partial x_2, \ldots$ for the $\partial/\partial t, \partial/\partial x,$ ... components of the ∇ operator in the previous equation. Dirac explained, "These two wave equations describe completely the relations between the two electrons and the field. No terms of the type of a Coulomb interaction energy are required in the operators of the wave equations."[53] As Dirac remarked, it may seem surprising that a theory with only plane waves could give the necessary electrostatic force between the electrons. But he proved that in the case of a one-dimensional, non-relativistic problem, this was indeed the case. Coulomb forces were contained implicitly in the theory and pictured as "vibrations of an intervening medium transmitted with a finite velocity."

Dirac's alternative quantum electrodynamics was no success in the physics community. It had little immediate impact and was soon shown to be formally equivalent to the Heisenberg–Pauli theory. Bohr reacted politely to Dirac's theory, but he was not especially interested in it; for, as he wrote to Dirac before ever seeing the completed theory, he was convinced of the soundness of the Heisenberg–Pauli formalism:[54]

I was very interested to hear from your letter that you hope soon to get a satisfactory theory of quantum electrodynamics. In the last few months, we have had here much discussion on the problem of the inner consistency of Heisenberg–Pauli's formalism and Rosenfeld and I hope soon to publish a paper in which we prove that the criticism of Landau and Peierls is wholly unfounded. In fact a closer study of the problem of the measurability of electromagnetic fields has revealed a complete agreement with the consequences of the formalism. Although this result was rather to be expected, the work, at least to ourselves, has been quite instructive, as we have been confronted with a lot of puzzles.

In April 1932, Dirac spent two weeks in Copenhagen, where he showed his new paper to Klein and asked him to read it. Klein recalled: "And when I turned the first page, Dirac said, 'You ought to read the paper more slowly; Heisenberg read it too fast.' And then I heard that Heisenberg had objected that this was just the old theory in a new form."[55] Not unexpectedly, the reaction from Germany was much sharper. Pauli, who was at the time Dirac's chief critic, rejected the theory totally. He believed, as he told Lise Meitner, that "it cannot be taken seriously; neither does it contain anything new, nor is it justified to speak of a 'theory.' "[56] When he addressed Dirac, his judgment was no less frank:[57]

Your remarks about quantum electrodynamics which appeared in the Proceedings of the Royal Society were, to put it gently, certainly no masterpiece. After a muddled introduction, which consists of sentences which are only half understandable because they are only half understood, you come at last, in an oversimplified one-dimensional example, to results which are identical to those obtained by applying Heisenberg's and my formalism to this example. (This equivalence is immediately recognizable and was then verified by Rosenfeld in an over-complicated way). This end of your work conflicts with your assertion, stated more or less clearly in the introduction, that you could somehow or other construct a better quantum electrodynamics than Heisenberg and I.

Having finished his work on the new quantum electrodynamics, Dirac went for a vacation to Norway, where he spent two weeks pursuing his favorite pastime, walking in the mountains. From Norway he went on to the USSR, where he discussed the problems of quantum field theory with the Russian physicists. While the physicists of the Copenhagen school reacted negatively to Dirac's theory, the Russian physicists found it interesting and promising. Vladimir Fock, Boris Podolsky, and K. Nikolsky developed aspects of the theory in several papers. For example, Fock and Podolsky extended Dirac's one-dimensional treatment of two interacting electrons to the more realistic case of three dimensions. They obtained the expected result, a Coulomb interaction term with the correct sign.[58]

At that time, Dirac knew that his new theory of quantum electrodynamics was mathematically equivalent to that of Pauli and Heisenberg, a fact that was made evident, for example, in Pauli's letter (quoted above). Even before Dirac's paper appeared in print, Bohr's associate Léon Rosenfeld had been able to prove the equivalence.[59] Dirac received this proof in April during his stay in Copenhagen, where he talked about his new theory and prepared a manuscript of it while also participating in a conference to celebrate the tenth anniversary of Bohr's institute.[60] Dirac must have felt disappointed after receiving Rosenfeld's equivalence proof. He was forced to accept it but still felt that his approach was physically superior to the standard quantum electrodynamics of Pauli and Heisenberg. The *conceptual* difference between the two theories – however *mathematically* equivalent they were – can be clearly seen in Dirac's reply to Rosenfeld:[61]

Dear Rosenfeld,

Thank you very much for the paper you sent me. I found it very interesting. The connection which you give between my new theory and the Heisenberg–Pauli theory is, of course, quite general and holds for any kind of field (not merely the Maxwell kind) in any number of dimensions. This is a very satisfactory state of affairs.

It does not seem certain to me that the singularities will cause an equal amount of trouble in both theories, because the factor which connects the two theories, namely the $e^{i\bar{H}_s x/hc}$ in your equation (31), will contain an infinity when there is a singularity in the field which makes $\bar{H}_s$ infinite. Could it not be so, that a mathematical process which is convergent for the one theory is divergent for the other? (Nur um zu lernen).

I have been studying the Heisenberg–Pauli theory again and find it difficult to understand why their formalism is invariant under a Lorentz transformation. I can follow all their arguments except the last sentence in § 5 on page 180 of Zeits.f.Phys. vol. 59. I should be very glad if you could explain the sentence a little more fully. . . . The paper that was in the library at Copenhagen is being published, together with the one-dimensional calculation. It will probably appear in this month's Roy.Soc.Proc. and you may expect to see it in a few days.

The paper, I believe, contains no mentioning of singularities and was merely intended to give a theory that is more closely connected with the results of observation than the preceding ones. I think you ought to publish your work.

Please give my kind regards to Prof. Bohr.

Yours sincerely,

P. A. M. Dirac

As one senses from the letter – and from the fact that he did not withdraw his paper – Dirac was not willing to admit that the mathematical equivalence implied a physical equivalence. He therefore continued to develop his approach, which half a year later resulted in a paper, co-authored by Fock and Podolsky, in the newly founded *Physikalische Zeitschrift der Sowietunion*. Dirac knew Fock and Podolsky from his travels to Russia, and Fock was an old acquaintance whom he had first met in the spring of 1927, during his stay in Göttingen. Podolsky was an American (but Russian born) theoretical physicist, who from 1931 to 1933 worked at the Ukrainian Physico-Technical Institute in Kharkov. The Dirac–Fock–Podolsky theory germinated in discussions Dirac had with Fock and Podolsky in September 1932, when they all attended a conference on the theory of metals held in Leningrad. After the conference Dirac went to the Crimea, where he vacationed with Kapitza, and on his way back to Moscow he stopped in Kharkov to discuss his new quantum electrodynamics with Podolsky. Fock and Podolsky had recently proposed a new formalism for the quantization of the electromagnetic field, which Dirac found more suitable for his purpose than the earlier formalism of Jordan and Pauli.[62] So he agreed to write a joint paper, which was completed in late October 1932.[63] The three authors derived the fundamental equations of quantum electrodynamics in a relativistically covariant way and proved that the equations yielded the Maxwell equations as conditions on the q-number wave function. In earlier formulations, such as the Heisenberg–Pauli theory, the covariance was far from clear, a result that Dirac traced back to a certain lack of symmetry between space and time coordinates in these theories; in the earlier formulations each electron was supplied with a separate space coordinate, but all particles were given the same time parameter. In the Dirac–Fock–Podolsky paper a lucid proof of Lorentz invariance was obtained by making use of the idea of multiple times: In addition to the common time for the entire system of particles and field (T), a separate field time (t) and separate times for each particle ($t_1, t_2, \ldots, t_n$) were introduced. The dependency of the wave function on the field time was stated as

$$\left(H_b - i\hbar \frac{\partial}{\partial t}\right)\psi = 0$$

where $\psi = \psi(t, t_1, t_2, \ldots t_n)$ and H_b is the field Hamiltonian in the absence of charges. In order to pass over to the usual one-time theory of Pauli and Heisenberg, Dirac and his co-authors considered the wave function in which the $n + 1$ times were equal to the common time, that is, $t = t_1 = t_2 = \ldots t_n = T$. They furthermore gave a proof of the equivalence between the Heisenberg–Pauli theory and Dirac's alternative theory that was more general and much more elegant than the "obscure" proof given by Rosenfeld. In this new proof Dirac, Fock, and Podolsky made use of a representation intermediate between the Schrödinger and Heisenberg representations. This representation was introduced by means of the canonical transformations

$$\psi^* = e^{(i/\hbar)H_bT}\psi$$
$$F^* = e^{(i/\hbar)H_bT}Fe^{-(i/\hbar)H_bT}$$

where ψ is a wave function, F an operator (in the Schrödinger representation), H_b the field Hamiltonian, and T the common time. The representation used by Dirac, Fock, and Podolsky later proved to be useful in cases of interaction in general, and today it is known as the "interaction representation" or "interaction picture," sometimes called the "Dirac picture."

While Pauli, as mentioned, did not think much of Dirac's 1932 theory, he was very interested in the Dirac–Fock–Podolsky theory, which he considered "a great improvement."[64] What impressed Pauli was not so much the physics of the paper as the mathematical elegance with which a manifestly covariant formulation of quantum electrodynamics was achieved. The many-time theory was regarded by the few specialists who took an interest in it, Pauli included, as merely a formal improvement of the standard Heisenberg–Pauli theory. Later in 1933, Pauli's former assistant Felix Bloch proved that, as far as physics is concerned, the Dirac–Fock–Podolsky theory was indeed equivalent to the Heisenberg–Pauli theory.[65] Bloch showed that if the many-time wave function was subjected to certain conditions, then the quantity $\psi\psi^*$ in the new theory could be interpreted in the same way as in usual quantum mechanics, that is, as a probability density. In effect, what Dirac had intended to be a radically new approach to quantum electrodynamics was shown, by the equivalence proofs of Rosenfeld and Bloch, to be just a reformulation.

Yet the formal innovations contained in Dirac's program proved to be important for the later development of quantum electrodynamics. When the emergence of modern renormalization techniques finally provided a breakthrough for the theory in 1947–8, Dirac's papers served as an important source of inspiration. Julian Schwinger, one of the architects of the new theory, was inspired by the Dirac–Podolsky–Fock formulation; he developed it greatly and also coined the term "interaction representation." Sin-Itoro Tomonaga, another of the fathers of modern quantum electrodynamics, was fascinated by Dirac's 1932 paper, which "attracted my interest because of the novelty of its philosophy and the beauty of its form."[66] The long and troublesome road toward renormalization thus took its start in aspects of Dirac's work – an irony, in the light of his later dislike of renormalization theory. Regarding the later development of quantum electrodynamics, another of his papers also deserves mention here, although its inspiration was somewhat delayed. In an investigation in 1933 of the formal quantum mechanical analogue of classical Lagrangian theory, Dirac argued that the Lagrangian method was in some respects more fundamental than the standard Hamiltonian method; but he did not apply his Lagrangian formulation of quantum mechanics to concrete physical problems.[67] Buried in the pages of the *Physikalische Zeitschrift der Sowietunion,* Dirac's Lagrangian theory was not much noticed in the thirties, but years later it was studied by the young Richard Feynman, who developed it into the space–time approach to quantum field theory for which he received the Nobel Prize.[68]

Dirac's occupation with the problems of quantum electrodynamics may be further illustrated by two papers of a mainly mathematical character written in the spring of 1933. Both were concerned with the possibility of generalizing the mathematical basis of quantum mechanics in order to overcome the difficulties of the existing quantum electrodynamics and secure its connection with classical dynamics. Dirac pointed out that in exceptional cases, which turned up in the theory of Heisenberg and Pauli, ordinary Hamiltonian methods were not applicable. In order to remove such "irregularities" and obtain "greater elegance," he proposed a reformulation of classical Hamiltonian dynamics by introducing homogeneous variables.[69] In another paper, read to the London Mathematical Society in June, Dirac argued that the existing quantum transformation theory was not suited for a relativistic treatment. He stated clearly his dissatisfaction with the existing Heisenberg–Pauli formulation of relativistic quantum mechanics, which, he wrote, "seems to be possible only if one introduces enormous complexity into the equations and sacrifices the directness of physical interpretation which was so satisfactory a feature of the non-relativistic theory."[70]

The fall of 1933 was occupied with travels and conferences. After his vacation in Norway, Dirac participated in a conference at Bohr's institute during September 14–20. Among other participants were Fermi, Heisenberg, Ehrenfest, and (or course) Bohr. This was the last time Ehrenfest would be among Dirac's colleagues in physics. Ehrenfest was seriously depressed, a fact of which Dirac was perceptively aware. He expressed his worry over Ehrenfest's mental state to Mrs. Bohr, but there was nothing they could do.[71] Five days after the conference ended, Ehrenfest committed suicide.

From Copenhagen Dirac traveled on to the First Soviet Conference on Nuclear Physics, which took place in Leningrad from September 24 to 30. There he reported on his latest work on his hole theory, which had by then become a theory of the positron. The hosts of the Leningrad conference were the Soviet physicists Fock, Iwanenko, and Tamm, all of whom Dirac knew well, and the cosmic ray specialist Dmitry Skobeltzyn; other participants included Francis Perrin from France, Guido Beck from Italy, and Victor Weisskopf and Otto Frisch from Germany. To Dirac's regret, Bohr was too busy to join the Leningrad conference. On August 20, Dirac had written to him:[72]

Dear Bohr,

Iwanenko has written to me to ask you to come to the Leningrad conference on nuclear physics beginning on September 25th. I expect you will already have heard of it directly from Leningrad. I am intending to go to it myself, after your conference in Copenhagen is finished, and I should like very much if we could travel there together. Probably you are very busy now, but it need not take up much more than about a week of your time. You may be sure of a warm welcome from the Russian physicists and I think you will find it interesting to see something of the modern Russia. (The economic situation there is *completely* different from everywhere else).

Peierls is working out the "polarisation" of the distribution of negative-energy electrons produced by a magnetic field. He thinks the results will turn out to be reasonable and in agreement with the electric field case.

I am looking forward very much to the Copenhagen conference. I am expecting first to take a holiday of about 2 weeks in Norway and then to arrive in Copenhagen about Sept. 14th.

With best wishes, and hoping you will find it possible to come to Leningrad.

Yours sincerely,

P. A. M. Dirac

After the conference ended, Dirac went on to Moscow, where he spent a few days. He was much impressed by what he saw in Russia, and noticed that the living standard and availability of consumer goods had greatly improved since his last visit. A few weeks after his return to England, Dirac left for Brussels to participate in his third Solvay Congress. This was the Seventh Solvay Congress, which took place from October 22 to 29. The main subject of the 1933 Solvay Congress was nuclear physics, which at that time had experienced revolutionary development: In 1930, Pauli had proposed the neutrino, at first believed to be a nuclear constituent. In late 1931, Harold Urey had discovered the deuteron, and shortly afterwards (in 1932) James Chadwick had discovered the neutron. Iwanenko had proposed in 1932 that Chadwick's neutron was a fundamental particle. In the same year, Heisenberg had published a proton-neutron theory of the atomic nucleus, John Cockcroft and Ernest Walton had reported nuclear transmutation with a high-voltage accelerator, and Ernest Lawrence and co-workers had constructed the cyclotron. And in 1933, Irène Joliot-Curie and Frederick Joliot had discovered artificial radioactivity. Dirac found the Seventh Solvay Conference interesting. He was, together with twelve other participants, invited to dine with the Belgian King Albert at the Palais de Laeken on October 27. Following the conference, he reported to Van Vleck:[73]

I have just been to another Solvay Conference, the subject being nuclei this time. Lawrence was there as representative of America, and he and Rutherford had some fine arguments together. I was one of those chosen to dine with the King this time. From where I was sitting I could not see whether he again drank two pitchers of water, but anyway I did (my own and a neighbour's). The conference was a most interesting one and there was rather more people there than usual. I have previously just been to Russia again, to a conference on nuclei in Leningrad, and I also visited Moscow again.

Although Dirac never jumped on the nuclear physics bandwagon, he showed at that time some interest in the field. After Chadwick's discovery of the neutron, there were several proposals about the constitution of the atomic nucleus. Some of them admitted α particles, protons, and neutrons as basic nuclear constituents; other included electrons and even neutrinos. According to Heisenberg, who lectured on his theory in Brussels, the nucleus contained only protons and neutrons. Dirac participated in the discussion following Heisenberg's report and defended the traditional view of nuclear electrons against the new ideas of Iwanenko and Heisenberg. He believed at the time that there were three kinds of elementary particles in the nucleus: protons, neutrons, and electrons. Having had to abandon his unitary "dream of philosophers" in 1930, he did not consider this a too crowded picture of the nucleus. "This number [three] may appear to be large, but, from this viewpoint, two is already a large number," he said.[74]

Dirac's insistence on nuclear electrons may seem old-fashioned for the fall of 1933, but it was not particularly so. It is only with hindsight that one can see Chadwick's discovery of the neutron in early 1932 to have made nuclear electrons obsolete. Chadwick himself believed in 1933 that what he had discovered was a composite particle, a proton–electron system. In Cambridge this view held considerable esteem, and at the Solvay Conference Chadwick had still not surrendered to the view of the neutron as a fundamental particle, which at the time was accepted by most physicists outside the Cavendish tradition.[75]

Another indication of Dirac's brief occupation with problems of nuclear physics can be seen in his unpublished argument that there might exist negative protons in nuclei. At that time the stability of the beryllium-9 isotope presented a puzzle, for its mass apparently exceeded that of two alpha particles and one neutron; therefore it would be expected to decay spontaneously. Gamow, who gave a talk on negative protons to the Kapitza Club in June 1934, wrote to Bohr that Dirac had suggested that instead of consisting of $(4p^+, 5n)$, the nucleus might consist of $(5p^+, 1p^-, 3n)$: "I spoke with Dirac about the stability of Berilium [sic] and he proposed the same point of view as Pauli in the discussion in Copenhagen: Be is stable because there is not enough neutrons to form two α-particles."[76] Remarkably, Dirac seems not to have thought of the negative proton as an anti-proton (in which case it would annihilate with an ordinary proton). Perhaps he then believed that there might be anti-protons as well as negative nuclear protons. The "beryllium anomaly" was resolved a year later when improved determinations of the nuclear masses proved that the decay process was, in fact, energetically forbidden.

For some years, Dirac continued to show a distant interest in problems of nuclear theory and also in other mainstream branches of physics. From

Dirac's publications alone, one might be tempted to infer that he was interested only in his own narrow field of specialization. But this was not really the case.[77] He just did not consider the new mainstream branches of physics, such as solid-state and nuclear physics, to be truly fundamental. And he did not want to engage himself in fields that he did not judge fundamental. Fashions never appealed to him.

Dirac had a predilection for general theories and fundamental problems, but he was not blind to the value of experimental physics. On one occasion he even involved himself in experiments. In March 1933, he discussed in a paper with Kapitza the possibility of obtaining electron diffraction from a grating of standing light waves, a quantum mechanical dual of the diffraction of light waves by a matter grating.[78] In dealing with the problem theoretically, Dirac and Kapitza considered two beams of light waves with the same frequency and moving in opposite directions, the spectral intensities of the beams being I_ν and I'_ν. An electron passing the beams with velocity v would not only be subjected to ordinary Compton scattering but also to stimulated Compton scattering, in which a photon absorbed from the initial beam would be re-emitted with the same frequency by the existence of the other, stimulating beam. Such stimulated scattering, which must occur according to Einstein's 1917 theory of radiation, had never been verified experimentally. Dirac and Kapitza calculated the probability of the stimulated deflection process to be

$$\frac{e^4 l}{2m^2c^2h^2v^4v} \int I_\nu I'_\nu \, d\nu$$

where l is the length of the electron's path in the radiation field and ν is the frequency of the absorbed or re-emitted photon. They estimated that the probability would only be of the order of 10^{-14} under ordinary experimental conditions. Consequently, they concluded that it would hardly be possible to detect the phenomenon, and they did not try to realize the experiment.

The next year Kapitza went to Russia, where he was forced to remain for thirty years. When he was awarded the Rutherford medal in 1966, he was at least able to visit Cambridge again, and in honor of the presence of the old Kapitza Club's founder, a special meeting of the club was arranged on May 10. Several of his old Cambridge friends participated, including Dirac. At this nostalgic meeting, Kapitza and Dirac reviewed their paper of 1933 in the light of new technological advances and concluded that the project was no longer unrealistic. Dirac met Kapitza for the last time in June 1979 at the Lindau meeting of Nobel Prize winners. Kapitza died one year later, at the age of eighty-six, and was paid tribute by Dirac in *Physics Today*.[79] As to the "Kapitza-Dirac effect," it was in fact observed, but only after both Kapitza and Dirac had died. In 1986, physicists at MIT studied the scattering of sodium atoms (rather than electrons) by a standing-wave laser field and detected the phenomenon predicted by Dirac and Kapitza fifty-three years earlier.[80]

An experimental project also resulted from Dirac's interest in the problem of how to separate a gaseous mixture of isotopes, a problem he began to explore in 1933–4. With the assistance of Kapitza, he invented a centrifugal method of separation that he attempted to develop experimentally. However, the work stopped when Kapitza went to Moscow with most of his equipment, and Dirac never published the method. Naturally, this experimental work by an arch-theorist, although it was left unfinished, pleased Rutherford, who always felt that experiment, not theory, should be the core of physics. "You may be interested to hear," Rutherford wrote Fermi, "that professor Dirac is also doing some experiments. This seems to be a good augury for the future of theoretical physics."[81] Writing Dirac from Moscow, Tamm also called attention to Dirac's experimental work and to that of two other great theorists: "I heard that you are making experiments on the centrifugation of gases. Well, Fermi and Bohr begin also to conduct experimental researches."[82]

Forays into nuclear physics and experiments aside, the problems of quantum electrodynamics contined to dominate Dirac's thoughts. He soon reached the conclusion that the existing theory was not satisfactory because it was haunted by infinities, a view he continued to hold throughout his life. Since the infinities played such an imporant role in the development of quantum electrodynamics, and in Dirac's research in particular, it is worthwhile to digress a little on this problem.

Basically, two related divergence problems turned up in quantum electrodynamics, one referring to the mass or energy of the electron, and the other to its charge. Of these, the mass divergence problem can be viewed as analogous to the old problem of the classical electron's electromagnetic mass, while the charge problem turned up only in 1933.

According to classical electrodynamics, the total field energy of an electron is

$$W = \frac{1}{8\pi} \int \vec{E}^2 dV$$

where $\vec{E}$ is the field strength and the integration is taken over the entire space outside the electron. The classical problem is that W involves a linearly divergent integral for the point electron (for which $E = e/r^2$ for $0 \leq r \leq \infty$). Using the classical theory, one can introduce various models for the structure of the extended electron in order to avoid the infinity; for example, if the charge is supposed to reside on the surface of a spherical electron with radius R, one gets

$$W = \frac{e^2}{2} \int_R^\infty \frac{dr}{R} = \frac{e^2}{2R}$$

In the framework of the theory of relativity, this contributes to the electron an electromagnetic mass, m_{EM}, in addition to its "mechanical mass," m_0:

$$m = m_0 + m_{EM} = m_0 + \frac{e^2}{2c^2R}$$

Here m is the experimentally determined mass. In pre-quantum theory the structure of the electron was widely discussed, within and without the theory of relativity, and there were many proposals about how much the electromagnetic mass might contribute to the observed mass. According to some theories, such as Abraham's, the entire mass was of electromagnetic origin.[83]

In quantum theory the notion of an extended electron became untenable, and hence the infinite self-mass became a serious, though at first not much noticed, difficulty. In early quantum mechanics the electron was usually conceived as a pointlike particle. For example, in 1925 Yakov Frenkel emphasized that "the electrons are not only indivisible physically, but also geometrically. They have no extension in space at all." He concluded that the mass could not possibly be interpreted electromagnetically and that the entire problem of the electron's inner structure was "scholastic."[84] This point of view, expressed shortly before the advent of Heisenberg's quantum mechanics, became a doctrine in quantum theory. The electron was considered to be pointlike because, as Dirac remarked in 1928, why should nature have chosen otherwise?[85] When the problem of self-energy came up in quantum theory, it was therefore formulated differently than in classical theory. But the problem itself remained essentially the same: The zero-point fluctuations of the electromagnetic field contribute to the electron an amount of energy. Since even in a vacuum the electromagnetic field fluctuates, it will make the electron perform forced oscillations. The energy possessed by the point electron by virtue of these oscillations turns out to be given by divergent integrals, and this leads to an infinite mass for the electron, which is of course absurd. Even the amplitudes of the zero-point oscillations become infinite. Formally the problem can be stated by analogy with that occurring in classical theory: The observed mass consists of two terms, that is,

$$m_{obs} = m + \delta m$$

where δm is infinite if the electron is a point particle.

In the early versions of quantum field theory, the infinite self-mass of the point electron survived. It was for some time believed that the self-energy (or self-mass) problem belonged to the classical electron and could be avoided in quantum theory. This view was held by Jordan, who, in his work with Klein from 1927, proved that in the case of non-relativistic interactions between electrons the self-energy terms could be made to disappear by means of a subtle mathematical trick. However, it was soon realized that the Klein–Jordan approach was not the answer sought for in the fight against the infinities; this was shown by Pauli and Heisenberg in their theory of 1929 and was confirmed by other physicists. In 1930, Oppenheimer published calculations, originally intended to form a sequel to the two Heisenberg–Pauli papers, that clearly demonstrated the acuteness of the problem.[86] He proved that in addition to the classical electrostatic self-energy, a new quantum effect turned up that contributed with a quadratically diverging term to the self-energy. Furthermore, as Oppenheimer pointed out, the divergences were not mathematical artifacts without observable consequences: If quantum electrodynamics anno 1930 was taken seriously, the spectral lines would be displaced infinitely.

When the positron was incorporated into quantum field theory, the infinities remained. According to Dirac's new picture of the vacuum, the field oscillations interacted with the "latent" electron pairs in the vacuum, resulting in induced charge and current fluctuations in the vacuum. These fluctuations would interfere with the induced fluctuations of the electron's field. Although the amplitudes of the zero-point oscillations would then be finite, the total effect on the electron's self-energy would still be infinite. Calculations of the self-energy in the positron theory were performed in 1934 by Weisskopf, who obtained a logarithmically divergent result.[87] If R is the radius of the electron, Weisskopf's result can be written as

$$\delta m = m\left\{ 1 + \frac{3e^2}{2\pi hc} \log\left(\frac{h}{mc}\frac{1}{R}\right)\right\}$$

Thus the self-mass became infinite for the point electron, although the divergence was much weaker than the quadratic divergence in earlier theories. Positron theory made the self-energy problem a little less troublesome, but it also introduced a new effect, vacuum polarization, which carried with it new difficulties. This effect was found by Dirac in the fall of 1933, and independently a little later by Oppenheimer and Furry.[88] This work of Dirac marked, according to Pais, the beginning of positron theory as a serious discipline and also Dirac's last important contribution to particle physics.[89] The latter point may be debated.

Dirac presented his new theory at the 1933 Solvay Congress in an address entitled "Theory of the Positron." Originally he was not invited to give a lecture at the conference. It was, after all, a conference on nuclear physics, a field in which Dirac was at most a peripheral figure. However, because of Pauli's intervention, an additional lecture on positron theory was included with short notice. Pauli's argument for the extra talk, as expressed to Paul Langevin, organizer of the conference, is interesting, not least because of his critical attitude toward Dirac's hole theory:[90]

The discovery of the positive electron once again reactualized Dirac's old idea of so-called holes, according to which gaps in the continuum generally occupied by the electron's negative energy states are re-interpreted as states of the positive electron with negative energy. Calculations not yet published have recently been made on this in Cambridge by Peierls and others (and overseas by Oppenheimer in America): Calculations which are very important for the general discussion of the theory of nuclei. That is why it would be very desirable to compose for the Solvay Congress, as a supplement to the theoretical report of Heisenberg, another shorter report on the development of the hole theory and its relationship with the positive electron. (Heisenberg has written to tell me himself that he finds the complementary report desirable.) Now, from what I learn from Cambridge, the circumstances surrounding the obtaining of such a report for the Solvay Congress are extraordinarily favourable owing to the very fact that Dirac is, in any case, already composing a similar report for a Russian congress in September; also, I would like to strongly suggest to you to *request Dirac to compose the report in question* for the Solvay Congress, and I consider it very probable that you receive from him an affirmative answer. The fact that there is a written document on this subject sent in advance to all participants would greatly facilitate the discussions at the congress.

A few months before the Solvay Conference, Dirac had informed Bohr about the ideas to which Pauli referred in his letter to Langevin:[91]

Dear Bohr,

Peierls and I have been looking into the question of the change in the distribution of negative-energy electrons produced by a static electric field. We find that this changed distribution causes a partial neutralisation of the charge producing the field. If it is assumed that the relativistic wave equation is exact, for all energies of the electron, then the neutralisation would be complete and electric charges would never be observable. A more reasonable assumption to make is that the relativistic wave equation fails for energies of the order $137mc^2$. If we neglect altogether the disturbance that the field produces in negative-energy electrons with energies less than $-137mc^2$, then the neutralisation of charge produced by the other negative-energy electrons is small and of the order 1/137. We then have a picture in which all the charged particles of physics – electrons, atomic nuclei etc. have effective charges slightly less than their real charges, the ratio being about 136/137. The effective charges are what one measures in *all* low-energy experiments, and the experimentally determined value for e must be the effective charge on an electron, the real value being slightly bigger. In experiments involving energies of the order mc^2 it would be the real charge, or some intermediate value of the charge, which comes into play since the "polarisation" of the negative-energy distribution will not have time to take on its full value. Thus one would expect some small alterations in the Rutherford scattering formula, the Klein–Nishina formula, etc. when energies of the order mc^2 come into play. It should be possible to calculate these alterations approximately, since, although the ratio (effective charge)/(real charge) depends on the energy at which we assume the relativistic wave equation to break down, it does so only logarithmically, and varies by only about 12% when we double or halve this energy. If the experiments could get sufficiently accurate data concerning these formulae, one would then have a means of verifying whether the theory of negative-energy electrons is valid for energies of the order mc^2.

I have not yet worked out the effect of magnetic fields on the negative-energy distribution. They seem to be rather more troublesome than electric fields.

With best wishes, and hoping to meet you in September,

Yours sincerely

P. A. M. Dirac

The exposition Dirac gave in Brussels was further elaborated three months later when he subjected the theory to a detailed mathematical analysis. In this paper Dirac defined his problem as follows:[92]

We now have a picture of the world in which there are an infinite number of negative-energy electrons (in fact an infinite number per unit volume) having energies extending continuously from $-mc^2$ to $-\infty$. The problem we have to consider is the way this infinity can be handled mathematically and the physical effects it produces. In particular, we must set up some assumptions for the production of electromagnetic field by the electron distribution, which assumptions must be such that any finite change in the distribution produces a change in the field in agreement with Maxwell's equations, but such that the infinite field which would be required by Maxwell's equations from an infinite density of electrons is in some way cut out.

In what follows we shall mainly deal with the Solvay address itself, which, though less mature, was the clearer and more simple presentation.

In his talk Dirac considered a weak electromagnetic field, which he treated as a perturbation. He then introduced the density matrix given by

$$(q'|R|q'') = \sum_r \bar{\psi}_r(q')\psi_r(q'')$$

where $\psi(q)$ is the four-component Dirac wave function, the summation is taken over all the occupied states, and R is the density operator that satisfies the equation of motion

$$i\hbar \frac{dR}{dt} = HR - RH$$

where H is the relativistic Dirac Hamiltonian for an electron moving in the field. The use of the density matrix was based on his earlier work of 1931 in which he had developed methods first introduced by von Neumann.[93] Dirac proved that R could be divided in two parts, $R = R_0 + R_1$, where R_0 is the charge distribution that produces no field and R_1 is the field-producing distribution. The physical behavior of an electron distribution is given only by R_1. Dirac calculated the electric density of R_1 and obtained a logarithmically divergent integral corresponding to an infinite electric charge and current density. About the divergence that thus appeared in the theory, he commented:[94]

One may think at first sight that the presence of this infinite charge makes the theory unacceptable. However, we cannot assume that the theory is applicable when energies larger than the order of magnitude of $137mc^2$ are in question, and the most reasonable way to proceed seems to be to arbitrarily limit the domain of integration to a value of the quantity $\frac{1}{2}(p' + p'')$ corresponding to electron energies of the mentioned order. This amounts, physically, to admitting that the distribution of negative energy electrons which are below a level of around $-137mc^2$ does not give cause to a polarization by the electric field in the manner indicated by our theory.

Dirac thus proposed to apply what is known as a cutoff procedure, and with it he obtained, after a complicated integration, the following result for the electric density:[95]

$$-\frac{e^2}{\hbar c}\frac{2}{3\pi}\left[\log\left(\frac{2P}{mc}\right) - \frac{5}{6}\rho\right] - \frac{4}{15\pi}\frac{e^2}{\hbar c}\left(\frac{\hbar}{mc}\right)^2 \nabla^2\rho$$

Here P is the magnitude of the momentum, and ρ is the electric density, which is related to the potential V by $\nabla^2 V = -4\pi\rho$, that is, the "external" charge density. The important part of the expression is the first term, which Dirac interpreted as "the electric density originating from the polarization produced by the action of the field on the distribution of electrons with negative energy."[96] If P is about $137mc$, the term becomes approximately $-\rho/137$, which signifies "that there is no density produced by polarization other than in the areas where the density ρ produced by the field happens to be situated, and that the induced energy neutralizes in it a fraction of around $1/137$ of the density produced by the field."[97] In effect, Dirac concluded that the induced electric charge contributed to the "real" charge of the electron an amount equal to $-136e/137$, so that

$$e_{obs} = e + \delta e = 136e/137$$

as stated in the letter to Bohr.

At the 1933 Solvay Congress, Bohr and Pauli commented on Dirac's address. They pointed out the mathematical difficulties the new theory raised, and objected that it was not valid when applied to distances of the dimensions of the classical electron. Dirac's more elaborate theory of 1934 was a subject of intense interest for the German physicists, who knew about it months before publication. Referring to his new ideas, Dirac told Bohr in November 1933 that "I shall write to Pauli about this and hope it will satisfy his objections to the theory of holes."[98] Whatever arguments Dirac made, he did not succeed in convincing the professor in Zürich. Pauli found Dirac's ideas artificial, mathematically complicated, and physically nonsensical. After studying Dirac's manuscript, he reported to Heisenberg:[99]

At present I am about ready to faint, having tried also to make practical calculations with his [Dirac's] formulae. The whole thing seems so artificial to me! . . . Thus Dirac proclaims natural law from Mount Sinai. Mathematically, of course, everything is very elegantly calculated. But it does not convince me as physics! Why should *this particular* formula be the true and correct one? What use is it that the electric polarization of a vacuum is finite, if the self-energy is infinite? And what use is anything, if pair production at high energies is too frequent according to the theory?

Heisenberg largely agreed, calling Dirac's theory "erudite nonsense, which no human being can take seriously."[100] At the end of April 1934, Heisenberg went to Cambridge, where he discussed the theory with Dirac, though without much profit.[101] Dirac's paper had not yet appeared in print, but the manuscript had for some time circulated among Heisenberg, Pauli, and Born.

In spite of the objections that Dirac's new theory of the positron met at first, the idea of vacuum polarization was at once taken up by other physicists. During the mid-1930s, the theory was developed by Furry and Oppenheimer, Edwin Uehling, Rudolf Peierls, Robert Serber, and others. For some time Pauli continued to use every occasion to criticize Dirac's conception of holes and his use of "limit-acrobatics" and "subtraction physics."[102] At first he joined Heisenberg in an attempt to develop Dirac's theory into a more consistent and complete scheme, but he soon resigned from hole theory, disgusted by what he felt was its lack of coherence. Heisenberg, on the other hand, engaged himself in positron theory and "subtraction physics" à la Dirac. In 1934, he presented an alternative formulation that was based on Dirac's "visualizable theory" but was more general.[103] Heisenberg managed to avoid some of the objectionable approximations used by Dirac and to treat electrons and positrons in a symmetric way. But as to the infinities, there was no real progress.

CHAPTER 7

FIFTY YEARS OF A PHYSICIST'S LIFE

IN the following chapters we shall be concerned with various aspects of Dirac's life and science during the period from 1934 to 1984. The present chapter primarily surveys biographical data, and Chapters 8 to 11 deal with Dirac's scientific contributions in the period.

In early 1934, Dirac made arrangements to spend the summer in Russia, where he hoped to hike in the Caucasus mountains with Tamm. Dirac also planned to visit Leningrad, Moscow, and Kharkov together with Niels Bohr, a journey he had long looked forward to.[1] However, these arrangements were canceled when Dirac decided to accept an appointment at Princeton's Institute for Advanced Study, where Einstein had settled the year before, after his emigration from Nazi Germany. Dirac left England on August 10 on the steamer *Britannica* after arranging with Van Vleck to resume, after three years, their travels together in America. "Dear Van," he had written at the end of July,[2]

I was very glad to get your letter (which crossed with mine to you) and to hear that you may be able to join me out west after all. It also looks as though we may be able to travel to the west together, as I shall be arriving in New York on Aug 18th and would like to leave for the west just after. With regard to where we go, my preference of one place over another is very small and I do not mind leaving all the decisions to you, as I know your likes and dislikes of places coincide very closely with my own. I should be quite glad to go to Glacier Park, in spite of my previous letter. I think that you would like Zion – there are enough trails radiating out from the camp for me to be able to spend a week there. I should like to try some new places during some of the time I am out west.

Dirac and Van Vleck met in Kansas City and proceeded to a small town in Colorado called Lake City.[3] There Dirac was again interviewed by a reporter from a local newspaper, as he had been in Wisconsin three years earlier.[4] Dirac and Van Vleck hiked and walked in the beautiful mountains and, among other things, journeyed to the top of the 4,360-meter

Uncompagrhe Peak. Traveling in that remote part of America was not comfortable, but, as we noted previously, Dirac never cared much for comfort. At one time he and Van Vleck rode inside a mail truck over rough roads, shut in with the mail sacks. After parting from Van Vleck at Silverton, Dirac made his way back east to Princeton. From there he reported to Van Vleck: "I stopped over a third day in Bryce and did a long new trail, which was still under construction at the far end, and then went back to Boulder Dam. On the way back east I was twice in a train following just behind a train which had an accident. (Is this a kind of Pauli effect?)"[5]

Dirac stayed for two terms at Princeton, attached to The School of Mathematics, during which period he also gave lectures at the University of Minnesota and at Harvard. Among the people at Princeton at the time were the physicists E. U. Condon and E. P. Wigner, the cosmologist H. P. Robertson, and the mathematicians L. Eisenhart, H. Weyl, A. H. Taub, and O. Veblen. And there was also Einstein, of course.

When Dirac went to Princeton in the fall of 1934, he also met his future wife, Margit Wigner Balasz, for the first time. She was visiting her brother, the Hungarian physicist Eugene Wigner, who had recently fled from Germany to the United States. Margit was divorced and had two children, Gabriel and Judith, who had stayed home in Budapest. Wigner, like Dirac, had been trained as an engineer and worked in fundamental problems of quantum physics. Dirac knew him superficially from his (Dirac's) stay in Göttingen in 1927 and came to know him well in Princeton, where they had adjoining rooms in Princeton's Fine Hall. Einstein's room was also next to Wigner's, but Dirac seems to have had very little, if any, contact with the founder of relativity. In an important paper in 1939, Wigner acknowledged Dirac's influence during the Princeton period: "The subject of this paper was suggested to me as early as 1928 by P. A. M Dirac. . . . I am greatly indebted to him also for many fruitful conversations about this subject, especially during the years 1934/35, the outgrowth of which the present paper is."[6] Much later Wigner recalled, with regard to their stay in Princeton: "I remember well that when we were in Princeton, both of us had visitors in turn. Then we usually went to lunch together to a restaurant. During the lunch there was very little discussion. He [Dirac] does not like to engage in conversation."[7]

After his appointment at Princeton ended, Dirac decided to make another tour around the world, this time alone. He left Princeton in mid-May and spent a week in Pasadena, where Oppenheimer was his host. In early June he left on the Japanese steamer *Asama Maru* for Tokyo, where he stayed with Nishina. "I have been two months in Tokyo," he wrote to Veblen from the Japanese capital. "I have been taken on many excursions in the neighbourhood by the physicists here. I am leaving tonight for

Kyoto and after a few days there will sail from Kobe for Tientsin and then go on to Peiping."[8] From the Chinese capital he went to Irkutsk and then on the Trans-Siberian Railway to Moscow, where he arrived on July 28. The route followed on this world tour was, except for his visit to China, largely the same as that taken in his 1929 trip.

In Moscow, Dirac met Tamm, with whom he had planned to climb in the Caucasus. However, once again the arrangement was canceled. Instead, Dirac spent much of his time in company with Kapitza, who in 1934 had been prevented by the Soviet authorities from returning to Cambridge.

Kapitza had wanted to go back to England, where his family and scientific equipment were, and was supported by Rutherford and other Western scientists. When it became clear that Kapitza would not be allowed to leave Russia, Rutherford proposed that a British scientist visit him and help him in resuming his work. Kapitza proposed various candidates for the visit, including Dirac, who visited him in Moscow along with the distinguished physiologist Edgar Douglas Adrian, who was, like Dirac, a Nobel laureate and was in Moscow during the summer of 1935 to attend a physiology congress. Dirac and Adrian had frank discussions with Kapitza, whom they found depressed but capable of resuming work if given sufficient support from England. Dirac wrote that he "stayed for several weeks in Bolshevo [outside Moscow] with Kapitza, as he was so lonely."[9] Dirac and Adrian's report to Rutherford convinced Anna Kapitza that she ought to be with her husband, and she left for Moscow soon after Dirac returned to Cambridge.[10]

On his way home from Moscow in August 1935, Dirac visited Budapest, to see the woman he wanted to marry. Back in Cambridge he reported to Van Vleck about his travel:[11]

I had a nice time on my travels, three weeks in Japan and one week in China, at Peiping. I was surprised to find so much difference between China and Japan – the Chinese houses are made of bricks and mud, while the Japanese are of paper, and their food and clothes are all different. (They never eat Chop Suey, except perhaps at foreign restaurants.) I am sorry about September 1936, but I still feel disinclined to come. The fact is I find I like conferences and lectures less and less as I get older, and I never did like celebrations. Of course I hope to see you again some time, but I would rather hike with you in the mountains than attend a conference. I am hoping to see the eclipse in Russia in June.

The conference referred to at the end of the letter was probably one associated with the tercentenary celebration of Harvard University. Through Van Vleck, who was appointed professor at Harvard in 1934, Dirac was invited to give a lecture and receive an honorary degree.

Dirac continued to see Margit Wigner. During the Christmas of 1935, she stayed with her children in Mariacell, near Vienna, and Dirac visited them there. The later Mrs. Dirac recalled of this visit: "He used to go off for long walks; he knew no fatigue, meals were unimportant to him, but not to me. . . . I often accompanied Paul, but usually regretted it. His enduring capacity would have been too much for most mortals."[12] The following year Margit again went to the United States to visit her brother, first sailing from England where she met with Paul.

As he mentioned in his letter to Bohr (to be quoted in Chapter 8), Dirac spent most of the summer of 1936 in the USSR. With Tamm and other physicists he went to the Caucasus, where there was to be a fine view of the solar eclipse. But Dirac missed the eclipse by a few days. While in the Caucasus he was informed that his father was dying, which prompted him to return to England for a short time. On June 17, he wrote to Tamm: "My father had died before I got back. I do not think I need to stay here very long, and hope to return to USSR at the end of June and go to the Caucasus with you in July."[13] Dirac returned to Moscow on June 27, and with Tamm he hiked in the Caucasus mountains, as they had planned to do the two previous years. At that time, Tamm was director of the theoretical division of the Lebedev Institute of Physics under the USSR Academy of Science, which was an influential post in Soviet science. From the Caucasus Dirac went to Moscow, where he stayed with Kapitza, whom he had also visited the year before and whom he visited again in 1937, that time bringing his wife with him.

The 1937 trip marked the seventh time during a period of eight years that Dirac visited the Soviet Union. His close contact with the country included cooperation with Russian physicists (Tamm, Fock, Kapitza), publication in Russian journals, and status as a corresponding member of the USSR Academy of Science. In 1937, Dirac wrote a paper in which he analyzed the concept of time reversal in quantum mechanics; it was written specially for the commemoration of the October Revolution on its twentieth anniversary.[14] Such close contact was unusual for a top British scientist because the political relationship between England and the USSR was strained during the period. As was mentioned in Chapter 5, Soviet physics had flourished in the early thirties, when relations with Western science were encouraged. However, from about 1935 the political climate had changed in the USSR under the impact of what is generally known as Stalinism. The following years were marked by increased xenophobia and political suppression that deeply affected Soviet science. Many Soviet scientists were arrested, and foreign scientists were viewed with suspicion, if not seen as potential spies. "Paul would like very much to go to Russia, but everybody advises him not to," wrote Margit from Budapest in the summer of 1937.[15] Dirac went anyway, and he took his new wife with him. The following year he wanted to visit the USSR once again, but this time his application for a visa was refused. In that year, the British Embassy in Moscow stopped issuing visas to Russians going to England, and in retaliation the Soviet authorities issued a general order to stop the granting of visas to British citizens. It was only after Stalin's dictatorship had ended that Dirac visited Russia again. In 1955–6, he served as a visiting professor at the Lomonosov University in Moscow, where he was celebrated as one of the founders of quantum mechanics. Later he published several times in Soviet scientific journals. Dirac visited the Soviet Union for the last time in 1973.

On January 2, 1937 Paul Dirac and Margit Wigner became husband and wife. A month later Margit went back to Budapest to see her family, and later in the year Paul and Margit again visited her native city. The marriage came as a surprise even to Dirac's closest friends, who had become used to regarding him as an inveterate bachelor. According to an often told anecdote, one of Dirac's old friends went to see him some time after he had gotten married. The friend, who had not heard of the marriage, was surprised to find a woman in Dirac's house. When Dirac noticed his curiosity, he said, "Oh, I'm sorry. I forgot to introduce you. This is . . . this is Wigner's sister."[16] Margit and Paul had two daughters, Mary Elizabeth and Florence Monica, and Margit brought two children, Gabriel and Judith, with her from her first marriage.[17] Margit, known as Manci to friends of the family, was more interested in social life than was her husband. Under her influence Paul almost became a social being. Paul and Margit moved to a house on Cavendish Road, and Paul gave up his room at St. John's College, where he had lived so large a part of his scientific life. The Dirac family also included Paul's mother, a widow since 1936, who would visit the house in Cambridge whenever it could be arranged. She happened to be on one of her visits there when she died in 1941.

The marriage inevitably caused Dirac to spend less time at the university, and he now did much of his work at home.[18] At that time life in Cambridge was quiet, and the social relations of the Dirac family were mostly with other Cambridge physicists. These included Rudolf Peierls, Maurice Pryce, and also Max Born, who in 1933 had left Hitler's Germany with his wife Hedwig. Born stayed in Cambridge until 1937, when he was appointed Darwin's successor in the Tait Chair of Natural Philosophy at the University of Edinburgh. In the same year Maurice Pryce married Margareth Born, the daughter of Max and Hedwig. In October 1937, the great man of modern British physics, Ernest Rutherford, died. Dirac, together with Ralph Fowler (who was Rutherford's son-in-law) and other distinguished scientists, attended the interment in Westminster Abbey.

During the last years of the thirites, it became increasingly clear that a new European war was threatening. Many scientists, including some of the best physicists, fled from Germany as a result of the anti-Semitic excesses of the Third Reich. Physicists in England tried to help those of their German-refugee colleagues who wanted to come to England. In Cambridge, Blackett and Rutherford were among those who engaged in helping refugees, the latter as President of the Academic Assistance Council. Dirac seems not to have been involved in these activities and never participated in the political debate. But on occasions when the problems arose close at hand, he tried to help. In 1937, his sister, Beatrice, had married a Jewish merchant from Hungary, Josef Teszler, with whom she went to Germany and later to Holland. Josef used his connection with Paul Dirac to ask Heisenberg for help in order to escape the fate of the concentration camp. With or without Heisenberg's assistance, the Teszler family managed to escape internment.[19] In 1938, Pauli begged Dirac to help a young (Jewish) cousin of his, Felix Pauli, get into England. Felix had fled from Nazi-occupied Austria to Switzerland, where he could not stay because he had neither passport nor visa. Dirac requested the Home Office in London to grant Felix a petition as a refugee.[20]

In the spring of 1939, it was recognized that physicists would perhaps play an important role in the coming war. The discovery of neutron-induced fission in uranium had been reported by German scientists (Otto Hahn, Fritz Strassman, Lise Meitner, and Otto Frisch), and it was known that secondary neutrons were released in the process. The discovery created a stir in the physics community, including discussion at Cambridge. On June 24, 1939, the Kapitza Club had as its guests Otto Hahn and Niels Bohr, who discussed "Fission of Uranium by Neutrons." Incidentally, this was the last Kapitza Club meeting until after the war, in 1948.

A few physicists realized that this discovery in pure physics might be developed into a device of enormous destructive power. The experiments that proved the possibility of a chain reaction took place in February 1939, primarily in France and the United States. Leo Szilard, working at the time with Fermi at New York's Columbia University, proposed that publications on the delicate subject be withheld and that the knowledge be distributed only to selected laboratories outside Germany. Szilard's group took action to convince French, British, and Danish physicists that this unusual limitation on the free diffusion of scientific results was necessary because of the situation in Germany. At the end of March, Wigner explained the matter to his British brother-in-law and asked him to support the proposal:[21]

I am writing to you in a rather serious matter this time. . . . What we should like to ask you at this time is to get in touch with Blackett and to actively support him in his endeavours if you find our position to be the reasonable one. It is my impression that there is some urgency in the matter. Although there exists apparently a great willingness for cooperation here, it is realised that the interests of the scientific workers in the U.S. may be prejudiced to some extent if America abeyed alone by the proposed procedure.

Dirac, not very interested in the matter himself, got in touch with Blackett and Cockcroft, who decided to follow the Americans. At that time several British physicists were working on the fission problem. Peierls, in particular, worked out an important theory for the production of neutrons in uranium fission. It was also Peierls who, in collaboration with Frisch, produced the first report on the possibility of constructing a "super-bomb."[22] Shorly thereafter a committee for research on military atomic energy was established. The committee was known as Maud (or M.A.U.D.); later, in 1941, the project's code-name changed to Tube Alloys.[23] The leading scientists of the Maud team were Peierls, Chadwick, and Franz Simon, who was a German-refugee theoretical chemist and former student of Nernst. The group also included the theoretical physicists Kemmer, Pryce, and Klaus Fuchs; the latter, a German-refugee physicist, would later become known as a spy for the Soviet Union.

Dirac, without doubt Britain's most distinguished theoretical physicist, was not a member of the Maud committee but acted occasionally as a consultant for the project. His war-related work involved two areas, the calculation of various models of a uranium bomb and the separation of isotopes. In the first area he completed three (classified) reports in the summer and fall of 1942. He did this theoretical work at home in Cambridge but kept in contact with Peierls and his group in Birmingham. Dirac seems to have been almost as fascinated by this work, of a more technical nature, as he was with pure quantum theory. Concerning some work he did on neutron multiplication in a supercritical mass of uranium enclosed in a container, he commented to Peierls in August 1942: "I was rather surprised to see how the whole theory comes out without a detailed knowledge of the processes by which the energy is degraded. I have enclosed my work on the effect of a non-uniform density in the outer layers in changing the rate of neutron multiplication."[24] Later the same year, he was asked if his reports could be studied by American physicists involved in similar projects. He answered: "I have no objection to your sending any of my reports to America but it might perhaps be better to wait until the work on the efficiency is completed, as this work should not take long now and the subject will then appear in a more finished form."[25] In 1943, Dirac did further calculations on neutron multiplication in non-spherical lumps of uranium. This resulted in two reports, one written alone and the other jointly with Peierls, Fuchs, and P. Preston.

Dirac's papers were studied by physicists in the United States, including Hans Bethe, who criticized some of Dirac's calculations for not being sufficiently precise. When the bomb project was taken over by the Americans' Manhattan Project, many British physicists were transferred to Los Alamos. The American–British group wanted Dirac to join them there, but he refused. About this episode Fred Hoyle told the following story:[26]

The clarion call rang out in Whitehall to get Dirac involved as a bargaining-counter with the Americans. . . . the Minister concerned, Sir John Anderson, telephoned Dirac in Cambridge to ask if he would call at the Ministerial office when next in London. Dirac said that he would. Sir John then went on to ask as an afterthought how often Dirac was in London, to which Dirac replied: "Oh, about once a year."

Methods of separating uranium-235 were a crucial part of the bomb project, and at one phase Dirac's early proposal for separating gas mixtures of isotopes (see Chapter 6) was considered. Peierls has recalled:[27]

When this work was started, Dirac was invited to Oxford to discuss the method. I was present at the meeting in the Clarendon Laboratory (there may have been more than one meeting) and I remember that the experimentalists expected a highbrow and abstract mathematician who would know the kinetic theory of the effect, but would not know one end of an apparatus from another. They were most impressed by Dirac's eminently practical and helpful remarks.

Dirac's wartime work on isotope separation resulted in reports that included new, important knowledge in the field of centrifuge science. In 1941, he succeeded in calculating the maximum possible output of a centrifuge and stated the result in a way that was valid for any mode of centrifuge operation. Dirac's centrifuge equations were, according to a recent reviewer, "probably the most important theoretical result in centrifuge technology."[28] He was in contact with Simon's group in Oxford, which he visited early in 1942 (as noted in Peierls's recollection quoted above). At that time, he suggested a jet isotope separation method that was subsequently (1942–5) investigated by physicists at Oxford and also tested at the Manhattan Project. Although this method was considered interesting and promising enough to test experimentally, it was not developed on a larger scale. Its efficiency was too low, and other alternatives, gas diffusion methods and electromagnetic separation, were given high priority.[29]

Apparently Dirac enjoyed his work on practically related physics, which may have appealed to his early background in engineering. As pointed out by Dalitz, this work supplies another side to our picture of the scientist. As compared to the corresponding work of other physicists, Dirac's involvement in war-related physics was rather sporadic. But neither this nor, for example, his refusal to join the bomb project in the United States should be interpreted as a conscious opposition to applying physics for military purposes. After all, his work was part of Britain's war effort, a fact of which he was well aware. Dirac's prime interest was in doing physics, and he seems not to have given much thought to its possible relation to the war. As long as he was involved in work that was not explicitly political or military, he just considered it interesting physics.

Dirac was the prototype of an ivory-tower scientist and valued his independence highly. Only when pressed hard would he leave his ivory tower and engage in matters of public importance. At the height of the war, there were still islands of physics research in Great Britain that remained unaffected by the war. "I continue my work undisturbed," wrote Born from Edinburgh in April 1940. "Soon my department will be the only spot in Great Britain where theoretical work is still done."[30] This was, as Born knew, an exaggeration. In Cambridge Dirac continued doing theoretical physics, largely undisturbed by the fact that so many of his colleagues were being drawn into military research.

Being a famous physicist with close relations to the USSR, and being fluent in French, Dirac could not completely withstand the pressure to become engaged in matters outside pure physics. In early 1940, arrangements were made to establish a British–French organization for scientific cooperation.[31] In April, Dirac agreed to accept, for one year, the presidency of the British committee, which also included J. Crowther, J. D. Bernal, Blackett, and Cockcroft. However, because of the fall of France the organization never got started. In February 1945, Dirac was invited to join the "International Senate," a conservative and elitist organization formed by French scholars. The Frenchmen, including Louis de Broglie, believed that selected, highly gifted individuals should be brought together to discuss politics independently of national boundaries. Dirac rejected the invitation because he found that the members of the organization consisted mostly of reactionaries and supporters of the discredited Vichy regime.[32]

After the war, the Anglo–French Society of Science was established, and Dirac delivered a lecture in the Palais de la Découverte in Paris in December 1945. The lecture was probably the only occasion on which Dirac, much to his surprise and discomfort, attracted a crowd of more than two thousand listeners. The audience thought he was going to speak about the new atomic bomb, which was a sensational topic at the time. Instead Dirac lectured on current problems in quantum theory and reviewed his recent attempts to reformulate quantum electrodynamics. He distinguished between the "direct" method of standard (Heisenberg–

Pauli) quantum electrodynamics and the "indirect" method associated with the idea of a hypothetical world (for this idea, see Chapter 8). Dirac advocated the latter method from an aesthetic point of view: "With the method that works for the hypothetical world, one finds more simple and elegant calculations. And I think I rather prefer this method, because I attach very much importance to the mathematical beauty of a fundamental theory of physics."[33] The audience in Paris probably did not understand much of the lecture. They were disappointed to learn that atomic physics was not identical with the atomic bomb.

Dirac's relationship with the Soviet Union would have made him an ideal science liaison officer if he had had the qualities and ambitions of a politician. James Crowther, a science journalist and leading figure in British science policy during the war, wanted Dirac to go to the USSR in 1942 on a diplomatic mission of some sort. He wrote Dirac:[34]

Sir Archibald Clark-Kerr, our ambassador in the USSR, who is in England at this moment, has approved the proposal that you should make a short visit to the USSR. He would like you to accompany him on his return in the middle of January, if that can be arranged. It is doubtful whether permission can be secured from the Soviet government in so short a time, but every effort is being made.

Dirac did not make the visit to Russia planned by Crowther, probably because permission was not granted in time by the Soviet authorities. Dirac's interest in the USSR did not necessarily imply any particular political involvement. In the thirties, there existed in Great Britain an active group of left-wing scientists who urged science planning and improved relations with communist Russia. The group included Bernal, Hogben, Haldane, and many others, but not Dirac. Whatever political sympathies he had, he always refused to become openly involved in political matters or to make himself a public figure. He was outside the group of Marxist scientists and also did not become involved in postwar attempts to create a socialist federation of scientists. His personal characteristics accorded well with the attitudes expected to be upheld by Cambridge scholars. The values generated from within the Cambridge culture meant, according to a historian, that[35]

Political commitments were your own affair, as long as they did not impede your full participation in the activities of your chosen research community. (But politics were thought to be such an irrational enterprise that any overt preoccupation with them was bound to cast some doubt on your "soundness".) Whatever else might be said about the life-style of Cambridge scientists, it was not one likely to inspire or sustain socialist convictions and practices.

However, Dirac was probably associated with, or a member of, the Marxist-dominated Society for Cultural Relations with the USSR, a branch of which was revived in Cambridge in 1936. At the end of 1941, Vladimir Semenovich Kemenov, a Russian cultural civil servant and chairman of the board of the All-Union Society for Cultural Relations with Foreign Countries, sent a telegram to Dirac that read: "Society for Cultural Relations sends you hearty new year greetings. Confident this year will bring victory over Nazism for all friends [of] world democracy."[36] After the final victory over Germany in May 1945, the Soviet Academy of Science invited representatives of British science to Moscow to celebrate the victory. Blackett, Mott, and Dirac were among the physicists invited, but the British government did not allow them to go, most likely because of their affiliation with the atomic bomb program, however weak it was in Dirac's case.

In 1948, the Cold War between the United States and the USSR began. At first Dirac only experienced the climate of suspicion and fear indirectly, in connection with Frank Oppenheimer, a physicist and the younger brother of the famous J. Robert Oppenheimer. In hearings before the Committee for Un-American Activities in 1948, Frank admitted that he had been a member of the Communist Party between 1937 and 1941. He was at once dismissed from his position at the University of Minnesota and was unable to find another position. At the request of J. Robert Oppenheimer, Dirac tried to find a position for Frank in England and Ireland. Although both Cecil Powell in Bristol and Lanos Janossy in Dublin wanted to help the ostracized American physicist, in the end nothing came out of the efforts.[37]

The Cold War later affected Dirac more directly. His contacts with Soviet science may have been the main reason this frequent visitor to the United States was denied a visa in 1954 after accepting an invitation to visit Princeton's Institute for Advanced Study during the academic year 1954–5. This was the period of McCarthyism, when many European physicists were faced with problems of hysterical visa restrictions and security measures under the Immigration and Naturalization Act. Dirac's exclusion caused some controversy and provoked a set of rare statements to the press. "I was just turned down flat," he told the *New York Times* on May 26. "They just said I was ineligible under Regulation 212A – and that's five pages long and covers many reasons."[38] And the following day, to *The Times*, he said that he did not understand the American decision: "It is probably because I had been in Russia before the war to attend scientific conferences – but it is rather strange that I should be stopped going to the United States the fourth time [after the war]."[39] Princeton physicists John Wheeler, Walther Bleakney, and Milton White protested vehemently, stating that the decision was a symptom of what might become organized cultural suicide: "We are very strongly aware of the advantages to this country of Professor Dirac's proposed visit. We are aware of no disadvantages."[40] In June the State Department ordered a review of the decision, and in September Dirac finally obtained his visa. However, at that time he had already accepted an offer to go to the Tata Institute of Fundamental Research in Bombay.[41] In India he was received by Pandit Nehru, the Prime Minister of the young nation. Dirac lectured on quantum mechanics and in January 1955 gave an address to the India Science Congress in Baroda.[42] In 1958, he was elected an Honorary Fellow of the Tata Institute.

In the fifties and throughout the rest of his career, Dirac continued his earlier way of life so far as his health permitted; he traveled widely, published steadily, and kept out of the public light. After his stay at the Tata Institute in 1954–5, he went by ship to Canada via Japan. Upon arriving in Canada, Dirac became seriously ill with hepatitis, and it took several months of medical care and recuperation in Canada and the United States before he had recovered enough to take up his research work in Ottawa. Other travels were to France (1959), Warsaw (1962), Dublin (1963), Trieste (1968, 1972), Florida (1968–9), Moscow (1956, 1973), Switzerland (1973), Australia/New Zealand (1975), and Israel (1979). In 1959 and again in 1962–3, he was at Princeton's Institute for Advanced Study, and he spent the academic year of 1963–4 at New York's Yeshiva University. During all these travels his home base was Cambridge University and his private home at 7 Cavendish Avenue, where he preferred to work. Dirac was invited to participate in the first meeting for Nobel laureates in physics, held in Lindau (in South Germany) in July 1953. Initiated in 1951 by Count Lennart Bernadotte from Sweden, the Lindau meetings alternate between chemistry, medicine, and physics. In 1953, Dirac met with other Nobel Prize-winning scientists, including Heisenberg, von Laue, von Hevesy, and Soddy. He liked the meeting and attended every one of the following Lindau meetings, his last one being in 1982. At several of the meetings he gave talks on his ideas about the basic problems of physics.

When Dirac retired in September 1969 from the Lucasian Chair, he had held the position for thirty-seven years and had been at Cambridge University for forty-six years. He was followed in the chair by the hydrodynamicist James Lighthill and then, in 1980, by Stephen Hawking, an authority on astrophysics and cosmology. At the age of sixty-seven, Dirac felt that he still had much to give to physics and did not want to retire completely. In 1968, he had been invited to the University of Miami's Center of Theoretical Studies in Coral Gables, where he spent several months of the years 1968–71. He decided to move permanently to Florida, where he was given excellent research conditions and where the climate was more suitable for his health than that in England. In June 1971, he joined the physics department of Florida State University in Tallahassee, where he stayed until his death thirteen years later.

During the period in Florida he remained productive.[43] He wrote his last research paper at the age of seventy-seven and continued publishing on physics until shortly before his death. Unlike other great scientists who in their old age may take up history or philosophy of science or write

Paul Dirac about 1960. Reproduced with permission of AIP Niels Bohr Library.

on the political, ethical, or cultural aspects of science, Dirac mostly stuck to his physics. But some of his contributions from this period were recollections or semi-historical sketches. During the Einstein centennial in 1979, Dirac participated in several meetings, and he used the occasion to emphasize the role played by aesthetic considerations in the creation of the theory of relativity. In Florida his health gradually began to fail, and in 1982 he had to undergo a serious operation. His failing health prevented most travel and his participation in many conferences. It also put a stop to his life-long custom of walking in the countryside. In July 1984, Dirac went to Aarhus, Denmark, to attend the interment of his stepson Gabriel. Three months later he died in Tallahassee, on October 20, 1984.

CHAPTER 8

"THE SO-CALLED QUANTUM ELECTRODYNAMICS"

B Y 1934, the majority of physicists engaged in quantum field theory felt that the theory was in a state of crisis and that "for very high energies the theory becomes false."[1] The reason for the pessimism – not to say frustration – was rooted partly in the theory's lack of conceptual and mathematical consistency and partly in its inability to account for high-energy phenomena such as the absorption data of cosmic rays. In the mid-1930s, cosmic radiation moved to the forefront of physics and was studied extensively by physicists in Europe and America. Before the discovery of the "heavy electron" (the meson or, today, muon) in 1937, it was generally believed that the disagreement between theory and data was caused by the breakdown of quantum electrodynamics for high energies, and not, as it turned out, by inadequate knowledge of the nature of the cosmic radiation.

Oppenheimer, one of the most active contributors to quantum electrodynamics during the period, was convinced that the theory was in a mess. After having struggled to produce a workable theory of pair production, he reported to George Uhlenbeck:[2]

We are prepared to believe that the theory can be improved, but we are skeptical, and think that this will not be so on the basis of quantum theoretic field methods. This point should be settled by summer; either Pauli or Dirac will have found the improvement, or they will have come with us to share the belief that it does not exist.

A couple of months later, Oppenheimer concluded that not only his own theory, but the entire basis of theoretical physics, was in need of radical transformation. In a state of despair, he wrote to his brother:[3]

As you undoubtedly know, theoretical physics – what with the haunting ghosts of neutrinos, the Copenhagen conviction, against all evidence, that cosmic rays are protons, Born's absolutely unquantizable field theory, the divergence difficulties with the positron, and the utter impossibility of making a rigorous calculation of anything at all – is in a hell of a way.

Physicists responded differently to the frustrating situation in quantum field theory. Some chose to leave the field – a few left physics entirely. Thus Max Delbrück, a promising theorist of the younger generation who was deeply engaged in the complicated calculations of quantum electrodynamics, gradually turned from quantum theory to biology; the reason for his conversion was, at least in part, his dissatisfaction with the state of quantum theory. Jordan, too, changed to biology, and later to geophysics. Of course, there were also those who had never had a high opinion of quantum field theory and, consequently, rather welcomed the crisis. Such was the case with Einstein, and also with Schrödinger, who never felt at home with the turn taken by quantum theory in the thirties. But most theorists stayed in the field, trying to improve it either by means of piecemeal methods or by suggesting more radical alterations.

Bohr, Dirac, Pauli, Heisenberg, Born, Oppenheimer, Peierls, and Fock came to the conclusion, each in his own way, that the failure of quantum electrodynamics at high energies would require a revolutionary break with current theory. In the second half of the thirties, several such schemes were proposed, and were intended to serve as radically new entries into the problems that faced quantum electrodynamics. Among these proposals, some sought to revise classical electrodynamics or electron theory in order to obtain a more satisfactory quantization. Along these lines, Max Born and Leopold Infeld constructed a unified field theory in which the electrodynamical equations were nonlinear.[4] Dirac also believed that a new classical theory of the electron had to be established, and he pursued this idea for several years, although in a very different direction from that followed by Born and Infeld. Heisenberg cultivated still another idea and built up a theory based on the notion of a fundamental length.[5]

Dirac had created the main part of the foundation of quantum electrodynamics and naturally felt committed to the field's problems and development. But he was no less dissatisfied with the situation than were other physicists, and he did not want to devote his time to what eventually emerged as mainstream quantum electrodynamics. Like Pauli, he turned for some time to other areas of physics, not directly related to quantum electrodynamics, as a sort of ersatz work. Dirac's critical attitude toward quantum electrodynamics during the period from 1935 to 1947 was neither unique nor particularly remarkable. As mentioned, many other physicists of eminence felt the same way.

The "revolutionary" programs were for some time considered to be serious candidates for a new quantum field theory. However, they turned out to be blind alleys and did not contribute to what retrospectively can be seen as the main road of development. With the recognition of new particles (mesons) in the cosmic radiation, the existing theory – gradually improved in its details but not changed in its essence – proved to be quite workable after all. The empirical disagreements became less serious, and by the end of the thirties most of the young theorists had learned to live with the theory. They adapted themselves to the new situation without caring too much about the theory's lack of formal consistency and conceptual clarity. The pragmatic approach of the "quantum engineers," including Fermi, Bethe, Heitler, and a growing number of young American physicists, proved to be of significant value, but it did not eliminate the fundamental problems that continued to worry Dirac, Pauli, and others. In 1947, revelation came after many years of trouble; when the modern theory of renormalization was established after the war, the majority of physicists agreed that everything was fine and that the long-awaited revolution was unnecessary. But Dirac still did not adopt this attitude and kept criticizing quantum electrodynamics along the same lines as before the war. It was only at this time that Dirac's views began to become decidedly unorthodox.

In the following account we shall follow Dirac from his first dissatisfaction with quantum electrodynamics to his final position as an outsider to the mainstream development of theoretical physics.

In February 1935, while Dirac was in America, Van Vleck invited him to be his guest and to give a lecture at Harvard.[6] Dirac's lecture there, which dealt with quantum electrodynamics, indicated how he felt about the subject at that time. Quantum electrodynamics, according to his lecture notes, was "not really a satisfactory theory." For one thing, he explained, it was an elaborate theory which had not led to new results; for another thing, "one can get solutions only by ignoring certain infinities." Nevertheless, Dirac had not yet completely lost confidence in the theory, to which he attributed some aesthetic qualities: "Q Eld [quantum electrodynamics] has, however, some beautiful and remarkable features. It is relativistically invariant – a very surprising result."[7] In his lectures at Princeton he expressed the same ambivalence toward quantum electrodynamics, pointing out that, although the theory was better than classical electrodynamics, it was at present of limited value. He developed a formalism that removed some of the infinities, but the infinite self-energies of electrons and photons remained. Dirac pictured these self-energies as the result of "nascent" or virtual particles:[8]

Just as the self-energy of the electron can be regarded as due to many nascent light quanta surrounding it, so the theory gives around each photon many nascent electrons and positrons which give it self-energy; i.e. the Hamiltonian contains terms corresponding to such transitions as cause the creation of electrons and positrons.

Dirac's main concern while at Princeton was with the mathematical aspects of relativistic quantum theory, which he discussed with the Princeton mathematicians, in particular Oswald Veblen. These mathematicians were examining various generalizations of the Dirac equation, for example, its formulation in cosmological spaces and in projective relativity. Shortly before Dirac's arrival, Schrödinger had lectured at Princeton on the Dirac equation in expanding universes, a topic to which he later returned.[9] Dirac's stay at Princeton resulted in two papers, both of which were published in the *Annals of Mathematics,* signifying that they were contributions to mathematical physics rather than to physics proper.[10] Dirac examined two geometries, de Sitter space and conformal space, in which the equations of physics could be expressed with tensors that took on five and six values, respectively. These more general spaces had previously been applied to general relativity and electromagnetism, and Dirac now showed that the fundamental equations of quantum mechanics could also be formulated in de Sitter space and conformal space. However, from the point of view of physics, nothing new was added through these mathematical formulations. Dirac probably assumed that this kind of interesting mathematics, if suitably developed, would somehow yield interesting physics too. In his work on conformal space quantum mechanics, he showed that "by making a further generalization of the space . . . a still greater symmetry of these [wave] equations is shown up, and their invariance under a wider group is demonstrated."[11] This was in complete accordance with his preface to *Principles of Quantum Mechanics,* in which he had stated that "further progress lies in the direction of making our equations invariant under wider and still wider transformations."[12]

At the beginning of 1936, Dirac worked out a new generalization of his linear relativistic wave equation.[13] By the mid-1930s, the linear Dirac equation of 1928 was known to describe the behavior of particles with half-integral spin, such as the electron, the positron, and presumably also the proton. The emergence of new elementary particles (the positron and the neutron) made physicists more receptive to the idea that other new particles, perhaps with spin different from one half, might exist in nature. Such hypothetical particles attracted theoretical interest, which in 1936 resulted in three attempts to generalize the wave equation to cover particles with arbitrary spin. Theories were proposed by Dirac, by the Belgian physicist G. Petiau, and by the Rumanian-French physicist Alexandre Proca.[14] Dirac applied the spinor formalism, introduced by Van der Waerden in 1929, in order to construct a system of relativistic wave equations linear in the energy operator. He proved that these general equations were also able to describe particles with spin greater than one-half and with either zero or nonzero rest mass. He justified as follows what might have appeared to be merely a mathematical exercise:[15]

It is desirable to have the equations ready for a possible future discovery of an elementary particle with a spin greater than a half, or for approximate application to composite particles. Further, the underlying theory is of considerable mathematical interest.

Probably the latter motive meant most to Dirac, who seems not to have been very interested in whatever new particles might be discovered. In 1935, the meson had been predicted by Hideki Yukawa in Japan, but Dirac's theory had no relation to this particle. In 1936, Yukawa's work was virtually unknown outside Japan, and there is no reason to believe that Dirac had noticed it. Although Taketani Mituo and other Japanese theorists had argued that the Yukawa particle would have spin one, this was unknown in Europe at the time.[16] When the meson made its entry into Western physics in the late 1930s, the generalized wave equations became more interesting from a physical point of view. Dirac's theory was taken up by Yukawa and Shoichi Sakata in Japan and by Pauli and Markus Fierz in Switzerland.[17] With an eye on nuclear forces, other physicists, including Richard Duffin, Nicholas Kemmer, and Homi Bhabha, developed first-order equations for particles with arbitrary spin.[18] Dirac showed little interest in this development and returned to the subject only in 1970 (see the end of this chapter).

During his work with mathematical generalizations of wave equations in 1935–6, Dirac's chief concern was still with the riddles of quantum electrodynamics. In the spring of 1936, he expressed most clearly how he felt about the state of affairs. His pretext, so to speak, was an experiment that had been carried out in the fall of 1935 by the American physicist Robert Shankland in order to test then current theories of photon scattering.[19] Shankland's results seemed to imply that energy was not conserved in the sort of individual atomic processes he had examined. This rather sensational news was quickly – indeed, too quickly – accepted by Dirac, who used it to launch a sharp attack on quantum electrodynamics. He discussed the conclusions to be drawn from the Shankland experiment at a Kapitza Club meeting in December,[20] and a few days earlier he had written to Tamm about the supposed implications, which included energy nonconservation, no neutrinos, and the abandonment of quantum electrodynamics.[21] In a paper of February 1936, he argued that Shankland's measurements necessitated that current relativistic field theory, including conservation of energy and momentum, be replaced by a new theory of the Bohr–Kramers–Slater type. The short-lived but influential Bohr–Kramers–Slater theory, originally proposed in 1924 as a rather desperate answer to the growing crisis in quantum theory, contained only conservation of energy on the statistical level.[22] Dirac believed that "physics is now faced with the prospect of having to make a drastic change involving the giving up of some of its principles which have been most strongly relied on (conservation of energy and momentum), and the establishment in their place of the B.K.S. theory or something similar."[23]

Quite willing to sacrifice the relativistic theory, he wanted to retain the general, non-relativistic theory and what he called the "primitive theory of radiation in quantum mechanics." This theory, he argued, "loses most of its generality and beauty when one attempts to make it relativistic." As to quantum electrodynamics, he now had no confidence in it at all:[24]

The only important part that we give up is quantum electrodynamics. Since, however, the only purpose of quantum electrodynamics, apart from providing a unification of the assumptions of radiation theory, is to account for just such coincidences as are now disproved by Shankland's experiments, we may give it up without regrets – in fact, on account of its extreme complexity, most physicists will be very glad to see the end of it.

Dirac's objections to what he referred to as "the so-called quantum electrodynamics" were partly addressed to Fermi's theory of β-decay. This theory, based on Pauli's neutrino hypothesis in order to reconcile the experimental data with the principle of energy conservation, was the most spectacular success – and perhaps the only one – of quantum field theory. Consequently, much discussion centered on the theory of β-decay. In theories of the Bohr–Kramers–Slater type, which were advocated by Dirac, there was no need to postulate the existence of neutrinos. Dirac considered the elusive particle of Pauli and Fermi to have been introduced in an ad hoc fashion: "A new unobservable particle, the neutrino, has been specially postulated by some investigators, in an attempt formally to preserve conservation of energy by assuming the unobservable particle to carry off the balance."[25]

Dirac's unreserved acceptance of Shankland's results was clearly preconceived according to his aesthetically based dislike of quantum electrodynamics. Usually he would not admit that theories could be falsified by experiments in such a simple way, but in this case he was in need of results that could justify his preconceived conclusion. His proposal to discard the relativistic quantum theory was drastic and essentially negative. It was only a state of intellectual despair that led him to the proposal, which, if taken seriously, would ruin the very achievements on which so much of his reputation rested. The relativistic version of quantum mechanics was Dirac's Nobel Prize–rewarded brainchild. Now he felt forced to renounce it without being able to offer an alternative. No wonder most of his colleagues considered his proposal a retrograde step. Kramers was surprised at Dirac's readiness to give up a large part of quantum mechanics and believed that Dirac was too uncritical in accepting experimental reports: "I think one must say that he [Dirac] lacks the independence towards the experimentalists with which for example Heisenberg looks at things."[26] However, a lack of confidence in theory and a belief in the primacy of experimental results were certainly not characteristic features in Dirac; on the contrary, in general he emphasized the autonomy and sometimes the priority of theory over experiment. His attitude in 1936 was anomalous.

Dirac's negative attitude toward the neutrino may appear somewhat surprising when looked at in the context of his general philosophy of physics (which will be examined in Chapter 13). Pauli's neutrino hypothesis was in fact based on reasoning rather similar to that which made Dirac postulate the anti-electron and the magnetic monopole. Clearly, Dirac would not accept the neutrino only on the ground that it rescued energy conservation. He wanted a mathematically sound argument in order to postulate new physical entities, which should, he thought, grow out of the mathematical structure of the equations. But it can hardly have escaped Dirac's attention that in addition to the experimental reasons there were, in fact, good theoretical reasons in support of massless neutral particles with spin one-half. Pauli had constructed a variant of the Dirac equation that described the neutrino, and in this sense proved that it was allowed by relativity and quantum mechanics. As Oppenheimer and Franklin Carlson remarked in 1932, Pauli's neutrino hypothesis had been advanced "on the further ground that such a particle could be described by a wave function which satisfies all the requirements of quantum mechanics and relativity."[27] It would have been in good agreement with Dirac's philosophy to accept the existence of the neutrino on the basis of such an argument. But he did not. His sense of mathematical beauty, applied to the unappealing mathematical structure of quantum field theory, led him to discard the neutrino early in 1936. Only one and a half years earlier, Dirac's attitude toward Fermi's theory had been quite different: "I think Fermi's theory is satisfactory as a beginning on a very difficult subject," he had written to Tamm in 1934. "The neutrinos seem to provide the only escape from non-conservation of energy and until something else turns up one should not be unsympathetic to them."[28]

Although in 1936 many physicists felt uneasy about the situation in quantum theory, only a few were willing to follow Dirac all the way. In private correspondence Heisenberg simply wrote off Dirac's idea as "nonsense" *(Blödsinn)*.[29] The Manchester physicist Evan Williams argued that if Shankland's results were correct, they would invalidate not only detailed energy conservation and quantum electrodynamics but also Heisenberg's uncertainty principle, and thus the very essence of the non-relativistic quantum mechanics for material particles.[30] Williams's objections were known to Dirac before he sent his paper to *Nature,* but apparently they did not convince him that the non-relativistic theory too, would have to be scrapped.[31] Jordan and Bohr both criticized Dirac's proposal and defended the neutrino and Fermi's theory.[32] In March, Bohr wrote to Kramers in Leiden:[33]

I am not at all pleased with Dirac's latest paper in "Nature," since I am extremely skeptical about the correctness of the new experiments in Chicago. All that Dirac does, however, lies in a higher plane of objectivity than most others', and I was deeply moved to hear, on my last visit to England, with what great thoroughness and understanding of our efforts he again had studied our old paper with Slater.

Moreover, Bohr was able to support his arguments with new experimental facts. At Bohr's institute Shankland's experiment was repeated by Jacob Jacobsen, who obtained results in good agreement with energy-conserving quantum electrodynamics.[34] These results, together with supplementary evidence obtained by German and American experimentalists, destroyed the empirical basis for Dirac's argument and reduced it to a much less convincing aesthetic objection.

Until Jacobsen's result was known, it was still possible to support Dirac's stand, as did Peierls, who suggested that even statistical conservation of energy might have to be abandoned.[35] Physicists who opposed the Bohr–Heisenberg trend in quantum theory welcomed Dirac's conclusion, although it was not in fact a protest against the Copenhagen interpretation. In March, Einstein wrote to Schrödinger: "I guess you have seen the mentioned note by Dirac in Nature? I am very happy that one of the real adepts now argues for the abandonment of the awful 'quantum electrodynamics.'"[36] Schrödinger too believed that Dirac's opposition to quantum electrodynamics signaled that the anti-Copenhagen camp had acquired a new, prominent ally. In a state of intellectual isolation he wrote to Dirac:[37]

I am awfully glad that you too feel this rather unsatisfactory state of affairs as unsatisfactory. I am awfully glad because, in general, there is a rather discouraging parallelism: the less a man really understands our quantum-mechanics, the more he agrees with me with respect to my discomfort in front of that situation! (And vice-versa: v. Neumann, Pauli, Heisenberg, Bohr disagree with me!) Which, time by time, made me fear that very probably I don't understand the thing myself. I feel a great relief, if you consent in principle.

However, the majority of physicists remained skeptical toward Dirac's hypothesis. Remembering that he had earlier turned "mad" ideas into fruitful theories (e.g., the hole theory), they wanted to wait and see if he could do the trick again. But this time there was no wizardry. "I have not yet built a non-conservation theory," he wrote. "One cannot make a drastic new theory whenever one pleases. One must wait for the ideas to come."[38]

In 1936, Bohr and Dirac agreed that the state of affairs in quantum electrodynamics called for some deep-rooted change in physical theory. But they disagreed over the kind of change needed and over whether it would be necessary to abandon the main body of current theory in order to produce the revolution; in particular, they disagreed in their views concerning the theory of β-decay. The following exchange of letters, dating from after Jacobsen's result was known, illustrates how they viewed the situation:[39]

Thank you very much for your letter and the copy of your article to "Nature." I am not in complete agreement with the views you put forward in the article. Your remarks emphasize the beauty and self-consistency of the present scheme of quantum mechanics; but I do not think they provide an argument against the possible existence of a still more beautiful scheme in which, perhaps, the conservation laws play an entirely different role. The non-relativistic nature of the present quantum theory appeared to me most strongly when I was writing my book. In the 1st edition, where I tried to build up everything from a relativistic definition of "state" and "observable," I found many things which were extremely awkward to explain. But these difficulties all vanish when one makes free use of non-relativistic ideas. I think there must be something fundamental underlying this. However, the conservation question must be decided by experiment, and at present it looks rather bad for Shankland. I am very sorry I cannot come to the conference, but I am leaving tomorrow for Russia, and hope to see the eclipse on June 19th in the Caucasus. I would like very much to see you later in the summer, but I think you are going to America, and I shall probably stay in Russia till August (climbing in the Caucasus).

Bohr replied three days later:

I thank you for your kind and interesting letter of June 9. I understand of course the weight of your arguments regarding the present difficulties of relativistic quantum mechanics, but I am inclined to think that the only way to progress is to trace the consequences of the present methods as far as possible in the same spirit as your positron theory. Just in this connection it may interest you that Fermi's theory [of] β-rays as pointed out by Heisenberg at our conference leads at any rate quantitatively to an understanding of the remarkable shower phenomenon. Besides these new and most promising considerations of Heisenberg appear to me to offer a most important clue to the old problem of the limitation of the very ideas of space and time imposed by the atomistic structure of all measuring instruments. You may remember that we have often discussed such questions but hitherto it seemed most difficult to find an unambiguous starting point. It now appears, however, that any measurement of such short lengths and intervals where the conjugated momenta and energy will cause all matter to split into showers will be excluded in principle. I need not say how much we all missed you at the conference. My wife and I hope, however, very much to see you on your way back from Russia. I have given up my journey to America in September, but unfortunately I will be away to a conference in Finland the three weeks of August; after that I will be in Copenhagen and it will give us all great pleasure to see you here. With kindest regards and best wishes from us both to you and all common friends in Russia.

Bohr's optimistic mood regarding quantum electrodynamics was indebted to work done in Copenhagen by E. J. Williams, C. F. Weizsäcker, and V. Weisskopf. Williams, who was at the time one of Bohr's right hands, applied one of Bohr's early ideas in order to account for the scattering of high-energy charged particles. He argued that the laws of standard physics applied well to these phenomena and, consequently, that it was unnecessary to look for a breakdown of quantum electrodynamics. Williams's work helped to convince the Copenhageners that quantum electrodynamics was, after all, a promising theory.[40]

It is interesting to compare the positions of Bohr and Dirac in 1936 with those they held in 1929–31, when energy conservation in β-decay was also a central topic (see Chapter 5). In 1936, their roles had almost completely reversed: Whereas six years earlier Bohr had pessimistically argued for a breakdown of relativistic quantum mechanics and had dismissed energy conservation, he was now quite optimistic; on the other hand, Dirac's attitude had changed completely and now mirrored that held by Bohr in 1930.

In the years following 1934, Dirac's negative attitude toward quantum electrodynamics manifested itself so that he did not publish on the subject. Still, it was always on his mind, and most of his later work can be seen as at least indirectly related to the problems that frustrated him so deeply. Indeed, looking back in 1979 on his career in physics, Dirac wrote: "I really spent my life mainly trying to find better equations for quantum electrodynamics, and so far without success, but I continue to work on it."[41] In later chapters we shall deal with his work in cosmology (1937–8) and his theory of the classical electron (1938 and later). The latter theory, in particular, was worked out with an eye on quantum electrodynamics; Dirac believed that a possible solution to the riddles of the infinities might be found if he could develop a better classical theory.

In 1939 and the years to follow, Dirac extended his recently developed theory of the classical electron to the quantum domain. He first presented his new ideas at a conference held at the Institut Henri Poincaré in Paris in April 1939.[42] In order to eliminate the infinities of quantum electrodynamics, he introduced a mathematical technique involving a new kind of electromagnetic field that in some respects differed from the Maxwell field. He applied it in connection with a certain limiting process (the λ-limit), which, he argued, was needed in order to formulate the equations of motion of the classical electron in Hamiltonian form. The λ-limiting technique had been introduced by Gregor Wentzel in 1933, but the general idea to make use of an alternative field with suitable mathematical properties (a Wentzel field) went back to the Dirac–Fock–Podolsky paper of 1932.[43] By means of this technique Dirac was able to eliminate the infinities for a single electron up to the order of e^2. However, new infinities appeared in the quantized version of the theory (see further Chapter 9).

In the following years, Dirac continued to develop mathematically his theory of 1939 in the hope that he could make the infinities disappear completely. Some of the work was done in Dublin, where he spent the summer of 1942 at the recently founded Institute of Advanced Studies.[44] He also attended a conference at the institute in 1944, which dealt with problems of quantum electrodynamics. The Dublin institute's department of theoretical physics was headed by Schrödinger, who had fled from Austria in 1938 after the Nazi takeover. In the forties, many of the best European physicists stayed for some time at the institute. Besides Dirac, frequent visitors included Born, Pauli, Eddington, Peierls, Heitler, Lanczos, and the Chinese physicist H. W. Peng. In 1942, Dirac was awarded a doctorate *honoris causa* at the National University of Ireland, and in 1944 he was elected an honorary member of the Royal Irish Academy. After the war, in July 1952, he returned to Dublin to participate in a colloquium series at the Institute of Advanced Studies.

In June 1941, Dirac delivered the Bakerian Lecture in London. In this lecture he suggested some rather radical ideas in order to overcome the difficulties of quantum electrodynamics. The existing theory, he stated in the spirit of his *Nature* article of five years earlier, "leads to such complicated mathematics that one cannot solve even the simplest problems accurately, but must resort to crude and unreliable approximations. Such a theory is a most inconvenient one to have work with, and on general philosophical grounds one feels that it must be wrong."[45] Dirac introduced into quantum mechanics an indefinite metric that led to a generalization of the quantum mechanical probability concept; with this generalization, probabilities were no longer constrained to be numbers in the interval between zero and one. The existence of negative probabilities in Dirac's scheme was, of course, problematical, but Dirac's commitment to the power of mathematics made him look for a physical interpretation of these states rather than to dismiss them as nonexistent:[46]

Negative energies and probabilities should not be considered as nonsense. They are well-defined concepts mathematically, like a negative sum of money, since the equations which express the important properties of energies and probabilities can still be used when they are negative. Thus negative energies and probabilities should be considered simply as things which do not appear in experimental results. The physical interpretation of relativistic quantum mechanics that one gets by a natural development of the non-relativistic theory involves these things and is thus in contradiction with experiment. We therefore have to consider ways of modifying or supplementing this interpretation.

But if negative energies and probabilities did not appear in the real world, where would they exist? Dirac suggested the idea of a "hypothetical world," that is, a world in which nearly all the negative-energy states were unoccupied with positrons (in the real world nearly all the negative-energy states are occupied). The point of introducing such an artifact was that it would allow the instrumentalistic power of usual quantum mechanics also to be secured in the relativistic domain. In order to bridge the gap between the two worlds, Dirac argued that transition probabilities calculated for the hypothetical world were the same as those of the actual world; only the initial probabilities of certain states would be negative in the hypothetical world.

The theory that Dirac built up on his idea of a hypothetical world did not receive much immediate attention, published as it was during the midst of the war. His introduction of state spaces with indefinite metrics was later developed by other physicists, especially by Suraj Gupta and Konrad Bleuler, who in 1950 proposed a reformulation of quantum electrodynamics based on Dirac's idea.[47]

When Pauli, then in Princeton, received news of Dirac's theory, he was at first unusually sympathetic to it: "I just studied very carefully your 'Bakerian Lecture' and I am very enthousiastic [sic]. This time I am sure, that you are on the right way (you remember, that I was on the contrary very critical in Paris and I was also right with it)."[48] However, on closer inspection Pauli found Dirac's theory less satisfactory.[49] Eventually, he came to regard the "hypothetical world" with suspicion. In November 1942, he told Dirac that in his forthcoming review of the theory[50]

... I certainly shall emphasize the incomplete character of the theory. The lack of completeness seems to me mostly obvious in the *rules which serve to translate the mathematical results concerning the hypothetical world into physical results of the actual world.* This translation does not seem to me really satisfactory yet and needs improvements.

The following year Pauli had this to say:[51]

It is my present opinion that your rules of translation of results concerning the "hypothetical world" into those of the "actual world" lead to irreasonable consequences, if they are applied to this problem [of low-energy photons] and that for this reason these rules can not be correct in higher approximation in $e^2/\hbar c$.

And another year later he wrote: "Personally, however, I have now the greatest doubts whether the idea of the introduction of an hypothetical world with its indirect connection with physics is really the correct way to a further progress."[52] When Pauli received the Nobel Prize in 1946, he had reached the conclusion that none of Dirac's attempts to formulate a satisfactory theory of quantum electrodynamics were acceptable. He ended his Nobel Lecture with saying that "a correct theory should neither lead to infinite zero-point energies nor to infinite zero charges, . . . it should not use mathematical tricks to subtract infinities or singularities, nor should it invent a 'hypothetical world' which is only a mathematical fiction before it is able to formulate the correct interpretation of the actual world of physics."[53]

A peculiar feature of Dirac's physics was his interest in notation and his readiness to invent new terms and symbols. We have already met with several cases of this inventiveness: the words commutator, *q*-number, *c*-number, eigenfunction, fermion, and boson; and the symbols [*x,y*] (for a commutator or Poisson bracket), δ (−function), and $\hbar$ ($h/2\pi$, "Dirac's h"). Dirac regarded the matter of notation as a relatively important part of physics and devoted an entire chapter of *Principles* to the subject. In 1939, he wrote:[54]

In mathematical theories the question of notation, while not of primary importance, is yet worthy of careful consideration, since a good notation can be of great value in helping the development of a theory, by making it easy to write down those quantities or combinations of quantities that are important, and difficult or impossible to write down those that are unimportant

In this paper Dirac introduced one of his most important notational innovations, the bracket or "bra-ket" formalism. He proposed to denote vectors corresponding to quantum states by symbols such as $|a\rangle$, where the letter *a* labels the state. He called these vectors – which corresponded to the symbols ψ or ψ_a for usual Hilbert space vectors – ket-vectors. The conjugate imaginary vector ($\bar{\psi}$ in traditional notation) was called a bra-vector and assigned the symbol $\langle a |$. In this way Dirac was able to write the inner product of two Hilbert vectors as $\langle b | a \rangle$. He further showed that a linear operator α could operate on a ket or bra, yielding states symbolized by $\alpha | a \rangle$ and $\langle a | \alpha$, respectively, and he proved a number of theorems for his new bra-ket algebra. At first Dirac's new notation was not much noticed (he used it himself for the first time in 1943), but eventually it gained acceptance and is today a common, powerful, and much admired notation in quantum theory.

In this context it may be mentioned that still another notational novelty was introduced by Dirac in 1944, in an important study of the unitary Lorentz transformation group.[55] In this work he coined the word "expansors" for certain tensorlike quantities that appeared as coefficients in a series expansion. Dirac considered the mathematical properties of the expansors and also suggested that they might be useful in physics. Applying the expansor formalism, he argued that particles having no spin when at rest might acquire a spin when moving. Although this work turned out to be of limited physical significance, it was of considerable mathematical import. Its analysis of infinite-dimensional irreducible representations was later developed further by Harish-Chandra, Valentin Bargmann, and other mathematical physicists.

For some time after the war, Dirac continued to uphold his idea of a hypothetical world. He reached the conclusion that the Schrödinger wave equation had no solution that could be expressed as a power series in the electronic charge. If the wave function ψ was written as

$$\psi = \psi_0 + e\psi_1 + e^2\psi_2 + e^3\psi_3 + \cdots$$

the series would not converge. According to Dirac, this was a most serious difficulty that could not be circumvented and that demanded some drastic change in the mathematical formalism. He considered, of course, the idea that the wave equation might have solutions that were not in the form of a power series, but he discarded this idea with the following argument:[56]

If they [solutions not expressible as a power series] exist they are presumably very complicated. Thus even if they exist the theory would not be satisfactory, as we should require of a satisfactory theory that its equations have a simple solution for any simple physical problem.

The idea that simple physical problems should be describable by simple mathematics was central in Dirac's thinking, although he never explained

Paul Dirac about 1945. Reproduced with permission of AIP Niels Bohr Library.

what "a simple physical problem" was. In his Paris lecture he expressed the idea as follows: "In order that a physical theory be satisfactory one ought to have that the mathematics of a simple process is itself simple, and that calculations should only be complicated if one deals with a complicated process."[57]

One way of avoiding the infinities was the "cutoff method" in which divergent integrals were cut off at some high value of the frequency. Such methods were generally applied in quantum electrodynamics at the time and had been used by Dirac as early as 1934 (see Chapter 6). But in the forties, Dirac argued that in classical theory it was unjustified to alter the physical laws of interaction in order to make the integrals converge, which was the physical meaning of cutoff procedures. In 1941, he argued that "the correspondence between the quantum and classical theories is so close that one can infer that the corresponding divergent integrals in the quantum theory must also be due to an unsuitable mathematical method."[58] Adding to the unattractiveness of the cutoff technique was that although it would lead to finite results, it would destroy relativistic invariance, a consequence Dirac found unacceptable.

In 1946, at a conference in London on Low Temperatures and Fundamental Particles, Dirac briefly considered the possibility that the trouble of quantum electrodynamics might be a result of the assumption of pointlike electrons. But at the time he saw no advantage of building up a theory in which the electron was an extended particle.[59] Whatever would turn out to be the solution, Dirac was firmly convinced that it would result from drastic changes in the mathematical structure of the theory. The conviction that progress lay ultimately in mathematics, and not in changes in physical interpretation, was deep-rooted in Dirac. The following quotation from 1941 expresses well this persistent element in his thinking:[60]

To have a description of Nature is philosophically satisfying, though not logically necessary, and it is somewhat strange that the attempt to get such a description should meet with a partial success, namely, in the non-relativistic domain, but yet should fail completely in the later development. It seems to suggest that the present mathematical methods are not final. Any improvement in them would have to be of a very drastic character, because the source of all the trouble, the symmetry between positive and negative energies arising from the association of energies with the Fourier components of functions of the time, is a fundamental feature in them.

Dirac spent most of the period 1946–51 in North America. In the spring semester of 1946, he was at Princeton University, and from the fall of 1946 to the summer of 1948 he served for three semesters as a visiting professor at the Institute for Advanced Study. During this period, quantum electrodynamics experienced an important breakthrough, known as the "renormalization" program. At Shelter Island near New York, a conference organized by Oppenheimer took place from June 2 to 4, 1947, the participants being primarily young American physicists including Julian Schwinger, Richard Feynman, Willis Lamb, Hans Bethe, Victor Weisskopf, Abraham Pais, and David Bohm.[61] Dirac did not attend. At the conference Schwinger and Feynman proposed new ideas to reformulate relativistic quantum theory so as to get finite answers for measurable quantities. From the discussions emerged a formalism, renormalization theory, that allowed one to subtract infinities yet end up with an unambiguous answer. The new quantum electrodynamics was developed in close interaction with experiments, especially those dealing with the magnetic moment of the electron and the hydrogen fine structure. The magnetic moment of the electron was found to differ slightly from the value predicted by the Dirac equation, and Schwinger was able to use his theory to calculate (in fact, predict) the correct result in excellent agreement with experiment. At the Shelter Island conference Lamb presented measurements of a small splitting-up of a spectral line in hydrogen that according to the Dirac equation was degenerate, a phenomenon that became known as the "Lamb shift." Bethe was at once able to account for the shift by means of a simple form of renormalization theory, and shortly afterwards the Lamb shift was calculated with amazing accuracy. These triumphs were substantiated when shortly after the Shelter Island conference it became known that the Japanese physicist Sin-Itoro Tomonaga had independently developed a theory of quantum electrodynamics that gave the same results as the theories of Schwinger and Feynman. In 1948, Freeman Dyson showed that the theories of Schwinger, Feynman, and Tomonaga were equivalent. A new paradigm was born and, as Weisskopf has expressed it, the "war against infinities" was ended.[62]

The general idea of (charge) renormalization had been introduced by Dirac in his theory of vacuum polarization from 1933–4 and was later developed by Heisenberg and Weisskopf. The more important concept of mass renormalization had been implicitly argued by Kramers in 1938 but was only developed into a manageable theory by Schwinger, Feynman, and Tomonaga. The simplest way to present the mass renormalization approach is perhaps to consider two electrons in different states of binding – say, a bound electron (1) and a free electron (2). For these electrons the finite, observable mass includes an infinite self-mass; formally, one has $m_{obs} = m + \delta m$ where m is the "bare mass" of the electron. Although the δm's are given by divergent integrals, they can be subtracted to obtain

$$m_{obs}^{(1)} - m_{obs}^{(2)} = \delta m^{(1)} - \delta m^{(2)}$$

Suitably calculated, this yields a finite residue. Since electron 2 is free, the difference is the contribution of the self-mass to the binding energy of the bound electron. This contribution is measurable, and hence renormalization calculations may be compared with experiments. In general, subtractions of infinite quantities are entirely ambiguous, but unambiguous results can be obtained if the subtraction procedure is formulated in a

manifestly Lorentz-invariant way. The renormalization theory of
Schwinger, Feynman, and Tomonaga included such a procedure.

As noted, Dirac did not participate in the Shelter Island conference,
and it appears that he was not in close contact with what went on; for
example, he first learned of the experiments on the Lamb shift through a
clipping from the front page of the *New York Times*.[63] But of course,
Dirac was not unaware that the young American physicists were attacking
the problems of quantum electrodynamics in their own ways. Before the
Shelter Island gathering, at the bicentennial conference of Princeton University in 1946, he had met for the first time with the twenty-eight–year–
old Richard Feynman, a former student of John Wheeler. In addition to
Feynman and Wheeler, the conference was attended by many celebrities,
including A. H. Compton, Bohr, Fermi, Kramers, Oppenheimer, Weyl,
Tolman, and Van Vleck.[64] On September 24, Dirac presented a paper on
"Elementary Particles and Their Interactions," which, in spite of its title,
dealt almost exclusively with his theory of the classical electron and the
attempts to quantize it. Lew Kowarski recalled that this paper was "in
very beautiful English, very clearly given, . . . but, as it seemed to me,
hardly anybody could understand a word because the paper was strictly
on Dirac's own level."[65] The audience might have expected a talk dealing
with other elementary particles too, in particular with mesons, which at
that time were at the forefront of the then new elementary particle physics. But Dirac believed not only that a satisfactory quantum electrodynamics was a precondition for a physics of elementary particles but also
that "electrons are presumably the simplest kind of charged particle that
can exist."[66] Then, as later, he saw no prospect in engaging in the physics
of other particles until the theory of the electron was well understood.
Feynman, who was discussion leader and had studied Dirac's manuscript
in advance, was asked to criticize Dirac's presentation.[67] He was rather
hard in his criticism, arguing that his famous senior colleague was "on
the wrong track" and that this was rooted in Dirac's insistence on using
Hamiltonian methods. About a year later, after the Shelter Island conference, Dirac and Feynman met again, at a seminar at the Institute of
Advanced Study where Feynman discussed "Dirac's Electron from Several Points of View." One of these views was Feynman's recent path integral approach. Although this was not Dirac's approach, he recognized
Feynman's genius and was impressed by his presentation.[68]

From March 30 to April 2, 1948, a follow-up conference to the Shelter
Island conference was held at Pocono Manor, Pennsylvania, where
Schwinger and Feynman were able to present their fully developed the-
ories. Dirac participated in the Pocono conference together with N. Bohr,
W. Heitler, G. Wentzel, E. Wigner, and others, and there got an oppor-
tunity to discuss the new electrodynamics with its creators. Half a year

later, he attended the eighth Solvay Congress in Brussels, where Oppenheimer gave a survey of the present state of quantum electrodynamics in front of Bohr, Pauli, Bhabha, Casimir, Dirac, and others. In the discussion following Oppenheimer's report, Dirac expressed his dislike for renormalization methods and his hope of avoiding infinities by looking for solutions to the wave equations without using perturbation methods.[69] In August and September 1949, he attended the Canadian Mathematical Seminar in Vancouver. In April of the following year, he participated in a large conference in Paris together with, among others, Pauli, Rosenfeld, Kemmer, Heitler, de Broglie, Dyson, and Feynman; Dirac gave an address in which he suggested a new and general formulation of relativistic quantum mechanics.[70] Later in the year, he lectured at Turin University in Italy before going back to Canada, where he spent most of a year. In July 1951, Dirac attended a conference at Bohr's institute in Copenhagen. At all these places, renormalization quantum electrodynamics was a topic of discussion, but Dirac did not share the view of most physicists, that the theory was satisfactory. In 1952, he was awarded the Copley medal "for his remarkable contributions to the quantum theory of elementary particles and electromagnetic fields."[71] At that time, he had decided that the new quantum electrodynamics could not possibly be correct.

Although acknowledging the empirical success and social appeal of the theory, Dirac found it completely unacceptable. This hostile attitude toward (standard) quantum electrodynamics remained throughout the rest of his life, much of which he spent in searching for an alternative theory. He hoped to repeat the success of his youth, when quantum mechanics was discovered by people who did not really know what they searched for. In a talk given in Vancouver in September 1949, Dirac argued that the problems of quantum electrodynamics had not been solved at all with the new renormalization theory. "What we need and shall strive after," he said, "is a change in the fundamental concepts, analogous to the change in 1925 from Bohr to Heisenberg and Schrödinger, which will sweep away the present difficulties automatically."[72] Dirac was very fond of this historical analogy, and he used it again and again. "The present work of Lamb and Schwinger is, in my opinion, somewhat analogous to the work on the development of the Bohr Theory which people were making shortly before the discovery of Heisenberg and Schrödinger's quantum mechanics."[73] He never fully accepted that the problems of quantum electrodynamics had been solved by a new generation of young physicists whose approach was essentially conservative and instrumentalistic. Relativity and quantum mechanics had been established by revolutionary steps, involving major conceptual changes, and Dirac firmly believed that a new revolution was needed. His lack of sympathy for the new quantum electrodynamics involved a lack of appreciation for the values of the new generation of physicists. As expressed by Samuel Schweber:[74]

The workers of the 1930s, particularly Bohr and Dirac, had sought solutions to the problems in terms of revolutionary departures. . . . The solution advanced by Feynman, Schwinger, and Dyson was in its core conservative: it asked to take seriously the received formulation of quantum mechanics and special relativity and to explore the content of this synthesis. A generational conflict manifested itself in this contrast between the revolutionary and conservative stances of the pre- and post-World War II theoreticians.

What, then, were Dirac's objections to postwar quantum theory? Basically, they reflected his general view of theoretical physics, in particular his aesthetically founded conviction that physics must build on "beautiful" mathematics (see Chapter 13). When in 1950 Dyson asked Dirac what he thought of the new development in quantum electrodynamics, he answered, "I might have thought that the new ideas were correct if they had not been so ugly."[75] He strongly felt that quantum electrodynamics, whether in its 1936 version or its 1950 version, was intolerable because it was "complicated and ugly."[76] This was a view that he kept with few changes until his death and that he repeated monomaniacally. The following quotation, dating from a talk given in 1975, is another characteristic expression of this attitude:[77]

Most physicists are very satisfied with the situation. They say: "Quantum electrodynamics is a good theory, and we do not have to worry about it any more." I must say that I am very dissatisfied with the situation, because this so-called "good theory" does involve neglecting infinities which appear in its equations, neglecting them in an arbitrary way. This is just not sensible mathematics. Sensible mathematics involves neglecting a quantity when it is small – not neglecting it just because it is infinitely great and you do not want it!

Statements similar to this occur frequently in Dirac's later publications. The core of the argument was that he would not tolerate theories that departed from what he called "sensible" or "sound" mathematics. In his view, renormalization quantum electrodynamics involved "a drastic departure from logic. It changes the whole character of the theory, from logical deductions to a mere setting up of working rules."[78] Again, at an American Institute of Physics conference in Florida in 1981, the seventy-eight–year–old Dirac delivered an address entitled "Does Renormalization Make Sense?" His answer was an uncompromising "no": "Some physicists may be happy to have a set of working rules leading to results in agreement with observation. They may think that this is the goal of physics. But it is not enough. One wants to understand how Nature works."[79] But what about the impressive and not ad hoc agreement with experiment, so incisively shown by the calculations of the Lamb shift and the anomalous magnetic moment of the electron? Dirac admitted that from an instrumentalist point of view renormalization quantum electrodynamics was undeniably a success. "But the price one must pay for this success is to abandon logical deduction and replace it by working rules. This is a very heavy price and no physicist should be content to pay it."[80]

In the sixties, Dirac developed what he believed was an improved alternative to standard electrodynamics, and he continued to work on the new scheme for the rest of his life. Most of his work on revising the foundation of quantum mechanics took place in the years 1962–5 during his stay at Yeshiva University's Belfer School of Graduate Science. In the fall of 1965, he received the Belfer School of Science Award, and on the same occasion a Dirac Chair in physics was established at Belfer. I shall not deal with Dirac's theory in detail but shall merely outline its main features, which in some respects differed drastically from ordinary quantum theory.

Quantum mechanics has traditionally operated with two representations or "pictures," known as the Schrödinger picture and the Heisenberg picture. In the usual Schrödinger theory the dynamical state of a system is given by a state (ket) vector or wave function that varies with the time, $|u(t)\rangle$; the physical quantities are given by stationary observables A, and the equation of motion is given by the Schrödinger equation

$$i\hbar \frac{d}{dt} |u\rangle = H|u\rangle$$

In the Heisenberg picture, on the other hand, no use is made of wave functions, and the dynamic state of the system is given by a stationary state vector $|u\rangle$; the observables vary with the time according to the Heisenberg equation

$$i\hbar \frac{dA}{dt} = AH - HA$$

The two pictures are connected in the sense that the Schrödinger picture can be transformed into the Heisenberg picture, and vice versa, by means of a unitary transformation, and then the two pictures become equivalent.

In practice, in quantum field theory and elsewhere, the Schrödinger picture came to be commonly used. In 1965, Dirac argued that this orthodoxy was false, that the two pictures were not equivalent, and that the Schrödinger picture was unsuited to meet the demands of quantum field theory. "All references to Schrödinger wave functions must be cut out as dead wood," he said.[81] Quantum theory should be based solely on the Heisenberg picture, he argued, because only the Heisenberg equation had reasonable solutions. Although Dirac was "really very loath to give it [the Schrödinger picture] up," he felt forced to base his reconstruction of quantum theory on this strategy.[82] It had a touch of historical irony, since it was Dirac himself who had first, in his transformation theory of 1926, demonstrated the equivalence between the two pictures.

In ordinary quantum mechanics states are represented by vectors in Hilbert space, but Dirac believed that the abandonment of the Schrödinger picture implied that the Hilbert space formalism had to be abandoned too. Furthermore, this step implied changes in the usual probabilistic interpretation of quantum mechanics. In the new theory the variables describing physical states could not, in general, be interpreted in terms of probabilities. For example, if the variable corresponding to a Schrödinger wave function was normalized at one time, it would not remain normalized. For that reason Dirac used the word "intensity" rather than probability to denote the square of the coefficients of the dynamic terms in his theory. The new theory also involved a greater amount of indeterminacy than did ordinary quantum mechanics. While in the usual theory, where Schrödinger's equation of motion held good, $|u(T)\rangle$ was determined by $|u(t)\rangle$ when $T > t$, in Dirac's theory the corresponding variables were not determined by their values at an earlier time.

Dirac managed to calculate the Lamb shift as well as the anomalous magnetic moment of the electron in his alternative theory.[83] He got results in good agreement with experiments, thus apparently duplicating the empirical success of renormalization quantum electrodynamics. However, his new theory did not provide any results not already known. The recalculation of the Lamb shift and the magnetic moment hardly impressed other physicists who did not share Dirac's strong commitment to (what he felt was) logical methods. Comparing standard quantum electrodynamics with his alternative theory, Dirac wrote:[84]

There is much similarity in the details of calculations in the two theories. But there is the underlying difference that the present calculations all follow logically from certain general assumptions applied to a suitable Hamiltonian, while the previous calculations made use of working rules without a logical connection between them.

Dirac's consistent work with the Heisenberg picture eliminated many of the infinities, but not all of them. While in earlier works he had objected to the use of cutoff procedures as a means to avoid infinities, he now warmly advocated cutting off the high energy part of divergent integrals. This was not an ideal solution, he admitted, since Lorentz invariance was then destroyed for high energies. But he considered this blemish a lesser evil than abandoning the logic of mathematics. He argued as follows:[85]

With a cut-off we eliminate at once all the difficulties about divergent integrals which have been plaguing theoretical physics for decades. These difficulties arise only because people want to have strict Lorentz invariance in an imperfect theory. In doing so they are aiming for something which may very well be impossible. They are setting their sights too high. They should adopt a more modest attitude and not require strict Lorentz invariance, and then they need not be disturbed by a cut-off.

Dirac's proposal for reconstructing quantum theory was a failure. It was isolated from the mainstream of quantum physics, and in the absence of new results arising from the theory, nobody felt induced to take it up. Dirac himself eventually reached the conclusion that his new quantum theory was a failure and that "this method with the cutoff is not a method to be recommended, and I don't think there's any future in it."[86] However, this recognition did not imply an acceptance of standard quantum electrodynamics.

Dirac's endeavors to improve quantum theory took many directions, some of which may seem rather desperate. In accordance with one of his favorite ideas, he often argued that what was needed was a new mathematical basis on which to build quantum theory. In one of his last papers he proposed that "pathological" representations of the Lorentz group would probably be important in the physics of the future of which he dreamt.[87] Dirac realized that his research program was unorthodox and, on the whole, lacked coherence: it consisted of a series of theories and approaches whose only common thread was that they were all attempts to replace existing theory. In 1974, at a conference at the Argonne National Laboratory, he said, "My work has been concentrated on how to get an improved quantum electrodynamics and once I feel that a certain line of work is not going to help in that direction, I lose interest in it."[88]

In Dirac's program of reconstruction, the search for Hamiltonian functions in conformity with the Heisenberg equation played an important role. In 1970 he found such a Hamiltonian, which he used to propose a new relativistic wave equation. From a formal point of view the new equation was very similar to the celebrated linear wave equation he first wrote down forty-two years earlier.[89] In units where $\hbar = m = c = 1$, the new wave equation, written in the Schrödinger picture, took the form

$$\left(\frac{\partial}{\partial x_0} + \alpha_j \frac{\partial}{\partial x_j} + \beta\right) q\psi = 0 \qquad (j = 1,2,3)$$

where the α and β quantities are 4×4 matrices satisfying certain anti-commutation relations. However, contrary to the 1928 equation, the new wave function had only one component. Also in contrast to his 1928 theory, Dirac supplied the particles described by his new equation with an internal structure. In the equation above, the quantity q is a column matrix of four elements, each of which is a dynamical variable associated with the internal degrees of freedom of the particle. The one-component wave function depends on two commuting q-variables, for example

$$\psi = \psi(x_0, x_j, q_1, q_2)$$

The quantity $q\psi$, which formally corresponds to the wave function of the 1928 theory, is then a four-row column matrix. Dirac showed that the new wave equation allowed only solutions with positive energy, contrary to the original theory in which negative energy solutions were also required.[90] The equation did not describe electrons but some unknown particles with zero spin. Dirac's new wave equation was purely hypothetical, a fact emphasized by the result that the particles described by the equation would be unable to interact electromagnetically. Such particles without charge or other electromagnetic characteristics have never been observed.

The relativistic wave equation of the seventies shared the fate of other contributions from Dirac's later years: it was ignored. Apart from Ennackel Sudarshan and co-workers, who generalized the equation to include electromagnetic interactions, physicists showed no interest in it.[91] Still, Dirac followed his own way. In 1982, referring to his positive-energy wave equation, he remarked in a pathetic mood: "I have spent many years looking for a good hamiltonian to bring into this theory, and haven't yet found it. I shall continue to work on it as long as I can. . . ."[92]

CHAPTER 9

ELECTRONS AND ETHER

DIRAC'S scientific work often dealt with subjects that were far from mainstream physics. Typically for such an original mind, he preferred to cultivate new subjects according to his own tastes. He never cared about the fashions of the physics community and accepted his self-chosen isolation. One example is his work on the classical theory of the electron, which started in 1938 and, after a long period of rest, was further developed in the 1950s. His paper of 1938 was an important contribution to electron theory and is still considered to be a classic. On the other hand, Dirac's later attempts to formulate new and better classical theories have largely fallen into oblivion. In the present chapter we shall consider the 1938 theory in some detail and shall also look at Dirac's later works on classical electrons. In the fifties, Dirac explicitly proposed to reintroduce the "aether" into physics. His ether theory was an offspring of his work on classical theories of the electron, and it naturally belongs to the research program initiated in 1938.

At that time, as we have seen, Dirac was very dissatisfied with the state of the art in quantum electrodynamics and desperately searched for new ways to get rid of the infinities that plagued the theory. One strategy toward that end was to base quantum electrodynamics on an improved classical theory. This strategy had been considered earlier, for example, by Oppenheimer, who argued in 1934 that perhaps "the origin of the critique of the theoretical formulae [of absorption of cosmic rays] lies in classical electron theory."[1] But neither Oppenheimer nor others seriously developed the idea before it was taken up by Dirac at the beginning of 1938. Although Dirac's theory of the electron was a classical one, very much in the tradition of Lorentz, Poincaré, and Abraham, it was clearly motivated by his wish to solve the divergence problems of quantum electrodynamics.

By the 1930s, classical electron theory had long ceased to be the interesting field of frontier research that it was before World War I. The rapid

189

development of atomic and quantum theory had endowed the Lorentz theory with an air of the past, and it was only taken up on rare occasions. Contributions to the theory continued to be made by, among others, Y. Frenkel in 1925 and 1934, A. Fokker in 1929, W. Wessel in 1934, and M. Pryce in 1936.[2] But these works did not have the status they once would have had and met with little response from leading quantum physicists.

According to classical electron theory, as developed between 1890 and 1910 by H. A. Lorentz, H. Poincaré, M. Abraham, and others, the electron is a particle of finite size, say a spherical distribution of electricity. If it moves in an external electromagnetic field, it is subjected to the Lorentz force

$$m\frac{d\vec{v}}{dt} = e\left(\vec{E} + \frac{\vec{v}}{c} \times \vec{B}\right)$$

However, since each part of the extended electron repels every other part with a Coulomb force, there is also a self-force

$$\vec{F}_{\text{self}} = \int \left(\vec{E} + \frac{\vec{v}}{c} \times \vec{B}\right)\rho dV$$

where $\vec{E}$ and $\vec{B}$ are the fields produced by the electron itself. As Lorentz showed in 1906, the self-force can be written as an infinite series of the form

$$\vec{F}_{\text{self}} = -\alpha\frac{e^2}{ac^2}\dot{\vec{v}} + \frac{2}{3}\frac{e^2}{c^3}\ddot{\vec{v}} - \beta\frac{e^2 a}{c^4}\dddot{\vec{v}} + \cdots \tag{9.1}$$

where α and β are coefficients whose value depends on the assumed structure of the electron, for example, its radius and charge distribution. The dots represent differentiation with respect to time. For a uniformly charged sphere of radius a, the coefficient of the first (acceleration) term represents the electromagnetic mass, equal to $\alpha(e^2/ac^2)$.

The problems of classical electron theory were closely related to the self-force, which would seem to imply that the electron is unstable: For a point electron ($a = 0$) the third and higher terms vanish, but then the first term becomes infinite; if $1/a$ is kept finite, the equation of motion contains not only an acceleration term but also derivatives of the acceleration to all higher orders. Pre-relativistic theory operated with two kinds of masses, the inertial mass (m) and the electromagnetic mass (m'). With these the content of equation (9.1) could be expressed as

$$m\dot{\vec{v}} = -m'\dot{\vec{v}} + \frac{2}{3}\frac{e^2}{c^3}\ddot{\vec{v}} \tag{9.2}$$

in which third and higher derivatives of v are neglected. The structure-independent $\ddot{v}$ term acts as a resistance of motion due to energy loss by radiation. The relativistic generalization of this term was calculated by Max von Laue in 1909. In covariant notation it reads

$$\frac{2}{3}\frac{e^2}{c^3}(\ddot{v}_\mu - \dot{v}_\nu\dot{v}^\nu v^\mu)$$

where v_μ denotes the relativistic four-velocity $dx_\mu/d\tau$ and the dots mean differentiation with respect to the proper time τ. The corresponding generalized equation of motion is

$$(m + m')\dot{v}^\mu = \frac{2}{3}\frac{e^2}{c^3}(\ddot{v}_\mu - \dot{v}_\nu\dot{v}^\nu v^\mu) + F^\mu \qquad (9.3)$$

where an external four-force F^μ has been added to the self-force. Equation (9.3) is only approximate and still does not allow for point electrons, for m' would then become infinite.

When Dirac took up the matter, he followed an approach that earlier had been cultivated by the Dutch physicist Adriaan Fokker, a student of Lorentz.[3] In his generalization of classical electron theory Fokker realized that in problems of relativistic interaction between several particles it was not even possible to set up a classical Hamiltonian formalism. But, Fokker argued, in order to solve the corresponding quantum problem, the Hamiltonian had to be known, which would necessitate an extension of the classical formalism. Fokker thus applied what has been called the inverse principle of correspondence, letting the problem of quantum theory determine the development of classical theory.[4] This was also the philosophy adopted by Dirac in 1938. The Dirac electron was, as in earlier works on quantum theory, a point electron. As Dirac argued, "The electron is too simple a thing for the question of the laws governing its structure to arise."[5] In accordance with his general view of physics, he did not attempt to build up a new model of the electron but tried to "get a simple scheme of equations which can be used to calculate all the results that can be obtained from experiment."[6] It was always Dirac's ideal, in formulating a physical theory, to be able to work out a "reasonable mathematical scheme" and then interpret the equations in the most direct and natural way. He elaborated:[7]

The scheme must be mathematically well-defined and self-consistent, and in agreement with well-established principles, such as the principle of relativity and the conservation of energy and momentum. Provided these conditions are satisfied, it should not be considered an objection to the theory that it is not based on a model conforming to current physical ideas.

Dirac retained the usual Maxwellian electrodynamics, the basic equations of which are ($c = 1$)

$$\frac{\partial A_\mu}{\partial x^\mu} = 0, \qquad \Box A_\mu = 4\pi j_\mu \tag{9.4}$$

The corresponding field quantities can be derived from the potentials by

$$F^{\mu\nu} = \frac{\partial A^\nu}{\partial x_\mu} - \frac{\partial A^\mu}{\partial x_\nu} \tag{9.5}$$

Usually only the retarded fields are taken into account, and the advanced fields are dismissed as nonphysical solutions. But Dirac stressed that the advanced field should play a role symmetrical to that of the retarded field.

If a moving electron interacts with an electromagnetic field, the actual field can be written as

$$F^{\mu\nu} = F^{\mu\nu}_{\text{ret}} + F^{\mu\nu}_{\text{inc}} \tag{9.6}$$

where the last term is the field representing the incoming electromagnetic waves on the electron. Since the advanced field is a time-reversed retarded field, the relation can also be written as

$$F^{\mu\nu} = F^{\mu\nu}_{\text{adv}} + F^{\mu\nu}_{\text{out}} \tag{9.7}$$

where F_{out} is the field of outgoing radiation leaving the electron. The radiation produced by the electron is then

$$F^{\mu\nu}_{\text{rad}} = F^{\mu\nu}_{\text{out}} - F^{\mu\nu}_{\text{inc}} = F^{\mu\nu}_{\text{ret}} - F^{\mu\nu}_{\text{adv}} \tag{9.8}$$

Dirac also used the $F^{\mu\nu}$ to define another field

$$f^{\mu\nu} = F^{\mu\nu} - \tfrac{1}{2}\,(F^{\mu\nu}_{\text{ret}} + F^{\mu\nu}_{\text{adv}}) \tag{9.9}$$

By using the mean of the retarded and advanced fields, rather than just the retarded field, for the force with which the electron acts on itself, Dirac was able to avoid the divergent v term in the self-force. The technique he used corresponded to a simple form of mass renormalization [on the left side of equation (9.3)].

Dirac then imagined the world line of the point electron to be surrounded by a very thin tube of radius ϵ corresponding to the radius of an extended electron. By means of the laws of conservation of energy and momentum, he proved that

$$\frac{e^2}{2\epsilon} \dot{v}_\mu - ev_\nu f''_\mu = \dot{B}_\mu \qquad (9.10)$$

where B_μ is an undetermined vector and f''_μ is the field defined by equation (9.9). In order to fix B_μ, Dirac assumed for simplicity that

$$B_\mu = kv_\mu \qquad (9.11)$$

with k as a constant. There were other possibilities for B_μ, but since these were more complicated, "one would hardly expect them to apply to a simple thing like an electron." Substituting equation (9.11) into equation (9.10) yielded

$$\dot{v}_\mu \left(\frac{e^2}{2\epsilon} - k \right) = ev_\nu f''_\mu \qquad (9.12)$$

An infinity appears implicitly in this equation since ϵ tends to zero for the point electron. In order to get a definite limiting form, Dirac assumed that

$$k = \frac{e^2}{2\epsilon} - m$$

where m is a constant, the rest mass of the electron. Then equation (9.12) became

$$m\dot{v}_\mu = ev_\nu f''_\mu \qquad (9.13)$$

which is the usual form of the Lorentz equation if f''_μ plays the part of the external field. To get an equation involving F_{inc}, equation (9.13) was transformed by means of equations (9.6), (9.8), and (9.9). This gave

$$f''_\mu = F_{\text{inc}} + \frac{1}{2} (F_{\text{ret}} - F_{\text{adv}}) = F_{\text{inc}} + \frac{1}{2} F_{\text{rad}} \qquad (9.14)$$

The radiation field was shown to obey the exact expression

$$F_{\text{rad}} = \frac{2}{3} e(\ddot{v}_\mu v_\nu - \ddot{v}_\nu v_\mu) \qquad (9.15)$$

Multiplying this by $ev_\nu/2$, and applying the formulae $v_\nu v'' = -1$ and $\ddot{v}_\nu v'' = -\dot{v}_\nu \dot{v}''$, the corresponding radiation reaction became

$$\frac{1}{2} ev_\nu F^{\mu\nu}_{\text{rad}} = \frac{2}{3} e^2(\ddot{v}_\mu + \dot{v}_\nu \dot{v}_\nu v_\mu) \qquad (9.16)$$

This was the same as in the Lorentz theory, but in Dirac's version no higher terms were neglected. Finally, the equations (9.16), (9.13), and (9.14) yielded the fundamental formula

$$m\dot{v}_\mu - \tfrac{2}{3}\, e^2(\ddot{v}_\mu + \dot{v}_\nu\dot{v}_\nu v_\mu) = ev_\nu F^{\mu\nu}_{\text{inc}} \tag{9.17}$$

which is known as the Lorentz–Dirac equation. In contrast to the corresponding formula of the Lorentz theory, the Lorentz–Dirac equation was exact and involved neither infinities not structure-dependent terms.

In his analysis of equation (9.17) Dirac called attention to two peculiarities. For one thing, since the equation contains a third derivative in the position, the usual initial conditions (position and velocity) would not suffice to determine the motion. In the case of no incident field, the result would be a whole family of solutions describing the self-acceleration of the electron. Dirac showed that nearly all of these solutions were physically unacceptable because with them (as so-called runaway solutions) the velocity of the electron would tend very rapidly toward the velocity of light. In order to get solutions that occurred in nature, Dirac was led to what he called "the most beautiful feature in the theory," namely, imposing the condition that the *final* velocity of the free electron be constant; obviously, v_μ = constant is a solution to equation (9.17) if $F_{\text{inc}} = 0$. Dirac wrote: "We must obtain solutions of our equations of motion for which the initial position and velocity of the electron are prescribed, together with its final acceleration, instead of solutions with all the initial conditions prescribed."[8] Pointing out a second peculiarity, he further proved that if an external force acted on the electron at time $t = 0$, it would acquire an acceleration just *before* $t = 0$. "It would appear here that we have a contradiction with elementary ideas of causality. The electron seems to know about the pulse before it arrives and to get up an acceleration (as the equations of motion allow it to do), just sufficient to balance the effect of the pulse when it does arrive."[9] Dirac seemed to accept this pre-acceleration as a matter of fact, necessitated by the equations, and did not discuss it further.

However, Dirac explained that the strange behavior of electrons in his theory could be understood if the electron was thought of as an extended particle with a nonlocal interior. He suggested that the point electron, embedded in its own radiation field, be interpreted as a sphere of radius a, where a is the distance within which an incoming pulse must arrive before the electron accelerates appreciably. With this interpretation he showed that it was possible for a signal to be propagated faster than light through the interior of the electron. He wrote: *"The finite size of the electron now reappears in a new sense, the interior of the electron being a region of failure, not of the field equations of electromagnetic theory, but of some of the elementary properties of space–time.*"[10] In spite of the appearance of superluminal velocities, Dirac's theory was Lorentz-invariant.

The first occasion on which Dirac explained his new theory of classical electrons was in a talk given to the $\nabla^2 V$ Club, probably in March 1938. At that time, Maurice Pryce served as president of the club, and the newly elected Fred Hoyle was its secretary. It was the secretary's duty to find speakers, and Hoyle phoned Dirac to persuade him to give a talk. "When he had understood my request," Hoyle recalled, "Dirac made a remark which nobody else in my experience would have conceived of: 'I will put down the telephone for a minute and think, and then speak again,' he said."[11]

The road to quantization of Dirac's new electron theory turned out to be troublesome. Most quantum physicists were skeptical. Pauli lectured in Cambridge in March 1938 and discussed Dirac's new, still unpublished theory with him. In November, Pauli wrote to him: "I would be very much astonished, if you would have been able to make a progress in the problems of the quantization in the mean time."[12] But Dirac was optimistic and was confident that he was following the right track. In a letter to Bohr he explained how he was proceeding:[13]

Bhabha told me you were in London a few days ago. I was sorry not to meet you then. Schrödinger was in London at the same time and did not know you were there and was sorry not to meet you. He is now staying with us a few days before going to Belgium.

I have spent the whole term working on the quantization of my classical electron theory. The first problem is to express the classical equations in Hamiltonian form. This can be done with the help of a limiting procedure. One has the usual Hamiltonian, $(W/c + eA_0)^2 - (p + eA)^2 - m^2c^2$, one for each electron, and one assumes as the Poisson Bracket relations between the field quantities

$$[A_\mu(\vec{x}'), A_\nu(\vec{x}'')] = g_{\mu\nu} \frac{1}{2} \{D(\vec{x}' - \vec{x}'' + \lambda) + D(\vec{x}' - \vec{x}'' - \lambda)\}$$

the other Poisson Bracket relations being of the usual form. Here D is the usual "4-dimensional δ-function" of Heisenberg and Pauli, and λ is a 4-vector whose direction is arbitrary, provided it lies within the light-cone, and whose length is made to tend to zero. On account of λ, the theory is not relativistic invariant before the passage to the limit (as the direction of λ provides a preferred time-axis), but in the limit it gives exactly the equations of my classical theory, without the condition that the final acceleration is zero. The limiting procedure is effectively the same as Wentzel's (Zeits. für Phys., vols. 86, 87, 1933–34), but it is now put in exact Hamiltonian form.

With the classical theory in Hamiltonian form it is merely a mechanical matter to go over to the quantum theory. I have not yet satisfied myself, however, that the resulting quantum theory has no infinities. From the closeness of the analogy, between classical and quantum theory one would expect that any classical theory from which the infinities have been eliminated would go over into a quantum theory without infinities. To make sure, however, it would be necessary actually to get solutions of the Schrödinger equation and see that they are alright, and I have not yet been able to do this.

I hope you have a good Christmas. I expect I shall be going to the mountains in France or Switzerland but have not yet made detailed plans.

Dirac presented his quantized electron theory, including the λ-limiting procedure, at a conference in Paris in April 1939.[14] Pauli and Wentzel were in the audience. They were not impressed by "Dirac's lectures on subtraction tricks," as Pauli reported to Heisenberg.[15] In reply to a letter of June 29 from Dirac, Pauli made it clear that he did not believe in Dirac's idea of developing a divergence-free classical theory as a basis for subsequent quantization:[16]

I don't think that difficulties of the type in question [discussed by Dirac in his letter] can be removed by mathematical tricks alone, without new physical ideas and I am on the contrary inclined, to draw from your results again the conclusion that the quantum-mechanical formalism, when applied to classical theories with an infinite number of degrees, leads to infinities, even if the corresponding classical model is finite (free from singularities).

In spite of protracted efforts, Dirac was not able to eliminate all infinities in his new quantum theory, which he developed in various ways from 1939 to 1946 (see also Chapter 8). The approach of developing improved classical theories for the electron was at the time followed by a few other physicists in England. Independently of Dirac, his former research student Andrew Lees in 1939 presented a general relativistic theory without considering quantization.[17] Lees had been supervised by Dirac from 1935 to 1938, along with Paul Weiss, his first Ph.D. student. Dirac's theory of point-electrons was criticized in 1945 by T. Lewis, who received Dirac's response two years later.[18] In Cambridge Maurice Pryce made an early contribution to Dirac's theory, which was further developed by Christie Jayaratnam Eliezer and Homi Bhabha.[19] Eliezer, a Ceylonese, became a research student under Dirac in 1941 and took his Ph.D. in 1946; during the war, he returned to Sri Lanka, where he was later appointed professor at the University of Ceylon. Bhabha, who was from India, also went from Cambridge to India during the war, first to become a reader at the Indian Institute of Science in Bangalore and later to assume directorship of the Tata Institute of Fundamental Research in Bombay. In agreement with Dirac's view, Bhabha suggested in 1939 that the problems of quantum electrodynamics "are probably due to the fact that it is not the quantization of the correct classical equations for point particles."[20] While leading to interesting results, the theories of Pryce, Eliezer, and Bhabha did not manage to turn Dirac's classical theory into a satisfactory quantum theory.

In his lecture at Princeton University's bicentennial conference in 1946, Dirac returned to his electron theory. He developed it in a form that did away with the unphysical runaway solutions, and he stated optimistically that "we have now got classical electrodynamics in a satisfactory form." However, he recognized that there were still some problems. For example, Eliezer had shown that the theory, if applied to a proton-electron system, did not permit the two particles to come into contact but only allowed them to be scattered. Yet Dirac did not consider this a very serious objection; he stated: "This is a rather unexpected and disturbing result, but the analogy between classical and quantum mechanics is not sufficiently close for one to be able to infer from it that in the quantum theory also there will be no states for which the electron is bound to the proton."[21] As to the quantized version of the theory, Dirac admitted that he had still not found a satisfactory solution. He tentatively considered the possibility of making a physical change in the theory, namely, of making use of an extended model of the electron. But, as in 1938, he discarded this possibility.

With the advent of renormalization quantum electrodynamics in 1947–8, interest in the Dirac theory largely died out. The mathematical details of the classical theory continued to be studied (by, e.g., Bhabha, Eliezer, and Rohrlich), but as a motivation for a quantum theory it was abandoned. The pre-acceleration featured in the theory has been the subject of some interest among philosophers. Does Dirac's theory really violate causality? If it does, is it a true example of backwards causation?[22]

In 1951, Dirac resumed work on the classical theory of electrons.[23] The new theory he proposed was based on physical assumptions very different from those he had made in 1938. However, its aim was the same, that is, to put forward classical ideas that could be transformed into a quantum theoretic alternative to the then current quantum electrodynamics. Dirac believed it was necessary to put the classical-relativistic theory of interacting particles on a firmer and more general basis, and with this aim he examined generalized dynamics in several papers around 1950. In 1949, he wrote:[24]

The existing theories of the interaction of elementary particles and fields are all unsatisfactory in one way or another. The imperfections may well arise from the use of wrong dynamical systems to represent atomic phenomena, i.e., wrong Hamiltonians and wrong interaction energies. *It thus becomes a matter of great importance to set up new dynamical systems and see if they will better describe the atomic world.*

The dynamic formalism developed by Dirac was more general than the usual Lagrangian or Hamiltonian formalisms. In 1950, he worked out the principles of how to deal with "constrained" dynamic systems, systems in which the (generalized) momenta are not independent functions of the velocities.[25] For such systems, the classical dynamic formalism did not make it possible to choose independent position and velocity coordinates that described the state of the system. In his works on constraints and other aspects of generalized classical dynamics, Dirac made use of the "inverse correspondence principle," clearly stated in the above quotation. He was guided by the structural similarities between classical and quantum mechanics and was pleased to observe that ideas identified in quantum mechanics could appear from a deeper study of classical dynamics. Dirac continued to work on his theory of constrained dynamics throughout the fifties. Together with other authors, he developed it into a powerful framework for the analysis of field theories of a very general sort. The importance of Dirac's formalism was not recognized until the 1970s, when it became applied in a very wide range of areas. Today it forms an important part of such fields as string theory and geometrodynamics.[26]

The most drastic change made by his theory of 1951 was that he no longer considered individual electrons as the raw material in a theory of electrons. He started with the Maxwell equations in the absence of charges; only the ratio e/m figured in the new theory, not the electronic mass and charge separately. "The existence of e should be looked upon as a quantum effect, and it should appear in a theory only after quantization, and not be a property of classical electrons," he argued.[27] He hoped that his new approach would eventually determine the elementary charge in terms of Planck's constant and would then also supply a theoretical value for the fine structure constant, a problem with which theoretical physicists had been occupied for years. It was also an old problem for Dirac and had served, for example, as the background for his 1931 theory of magnetic monopoles.

In electrodynamics the potentials are not completely fixed by the fields appearing in the Maxwell equations. If A_μ coresponds to the field $F_{\mu\nu}$ by means of equation (9.5), the same field can be described by the gauge transformed potential

$$A'_\mu = A_\mu - \frac{\partial S}{\partial x^\mu} \tag{9.18}$$

where S is an arbitrary function with no physical significance. The existence of gauge transformations implies a surplus of mathematical variables compared with those physically necessary. In order to remove these superfluous variables, a definite gauge is introduced, usually through the condition

$$\frac{\partial A_\mu}{\partial x_\mu} = 0 \qquad\qquad (9.19)$$

In 1951, Dirac suggested that in order to get "a more interesting and more powerful mathematical theory" the standard condition (9.19) should be abandoned. The extra variables that would then occur would be "to a certain extent, *at our disposal*, and we shall see that they can be made to serve in the description of electrons, instead of remaining physically meaningless."[28] He suggested that the gauge transformation be destroyed by imposing the condition

$$A_\mu A^\mu = k^2 \qquad\qquad (9.20)$$

which then served as a substitute for equation (9.19). In order to get agreement with the Lorentz equation, the constant k was identified with m/e. The four-velocity v_μ of a stream of electrons was found to be related to A_μ by

$$v_\mu = \frac{1}{k} A_\mu \qquad\qquad (9.21)$$

Schrödinger, himself an outsider in postwar quantum theory, was one of the few physicists who showed an interest in Dirac's new classical theory of electrons. In a letter, now lost, he asked Dirac about some points in his paper of 1951. Dirac replied:[29]

Dear Erwin,

Thanks for your letter. It is not necessarily true that field equations are consistent if they come from a variation principle, (e.g. if one has just one field quantity V and one takes the action density $L = V$). I believe now that my equations are too restricted. It was pointed out to me by D. Gabor that they require the vector $kv_\mu - A_\mu^*$ to be irrotational, while in practice one can easily get this vector to be vortical.

I have not got a more general theory, taking

$$A_\mu = kv_\mu + \xi \frac{\partial \eta}{\partial x^\mu}$$

where ξ and η are two new field quantities. This still gives the Maxwell, Lorentz eqns. and still involves only the ratio e/m. I hope this will meet your objections.

In a series of papers written during 1952–5, Dirac extended the theory along the lines indicated in the letter to Schrödinger, and dealt also with vortical streams of electrons and with several interacting streams.[30] He emphasized that, unlike the 1938 theory, the new theory was not a theory of point charges, but of continuous streams of electricity. Commenting on the ambitious aim of establishing a quantum theory of electrons that would permit the deduction of the fine structure constant, he wrote: "It seems hopeless to attack this problem from the physical point of view, as one has no clue to what new physical ideas are needed. However, one can be sure that the new theory must incorporate some very pretty mathematics, and by seeking this mathematics one can have some hope of solving the problem."[31] The idea of posing classical electrodynamics, and subsequently quantum electrodynamics, in a form that did not make use of the gauge condition [equation (9.19)], or any other gauge, continued to occupy Dirac. In 1955, in lectures given to the National Research Council in Ottawa, he showed that electrodynamics could be formulated in a gauge-invariant way without imposing any specific conditions on the potentials.[32] In these lectures he suggested that in order to explain the elementarity of the electronic charge, one should make use of a quantized version of the old field concept of Faraday and J. J. Thomson:[33]

A reasonable attempt at an explanation could be founded on the assumption that quantization of the electromagnetic field results in the electric field being composed of discrete Faraday lines of force, each associated with the charge e, so that there is just one line ending on each charge e. The lines would then be the elementary concept in terms of which the whole theory of electrons and the electromagnetic field would have to be built up.

The most remarkable feature of the new theory of streams of electrons was that it, as interpreted by Dirac, allowed for the existence of a universal ether. As is well known, the concept of the ether played a predominant role in nineteenth-century physics and was the foundation of the pre-relativistic electron theories of Lorentz, Poincaré, Abraham, and Larmor. The acceptance of Einsteinian relativity killed the ether, which since the 1920s was virtually excluded from physics. Although a few physicists and philosophers still stuck to the ether and opposed the theory of relativity, the ether definitely became a concept beyond the fringe of scientific respectability. To maintain the existence of an ether became a sign of hopeless reaction, incompatible with the march of true physical progress.

Dirac's ether did not belong to the tradition of the anti-relativistic outsiders. He never dreamed of challenging the theory of relativity, which he always regarded as a most beautiful and perfect theory. But he believed that the ether might well be reconciled with relativity after all. Interestingly, a somewhat similar view had been suggested much earlier by no less than Einstein himself, the destroyer of the classical ether; for example, in an address of 1919 he said:[34]

The hypothesis of ether in itself is not in conflict with the special theory of relativity. Only we must be on our guard against ascribing a state of motion to the ether. . . . There is a weighty argument in favour of the ether hypothesis. To deny the ether is ultimately to assume that empty space has no physical qualities whatever. The fundamental facts of mechanics do not harmonize with this view.

The main motivation that led Dirac to suggest his ether hypothesis was clearly his dissatisfaction with quantum electrodynamics. In light of the severe criticism of the classical ether, such a hypothesis might seem desperate, but Dirac was willing to try any idea, however strange, in his search for a better quantum theory. If one idea did not prove feasible, he would leave it and think of another. He did not become particularly committed to the idea of an ether, as were the physicists at the turn of the century, but he merely considered it to be a possible ally in the fight against the fashions of current quantum theory. In one article, he wrote:[35]

I would be quite willing to give up all ideas of the aether if a satisfactory theory could be set up without it. It is only the failure of the world's physicists to find such a theory, after many years of intensive research, that leads me to think that the aetherless basis of physical theory may have reached the end of its capabilities and to see in the aether a new hope for the future.

Dirac's reintroduction of the ether relied not on relativistic arguments but on the new picture of the physical vacuum that quantum mechanics has provided. Already, in Dirac's "sea" of negative energy of 1930, a material vacuum resembling the ether had been implicitly introduced. In contrast to Einstein's rather cautious and largely unarticulated position about some "ether" and its properties, Dirac considered his ether as a real physical quantity characterized by a state of motion. In order to supply the ether with a velocity without violating the principle of relativity, he used arguments based on quantum mechanics.

Dirac's idea of a relativistic quantum-based ether first appeared in 1951 in a paper entitled "Is There an Aether?" It was elaborated on later occasions, especiallly in a lecture Dirac delivered on July 1, 1953, at the third Lindau meeting for Nobel Prize winners.[36] His basic argument was that relativity would only rule out the ether if it was endowed with a definite velocity at a definite point in space–time. However, Dirac argued, if the velocity of the ether was considered to be a dynamic variable, it would be governed by the laws of quantum mechanics and would thus be subject to the indeterminacy relations. The velocity would be distributed over various posssible values according to some probability law. The principle of relativity indeed forbade that there be any preferred direction of space–time, which is a perfect vacuum, but this requirement could be reconciled with the ether hypothesis if one assumed that in a vacuum all velocities of the ether would be equally probable and distributed in a Lorentz-invariant way. That is, in a perfect vacuum the four-velocities v_μ of the ether particles would be uniformly distributed on the hyperboloid

$$v_\mu v^\mu = 1 \qquad (9.22)$$

He explained:[37]

Let us imagine the aether to be in a state for which all values for the velocity of any bit of the aether, less than the velocity of light, are equally probable. . . . This state of the aether, combined with the absence of ordinary matter, may well represent the physical conditions which physicists call a perfect vacuum. In this way the existence of an aether can be brought into complete harmony with the principle of relativity.

Dirac identified the ether velocity with the stream velocity of his classical electron theory, satisfying equations (9.21) and (9.22). Of course, the ether velocity was not simply the velocity of electric charges, since the ether permeated even the regions of the vacuum. But it was the velocity with which small charges would flow if they were introduced. "It is natural to regard it [v_μ] as the velocity of some real physical thing. Thus with the new theory of electrodynamics we are rather forced to have an aether."[38] Dirac thus pictured the ether as a velocity field, although he also described it in more concrete terms as "a very light and tenuous form of matter."[39] At least from a qualitative point of view, Dirac's ether was not so very different from the ether of classical physics.

If the ether were accepted, there would seem to be no reason not to accept absolute time as well. Indeed, at the Lorentz–Kammerlingh Onnes conference held in Leiden in the spring of 1953, Dirac proposed that absolute time be reconsidered. Among his audience in Leiden were Bohr, Pauli, Heisenberg, Peierls, Fokker, Pais, and Lamb. Using arguments similar to those used in defending the ether, Dirac maintained that it was possible to reintroduce absolute simultaneity in a way that would not violate the principle of relativity. The ether, absolute simultaneity, and absolute time[40]

. . . can be incorporated into a Lorentz invariant theory with the help of quantum mechanics, so that there is no reason for rejecting them on the grounds of relativity. Whatever nature has made use of any or all of these devices can be decided only by detailed investigation.

The arguments for Dirac's ether hypothesis were qualitative and rested largely on aesthetic considerations. He believed that the new ether, if developed mathematically, would be able to restore to quantum mechanics the "inherent simplicity" that the current theory lacked. But he was unable to work out a satisfactory quantum theory with absolute time and had to rest content with the conclusion that "one can try to build up a more elaborate theory with absolute time involving electron spins, which one may hope will lead to an improvement in the existing quantum electrodynamics."[41]

Dirac's new ether hypothesis received attention in the newspapers, where it was described as an attempt to restore the pre-Einsteinian ether.[42] As one might expect, it had no impact on the community of physicists. Those who did show an interest did so because of objectives Dirac did not share. Astrophysicists Hermann Bondi and Thomas Gold sought to interpret Dirac's ether in terms of their own theory of continuous creation on cosmological matter. But Dirac denied that the ether velocity appearing in steady-state cosmology had anything to do with his electrodynamic ether.[43] In attempts to develop deterministic, hidden-variable theories of quantum mechanics, Dirac's ether has occasionally been considered.[44] However, Dirac was not attracted by hidden-variables theories and never joined forces with those physicists who criticized the Copenhagen interpretation of quantum mechanics. Although he never elaborated his ether theory beyond the vague state in which it was presented in the years 1951–3, he continued to entertain it. At the 1971 Lindau meeting, he reiterated his stand and added that an ether might prove necessary for the description of certain elementary particles.[45]

Dirac's search for new electron theories on which an alternative quantum theory could be based did not stop with the failure of the ether hypothesis. In 1960, he reexamined the nonlinear Born–Infeld electrodynamics of the thirties, in which the field equations differed slightly from the Maxwell equations.[46] As usual, he worked out a comprehensive action principle and then passed over to a Hamiltonian formalism. Although he found the Born–Infeld theory satisfactory because it avoided both infinities and runaway solutions, he had to conclude that this theory also was unsuited for development into an improved quantum theory.

In 1962, Dirac proposed yet another theory of the electron.[47] He now abandoned the point electron, the virtues of which he had praised since 1928, and replaced it with an electron of finite size. This step was justified, he believed, in view of the fact that the muon could be considered to be a heavy electron. Dirac pictured the electron as "a bubble in the electromagnetic field," a small sphere with a charged surface. The muon was considered to be an electron in its lowest excited state. This model, so similar to the classical models of the Lorentz age, had to be supplied with a non-Maxwellian surface force in order to keep the particle from exploding. This, too, was a familiar problem of pre-relativistic electron theories, where such a force was introduced by Poincaré in 1906. Dirac's new, extended model of the electron was, in spirit and content, curiously similar to the models in vogue before World War I. It would have been appreciated by people like J. J. Thomson, Larmor, Abraham, and Lorentz, but in 1962 it seemed hopelessly outdated. The theory was in some sense a model of the muon, and thus a contribution to particle physics, but it was completely out of harmony with the trend of elementary particle physics in the sixties. Nevertheless, Dirac's idea of picturing the electron as a "bubble" or "bag" turned out to be fruitful for certain aspects of elementary particle physics. In the mid-1970s, physicists reconsidered Dirac's model as a possible model – called the bag model – for hadronic particles. In this case the Coulomb attraction was replaced by the pressure from the quarks, which were assumed to be confined in the "bag."[48]

CHAPTER 10

JUST A DISAPPOINTMENT

I N May 1931, at a time when Dirac was deeply engaged in his new theory of holes, he submitted to the *Proceedings of the Royal Society* a remarkable paper entitled "Quantised Singularities in the Electromagnetic Field." With this paper three new hypothetical subatomic particles were introduced into physics: the anti-electron, the anti-proton, and the magnetic monopole.

A (magnetic) monopole, as the magnetic analogue of the electron, is a particle carrying an isolated magnetic charge. As far as experiments have been able to tell, such particles do not exist in nature. This was well known in the late nineteenth century, when the lack of symmetry between electricity and magnetism became codified in the Maxwell equations:

$$\nabla \cdot \vec{E} = 4\pi\rho, \quad \nabla \times \vec{E} = -\frac{1}{c}\frac{\partial \vec{B}}{\partial t}$$

$$\nabla \cdot \vec{B} = 0, \quad \nabla \times \vec{B} = \frac{4\pi}{c}\vec{j} + \frac{1}{c}\frac{\partial \vec{E}}{\partial t} \tag{10.1}$$

The equation $\nabla \cdot \vec{B} = 0$ expresses the empirical fact that monopoles do not exist. Only in a vacuum, where $\rho = \vec{j} = 0$, do the Maxwell equations appear in a symmetric form. If monopoles are assumed, equations (10.1) are relaced by

$$\nabla \cdot \vec{E} = 4\pi\rho, \quad \nabla \times \vec{E} = \frac{4\pi}{c}\vec{j}' - \frac{1}{c}\frac{\partial \vec{B}}{\partial t}$$

$$\nabla \cdot \vec{B} = 4\pi\rho', \quad \nabla \times \vec{B} = \frac{4\pi}{c}\vec{j} + \frac{1}{c}\frac{\partial \vec{E}}{\partial t} \tag{10.2}$$

where ρ' is the magnetic charge and $\vec{j}' = \vec{v}\rho'$ is the corresponding current density. The Maxwell equations can be written in relativistic notation by introducing the six-vector $F_{\mu\nu}$, as first done by Minkowski in 1908; the components of $F_{\mu\nu}$ are $\vec{E}$ and $\vec{B}$, and its dual six-vector $(F^\dagger)_{\mu\nu}$ is obtained by substituting $\vec{E}$ for $\vec{B}$ and $-\vec{B}$ for $\vec{E}$. Then equations (10.2) read

$$\frac{\partial F_{\mu\nu}}{\partial x_\nu} = -4\pi j_\mu \qquad (10.3)$$

and

$$\frac{\partial (F^\dagger)_{\mu\nu}}{\partial x_\nu} = -4\pi k_\mu \qquad (10.4)$$

The variable x_ν is a point in space–time ($\mu,\nu = 0,1,2,3$), and the velocity of light c is taken to be unity. The motion of an electrically charged particle is given by the Lorentz law, which can be written

$$m\frac{d^2 z_\mu}{d\sigma^2} = eF_{\mu\nu}\frac{dz^\nu}{d\sigma} \qquad (10.5)$$

where $z_\mu = z_\mu(\sigma)$ is the world-line of the particle and σ is the proper time. In analogy with this equation, the equation of motion for a monopole is

$$m\frac{d^2 z_\mu}{d\sigma^2} = g(F^\dagger)_{\mu\nu}\frac{dz^\nu}{d\sigma} \qquad (10.6)$$

where g is the magnetic charge of the pole.

All this was known long before Dirac's work, and monopoles or "magnetic matter" had occasionally been discussed by a few physicists. For example, at the turn of the century Poincaré and J. J. Thomson both briefly examined how electrically charged particles would move in a field generated by monopoles,[1] However, the monopole field discussed by these authors was conceived as the field from a long magnetic rod in which the contribution from the other pole could be ignored. In 1905, the French physicist Paul Villard, known as the discoverer of gamma radiation, believed that he had discovered magnetic monopoles, or, as he termed them, magnetons. Villard's discovery claim was inspired by the symmetry between electricity and magnetism that would then result, but it was soon shown that the "magnetons" were just cathode rays.[2]

More interesting in the present context, since 1885 Oliver Heaviside had preferred to present the electromagnetic equations in a form symmetric between electric and magnetic terms.[3] Heaviside did not believe in the existence of monopoles but found the equations to be simpler and mathematically more satisfactory if the symmetric form was admitted. Dirac knew Heaviside's system well from his time as an engineering student and may have received some inspiration from it, although the problem area which Dirac addressed in 1931 was, of course, widely different from that with which Heaviside worked.

If monopoles remained at the periphery of scientific inquiry in the twentieth century, it was not only because they were not found in nature. With the establishment of the Maxwell-Lorentz theory as a fundamental law of nature, monopoles seemed to be precluded on theoretical grounds. Specifically, the formulation of electrodynamics in terms of potentials (i.e., the vector potential $\vec{A}$ and the scalar potention ϕ), given by equation (9.5) or by

$$\vec{B} = \nabla \times \vec{A} \quad \text{and} \quad \vec{E} + \frac{1}{c}\frac{\partial \vec{A}}{\partial t} = -\nabla \phi \qquad (10.7)$$

would not be possible if monopoles were present. The identity $\nabla \cdot (\nabla \times \vec{A}) = 0$ proves that equations (10.7) are inconsistent with equations (10.2). When quantum theory emerged, the electromagnetic field and its interaction with matter were described in terms of Lagrangian or Hamiltonian formulations of electrodynamics, formulations that were based on potentials rather than field quantities. The usual way of transferring electrodynamics to quantum mechanics – used, for example, by Pauli and Heisenberg in their quantum electrodynamics of 1929 – was crucially based on the existence of a magnetic vector potential, and thus on the implicit denial of monopoles. At any rate, it is a fact that magnetic charges were not considered at all in quantum theory prior to 1931. There was, however, a "magnetic tradition" in England during the 1920s, when Samuel McLaren, Herbert Allen, and Edmund Whittaker, among others, advocated hypotheses involving discrete magnetic Faraday tubes, or, if one likes, monopoles.[4] Although Dirac was probably aware of this work, it was in a tradition very different from that followed by mainstream quantum theorists. As far as quantum theory is concerned, Dirac invented the monopole.

Dirac's original aim was not to work out a theory of monopoles, which he had probably not thought about at all before 1931. The monopole appeared rather as an accidental result, a by-product of a more general line of research concerned with explaining the existence of elementary electrical charges. Many years later, Dirac recalled: "It oftens happens in scientific research that when one is looking for one thing, one is led to discover something else that one wasn't expecting. This is what happened to me with the monopole concept. I was not searching for anything like monopoles at the time."[5] In a lengthy introduction to his paper of 1931,

Dirac stated his general methodology of physics, and this statement, together with the preface to *Principles of Quantum Mechanics,* was the nearest he came to expounding a philosophy of science at the time. Reminding his readers of the unexpected role that abstract mathematical fields such as non-Euclidean geometry and non-commutative algebra had come to play in recent physics (in the theory of relativity and in quantum mechanics, respectively), he wrote:[6]

It seems likely that this process of increasing abstraction will continue in the future and that advance in physics is to be associated with a continual modification and generalisation of the axioms at the base of the mathematics rather than with a logical development of any one mathematical scheme on a fixed foundation. There are at present fundamental problems in theoretical physics awaiting solution, *e.g.,* the relativistic formulation of quantum mechanics and the nature of atomic nuclei (to be followed by more difficult ones such as the problem of life), the solution of which problems will presumably require a more drastic revision of our fundamental concepts than any that have gone before. Quite likely these changes will be so great that it will be beyond the power of human intelligence to get the necessary new ideas by direct attempts to formulate the experimental data in mathematical terms. The theoretical worker in the future will therefore have to proceed in a more indirect way. The most powerful method of advance that can be suggested at present is to employ all the resources of pure mathematics in attempts to perfect and generalise the mathematical formalism that forms the existing basis of theoretical physics, and *after* each success in this direction, to try to interpret the new mathematical features in terms of physical entities (by a process like Eddington's Principle of Identification).

Dirac viewed his recently proposed theory of holes as "a small step according to this general scheme of advance." He used the occasion to revise his identification of negative-energy electrons with protons, and introduced as an alternative the anti-electron. It is not obvious what this had to do with monopoles, but according to Dirac the link was as follows:[7]

The object of the present paper is to put forward a new idea which is in many respects comparable with this one about negative energies. It will be concerned essentially, not with electrons and protons, but with the reason for the existence of a smallest electric charge. This smallest charge is known to exist experimentally and to have the value *e* given approximately by

$$hc/e^2 = 137$$

The theory of this paper, while it looks at first as though it will give a theoretical value for *e*, is found when worked out to give a connection between the smallest electric charge and the smallest magnetic pole.

Dirac's concern with explaining the "magical" number 137 was no doubt inspired by Eddington's recent attempts to deduce the value of the fine structure constant from fundamental theory.[8] Eddington's claim that the value was exactly 1/137 was fascinating but was received with skepticism by most physicists. Dirac was no exception, as intimated by his word "approximately" in the above quotation. But although Dirac rejected Eddington's approach, he did not reject his aspirations. He continued to hint that the value of the fine structure constant should be explainable by physical theory.[9] At some time during the thirties, Dirac apparently played with the idea that the fine structure constant might be related to the temperature concept. In 1935, Heisenberg wrote him:[10]

I don't believe at all any more in your conjecture that the Sommerfeld fine structure constant may have something to do with the concept of temperature; that is, neither do I any more believe in the Lewis value. Rather, I am firmly convinced that one must determine $e^2/\hbar c$ within the hole theory itself, in order that the theory may be formulated in a sensible way. As to the numerical value I suppose $\hbar c/e^2 = 2^4 3^3/\pi$, but that is of course in play.

Dirac showed in his 1931 paper that quantum mechanics allowed for magnetic poles. He noticed with satisfaction that this result was obtained by following the general approach outlined in the introduction to the paper:[11]

The present development of quantum mechanics, when developed naturally without the imposition of arbitrary restrictions, leads inevitably to wave equations whose only physical interpretation is the motion of an electron in the field of a single pole. This new development requires *no change whatever* in the formalism when expressed in terms of abstract symbols denoting states and observables, but is merely a generalization of the possibilities of representation of these abstract symbols by wave functions and matrices.

Dirac's "generalization" was essentially the introduction of a non-integrable phase factor into the wave function. He showed that under certain circumstances this was equivalent to introducing a magnetic field with a point (magnetic) charge as source. The main steps in his reasoning were as follows.

Consider the wave function $\psi = \psi(x_j, t)$, $j = 1,2,3$. This function is only determined to within multiplication by an arbitrary phase factor $\exp(i\gamma)$ since ψ and

$$\Psi = \psi e^{i\gamma} \tag{10.8}$$

give the same probability distribution. In general, the phase γ will be a function of position and time. If, as a special case, γ is a non-integrable function, then it does not have a definite value at each point but has a definite change in value from one point to a neighboring point; that is, it has definite derivatives

$$\partial\gamma/\partial x_j = \kappa_j \quad \text{or} \quad \nabla\gamma = \vec{\kappa}$$

For such a non-integrable phase, the change in phase around a closed loop

$$\Delta\gamma = \oint \kappa_j dx_j$$

will in general be different from zero. Applying Stokes's theorem, the change in phase can be written

$$\Delta\gamma = \int_s (\nabla \times \vec{\kappa})_n dS \tag{10.9}$$

where S is a surface whose boundary is the loop considered. The wave equation, whether relativistic or non-relativistic, involves the momentum operator $p_j = -i\hbar\partial/\partial x_j$ and the energy operator $E = i\hbar\partial/\partial t$; and then, from equation (10.8), one has

$$-i\hbar \frac{\partial\Psi}{\partial x_j} = e^{i\gamma}\left(-i\hbar\frac{\partial}{\partial x_j} + \kappa_j\hbar\right)\psi$$

$$\text{and} \quad i\hbar \frac{\partial\Psi}{\partial t} = e^{i\gamma}\left(i\hbar\frac{\partial}{\partial t} - \kappa_0\hbar\right)\psi \tag{10.10}$$

where $\kappa_0 = \partial\gamma/\partial t$. In spite of the fact that $|\psi|^2 = |\Psi|^2$, the two wave functions will thus satisfy different wave equations. A linear wave equation of the general form

$$H\left(-i\hbar\frac{\partial}{\partial x_j}, x_j\right)\Psi = i\hbar\frac{\partial\Psi}{\partial t}$$

can be transformed into an equation for ψ by means of equations (10.10), yielding the result

$$H\left(-i\hbar\frac{\partial}{\partial x_j} + \kappa_j\hbar, x_j\right)\psi = \left(i\hbar\frac{\partial}{\partial t} - \kappa_0\hbar\right)\psi$$

Thus, if Ψ satisfies any wave equation involving p_j and E, ψ will satisfy the corresponding equation in which p_j has been replaced by $p_j + \kappa_j\hbar$ and

E has been replaced by $E - \kappa_0 \hbar$. This situation is similar to the one in which the electromagnetic field is introduced: in the latter case the equation of motion is the same as in the field-free case if only the substitutions

$$-i\hbar \frac{\partial}{\partial x_j} \to -i\hbar \frac{\partial}{\partial x_j} + \frac{e}{c} A_j \quad \text{and} \quad i\hbar \frac{\partial}{\partial t} \to i\hbar \frac{\partial}{\partial t} - e\phi$$

are made. The analogy thus implies that the introduction of a non-integrable phase factor amounts, in effect, to introducing an electromagnetic field for which

$$\frac{e}{c} A_j = \kappa_j \hbar \quad \text{and} \quad e\phi = \kappa_0 \hbar$$

Then one has

$$\Delta\gamma = \frac{e}{\hbar c} \int_s (\nabla \times \vec{A})_n \, dS = \frac{e}{\hbar c} \int_s B_n \, dS \qquad (10.11)$$

that is, the magnetic flux going through the loop is related to the change in the phase $\Delta\gamma$ around a loop.

The connection between gauge transformations and phase factors of wave functions was well known in 1931. The more general connections between non-integrable phases and the electromagnetic potentials had been included in works by Weyl and by Fock in 1929.[12] Thus, the portion of Dirac's paper detailed above so far contained no new reasoning, and he emphasized that what was shown up to this point was merely that "non-integrable phases are perfectly compatible with all the general principles of quantum mechanics and do not in any way restrict their physical interpretation."[13]

But a new feature was brought in when Dirac considered the change in phase

$$\gamma \to \gamma + n2\pi, \qquad n = 1, 2, 3, \ldots$$

which leaves the wave function unaffected but does affect the result stated in equation (10.11): while the flux is determined, the value of $\Delta\gamma$ will depend on the value of n. This requires equation (10.11) to be generalized to

$$\Delta\gamma + n2\pi = \frac{e}{\hbar c} \int_s B_n \, dS \qquad (10.12)$$

where n is some definite but unknown integer. Usually n is zero in equation (10.12) for very small loops; the magnetic flux will be close to zero, and the change in phase of the continuous wave function must then also be very small. In this case there is no difference between equations (10.11) and (10.12). However, the argument is not valid for regions in space where ψ vanishes. If, for example, ψ is zero, the phase factor is completely undetermined; and if ψ is close to zero, even small changes in ψ may correspond to appreciable changes in γ, so that n has to be nonzero in equation (10.12), while the flux is still close to zero. Since ψ is a complex function, $\psi = \psi_1 + i\psi_2$, its vanishing would require two conditions, one for ψ_1 and one for ψ_2. In general, the points at which ψ vanishes will therefore lie along a line that Dirac called a nodal line. For small loops along a nodal line, $\Delta\gamma$ will be equal to $2\pi n$, with n undetermined but nonzero.

Dirac treated large loops by dividing them up into small loops lying on a surface whose boundary is the large loop. The flux passing through the large loop will be equal to $\Sigma\Delta\gamma$ for all the small loops plus a contribution of $\Sigma 2\pi n$ from the nodal lines cutting the surface,

$$\Delta\gamma = 2\pi\Sigma n + \frac{e}{\hbar c}\int_s B_n\, dS$$

The summation is over all the nodal lines, with one term for each line. For any closed surface, $\Delta\gamma$ must vanish, because the boundary perimeter shrinks to zero; that is,

$$2\pi\Sigma n = -\frac{e}{\hbar c}\int_s B_n\, dS$$

Finally, if one or more nodal lines have their end points inside the closed surface, Σn will not vanish and there will be a net flux crossing the surface,

$$\int_s B_n\, dS = \frac{2\pi\hbar c}{e} n \qquad (10.13)$$

which is the magnetic analogy of Gauss's law for electric flux. Dirac wrote: "Since this result applies to *any* closed surface, it follows that the end points of nodal lines must be the same for all wave functions. These end points are then points of singularity in the electromagnetic field."[14] And he concluded that at the end points of a nodal line there will always be a magnetic monopole the strength of which is given by

$$g = \frac{\hbar c}{2e} n \qquad (10.14)$$

It should be noted that equations (10.13) and (10.14) express a kind of flux quantization, a concept that was only introduced explicitly by Fritz London in 1950 in his theory of superfluidity.[15] From 1934, London based his works on superconductivity on mathematical arguments strikingly similar to those contained in Dirac's 1931 paper, and in 1950 he suggested that the magnetic flux was quantized in multiples of the universal unit $\hbar c/e$, which is the same expression derived by Dirac for magnetic poles. However, in spite of the formal identity of the two results, there seems to be no generic connection between Dirac's monopoles and London's theory. If London was inspired by Dirac, he did not refer to him.

Dirac's argument above showed not only that quantum mechanics would accommodate monopoles but also that magnetic and electric charges were interconnected. Dirac emphasized that if monopoles (or just one monopole) existed, this would amount to an explanation of why electricity is quantized, although then the quantization of magnetic charge would remain unexplained. Dirac found it "rather disappointing to find this reciprocity between electricity and magnetism, instead of a purely quantum condition."[16] He did not change his view. Fifty years later, at a time when monopoles had become highly interesting objects for physical research, Dirac tersely referred to his entire monopole theory as "just a disappointment."[17]

In his work of 1931, Dirac recognized, of course, that for his monopole theory to be consistent, some constraint had to be put on the equation $\vec{B} = \nabla \times \vec{A}$: "There must be some singular line radiating out from the pole along which these equations are not satisfied, but this line may be chosen arbitrarily."[18] Using this idea, later known as the Dirac string, Dirac treated the case of an electron moving in the magnetic field of a monopole. In polar coordinates the equation of motion was found to be

$$-\frac{\hbar^2}{2m}\left\{\nabla^2 + \frac{i}{2r^2}\sec^2\left(\frac{\theta}{2}\right)\frac{\partial}{\partial\phi} - \frac{1}{4\pi r^2}\tan^2\left(\frac{\theta}{2}\right)\right\}\psi = W\psi$$

where W is the energy. Dirac showed that the equation did not allow for stable bound states of the monopole–electron system, but he did not solve the equation. A general solution to the monopole wave equation was, at Dirac's instigation, worked out by Tamm, who at the time was living in Cambridge.[19] Tamm read Dirac's manuscript before publication, and his paper appeared simultaneously with Dirac's. Reporting to someone in Moscow, he told about his work in Cambridge:[20]

My contacts with Dirac were of special value to me. Dirac is one of the leading physicists of our time, and my discussions with him on fundamental matters of quantum theory and the road ahead in it were very valuable. As a follow-up to his latest investigations I did a piece of research in Cambridge – Generalised Spherical Functions and so forth. I was attempting a mathematical investigation of the motion of the electron in an isolated magnetic pole field, the existence of which is ruled out in classical physics theory. However, of late the idea has gained new ground: Dirac's new research has proved the existence of such poles not to be at variance with quantum theory. Dirac thinks that such poles may be important to the structure of the nuclei of heavy atoms . . .

The Dirac monopole was a hypothetical particle, justified only in the sense that it was proved not to be forbidden by quantum mechanics. Dirac realized, of course, that this did not secure the actual existence of monopoles in nature, a question that could only be settled experimentally. However, adopting a plenitude principle, he tended to believe that since there was no theoretical reason barring the existence of monopoles, they would probably exist somewhere in nature. "Under these circumstances," he concluded, "one would be surprised if Nature had made no use of it."[21] But if monopoles existed, why had they not been observed? Dirac suggested that the coupling between a north pole and a south pole was very much larger than the corresponding electrical coupling and that a very large energy would therefore be required to break up a system of bound monopoles. This idea – that oppositely charged monopoles would be very hard to keep apart – was featured a year later in the *Faust* parody performed at the Bohr Institute. At the "Quantum theoretical Walpurgisnight" the monopole enters, singing:[22]

> Two monopoles worshiped each other,
> And all of their sentiments clicked.
> Still, neither could get to his brother,
> Dirac was so fearfully strict!

Dirac's audacious theory received attention in the press but was politely ignored by the physics community.[23] It was presented to the Kapitza Club in a lecture given in July 1931, on which occasion Dirac apparently proposed the name magnon for the magnetic monopole.[24] In 1933, Blackett for a short time may have believed that he had found a monopole track in cloud chamber pictures. According to S. A. Altschuler, Dirac wrote Tamm about this possible discovery but informed him shortly afterwards that it was based on a mistake.[25] Since Blackett had recently confirmed Dirac's anti-electron, it would have been natural for him to look for the monopole also. However, Blackett's observations, whatever they were, never reached publication.

In 1931, Dirac sent a copy of his manuscript to Bohr in Copenhagen, where it was read by the physicists before publication. In the concluding remark, "Under these circumstances . . ." (quoted above), the typist in

Cambridge had inadvertently omitted the word "no," thus reversing the meaning of Dirac's statement. According to Delbrück, who was in Copenhagen at the time, "everybody had overlooked this amusing omission, and when I noticed it and showed it to Bohr he was very happy about it, showing it to everyone and remarking that this version exactly expressed his own attitude to this paper, and to much of Dirac's approach in general."[26] While Bohr was very skeptical toward both Dirac's monopoles and his anti-particles, Delbrück was fascinated by these entities and was quite willing to admit them as real. The ending phrase of Dirac's paper particularly fascinated Delbrück as an example of the principle of plenitude. When he later changed to molecular biology, Delbrück was guided by a similar attitude and often quoted Dirac's words in discussions with biologists.[27]

When Dirac lectured in Princeton in the fall of 1931, he discussed his recent monopole theory at a colloquium on October 1 where Pauli talked of neutrinos (see also Chapter 5). A few days earlier, Pauli had mentioned the forthcoming colloquium in a letter to Peierls, describing it ironically as "a first national attraction" and referring briefly to the new pair of hypothetical particles: "From a purely logical point of view the magnetic poles seem to be more satisfactory than the magnetic dipoles, but then one cannot understand the false statistics of the N-nucleus. It is also perfectly possible that there are neither magnetic poles nor dipole-neutrons."[28] In the early 1930s the neutrino and the monopole were considered in much the same light by the physics community. Both entities were the result of bold hypotheses, if not speculations. Since they broke with the two-particle orthodoxy, they were received skeptically if not outright rejected. However, whereas the neutrino soon found support in Fermi's theory of a β-decay and achieved increasing respectability, the magnetic monopole remained at the fringe of physics for almost three decades. The view of Carl Wilhelm Oseen in his evaluation of Dirac's work for the Nobel Committee was no doubt shared by most physicists: "[The monopole paper] is strongly speculative but of pretty much interest to those who are not afraid of speculation."[29]

The lack of success experienced by the monopole is emphasized by the literature on the subject: During the thirties, it was discussed only nine times, Dirac's original paper included.[30] The only quantum physicist of eminence who seems to have taken Dirac's idea seriously was Jordan, who discussed it twice, in 1935 and 1938. Jordan introduced his 1938 paper on monopoles as follows: "The assumption that such magnetic poles exist has been received with much doubt. In the meanwhile, however, the number of known elementary particles has increased considerably, that one would now rather be inclined to regard the Dirac pole as a possibility worthy of serious investigation."[31]

The British physicist Owen Richardson, a pioneer in thermionic theory and a Nobel laureate of 1928, responded to Dirac's paper by suggesting the existence of "magnetic atoms" as counterparts to the usual electric atoms. Speculating freely, Richardson calculated the spectra and dimensions of such atoms. He suggested that monopoles might exist in the cosmic radiation and also have a significance for cosmology: "The possible existence of such isolated magnetic poles, with properties so very different from those of electrons and protons, obviously changes the basis for discussion of a good many cosmological questions."[32] Also, Frederick Kestler of the University of Kansas thought that monopoles might be of extraterrestrial origin; he suggested privately to Dirac that the high energies of cosmic ray particles might be the result of monopoles accelerated in the earth's magnetic field.[33] Another American physicist, Rudolph Langer at Caltech, speculated in 1932 that the recently discovered neutron might be a combination of two oppositely charged magnetic poles.[34] But all this was speculation that was hardly noticed by leading quantum physicists.

Dirac himself remained silent on the subject for seventeen years. It was only in 1948 that he returned to monopole theory with a substantially improved version of his old theory. At the Pocono conference in April 1948, he lectured on his new theory and shortly thereafter submitted it for publication. What prompted Dirac to take up his old, half-forgotten theory is uncertain, but probably he saw it as a possible answer to the problems of quantum electrodynamics. A few years later, he admitted that "it [the monopole theory] is not satisfactory because it does not enable us to calculate the magnitude of the electronic charge."[35]

Whatever Dirac's reason for resuming work on the monopole in 1948, it was not because the particle had become more respectable or was considered closer to experimental detection. The only one who had searched for Dirac's pole during the preceding years (except perhaps for Blackett, as noted previously) seems to have been Merle Tuve at the Carnegie Institute in Washington, Tuve, a pioneer in high-energy accelerator physics, believed that the discovery of the positron justified a search also for the monopole, and he undertook some experiments with this purpose in mind.[36] No monopoles were detected.

The controversial Austrian physicist Felix Ehrenhaft not only searched for but claimed to have actually found magnetic poles, although these were not necessarily Dirac poles. Ehrenhaft had at that time a low reputation because of his old controversy with Millikan over sub-electronic charges, the existence of which Ehrenhaft continued to defend. In investigations made in the thirties and forties, Ehrenhaft detected particles that behaved like single magnetic charges and later thought that they were perhaps a kind of Dirac pole.[37] He tried in vain to obtain Dirac's support for his claimed discovery, which was either ignored or rejected by the physics community. In 1937, he invited Dirac to give a talk in Vienna, but Dirac declined, with the result that Ehrenhaft felt insulted.[38] During the war, when Ehrenhaft lived in New York as a political refugee and scientific outcast, he persistently but unsuccessfully tried to interest Dirac and others in his discoveries. Dirac received several manuscripts and letters from Ehrenhaft but did not want to get involved with the increasingly frustrated Austrian professor. An arch-experimentalist with little respect for theory, Ehrenhaft expressed his personal frustration and methodological views in a letter to Dirac in 1944:[39]

It is a very sad thing that there are physicists that are called experimental physicists, who are not quite able to understand theories, and there are theoretical physicists who admit they are not able to understand the experiments. As a result, physics seems to be developing in a very wrong way. You say that you as a theoretical man, can not give me help in my problem. It seems to me that since theory can only be built on facts, if one does not include all the facts the theory can not be valid. . . . What kind of physical theory will you offer the young generation returning from the war? If the theory ignores important facts, it is worthless, and I hope you will keep this thought in mind in your speculations on the construction of the atom and matter itself.

Dirac, an arch-theorist with little respect for "facts," was not convinced by Ehrenhaft's argument. He later recalled:[40]

I met Ehrenhaft several times on later occasions, at meetings of the American Physical Society. Ehrenhaft was not allowed by the secretaries to speak at these meetings. His reputation had sunk so low, everybody believed him to be just a crank. All he could do was to buttonhole people in the corridors and pour out his woes. He often talked to me like that in the corridors. I formed the opinion that he was in any case sincere and honest, but he must have given the wrong interpretation to his experiments. He kept saying that he had these experimental results and nobody would listen to him.

In his new paper of 1948, Dirac referred briefly to Ehrenhaft's claim, but only to deny its relevance to monopole theory: "This [Ehrenhaft's result] is not a confirmation of the present theory, since Ehrenhaft does not use high energies and the theory does not lead one to expect single poles to occur under the conditions of Ehrenhaft's experiments."[41]

In accordance with his view of 1931, Dirac pointed out that any search for free monopoles should consider high-energy atomic processes because of the large binding energy for pairs of monopoles. He further pointed out that a monopole would probably leave a strongly ionizing track in a photographic emulsion, its ionization roughly remaining constant as it slowed down (in contrast to ordinary charged particles, which would produce an increasing ionization). This method of detection was to become the standard for monopole hunters later on. As to the way monopoles would exist in nature, if they did, Dirac mentioned casually that they might possibly be constituents of the proton.[42] He also considered the possibility that monopoles might possess electrical charge in addition to their magnetic charge, but he left the question undecided.[43]

Dirac's aim in his 1948 paper was to set up a complete dynamical theory for the interaction of charged particles and monopoles. In order to express the monopole theory in the language of quantum electrodynamics, he had to put the classical theory in Hamiltonian form, for which purpose he needed an action principle. The action principle involves the electromagnetic potentials, and in accordance with his "primitive" theory of 1931 Dirac proposed to change equation (9.5) so that it would fail at just one point on every surface surrounding the monopole. He called the line of these points, extending outward from the pole, a string. A monopole, he argued, could exist only if it was attached to the end of a string. The string idea called for revision of equation (9.5), which Dirac replaced with

$$ F_{\mu\nu} = \frac{\partial A_\nu}{\partial x^\mu} - \frac{\partial A_\mu}{\partial x^\nu} + 4\pi \sum_g (G^\dagger)_{\mu\nu} $$

where $(G^\dagger)_{\mu\nu}$ is a new field quantity that vanishes everywhere except on the string. The $G_{\mu\nu}$ field was fictitious in that it belonged to the string and not to the monopole. Dirac called it an unphysical variable because it did not correspond to anything observable.

Through a series of complicated calculations, Dirac managed to derive electrodynamic action integrals that accommodated the monopole. Working out the variation of the new action integrals, he showed that they gave the correct result, that is, the modified Maxwell–Lorentz equations (10.3), (10.4), (10.5), and (10.6). In order to get the equations, Dirac had to impose the condition that a string could never pass through a charged particle, a condition that later became known as the Dirac veto.[44] Dirac's purpose in establishing the action principle of monopole electrodynamics was to use it as a step for a subsequent quantization. He found the Hamiltonian equation of motion for a monopole and, assuming for simplicity that monopoles had spin one-half, formulated the corresponding quantum wave equation. In this equation there were no terms arising from the effect of the field on the pole. However, this did not imply that the field would not affect the motion of poles. Dirac pointed out that the field would affect the motion of monopoles only through the strings at the ends of which monopoles were constrained.

At the end of his paper Dirac deduced in a new way the result of his 1932 theory, that quantization of the equations of motion was possible only for those values of e and g that satisfied the Dirac condition (10.13). In summing up his theory, Dirac admitted that it was not perfect because it still contained infinite quantities. However, this was a general difficulty of quantum electrodynamics, with or without monopoles. "The occurrence of these difficulties," he stated, "does not provide an argument against the existence of magnetic poles."[45]

Dirac's improved theory of magnetic monopoles placed the subject within the tradition of the then current research front of theoretical physics. Together with his theory of 1931, it became the work on which all later theories of monopoles were based. In a letter to Bethe, Pauli referred to Dirac as "Monopoleon," and added that his attitude to the theory of monopoles was "not unfriendly."[46] In October 1949, Fermi gave a series of lectures in Italy, one of them dealing with Dirac's theory of monopoles. Fermi mentioned that he and Teller were examining the possibility that unidentified cosmic ray tracks were caused by magnetic poles.[47] However, Dirac's theory did not really arouse a positive response from the physics community until much later, and monopole theory remained a fringe subject for another two decades. Dirac himself only returned to the subject in 1975, at which time it had been transformed into an almost fashionable field of research. As shown by the citation structure in Appendix I, monopole theory became a major research area in the mid-1970s.

After 1974, when Gerardus t'Hooft and Alexander M. Polyakov showed that certain gauge field theories like SU(5) predicted magnetic monopoles, much interest was focused on these poles rather than on those originally described by Dirac. The so-called Grand Unification Monopoles (GUMs) in some respects differ markedly from the Dirac poles. They are exceedingly massive and endowed with an internal structure. Also in contrast to the Dirac poles, they do not require a string. GUMs are believed to have been produced in copious number in the very early phases of the Big Bang and to act as catalyzers for the proton decay predicted by Grand Unified Theories. A year after GUMs were introduced, Paul Buford Price and co-workers claimed to have discovered in the cosmic radiation a magnetic monopole of strength $137e$. The claim caused a sensation and was criticized by many physicists, including Luis Alvarez, until consensus decided a year later that the monopole announced by Price was a mistake.[48]

Dirac seems not to have been much interested in these widely announced developments. In August and September 1975, he delivered a series of lectures in different places in Australia and New Zealand. He talked about the monopole theory in Christchurch, New Zealand, on September 12, which was only a few days after Price's discovery claim had

appeared in print, promising to vindicate Dirac's old theory. But Dirac
was cautious about accepting Price's conclusion and stated reservedly
that the possible existence of monopoles was still an open question. He
emphasized that his theory merely proved that monopoles were consistent with quantum mechanics and therefore could exist, and he further
stated: "There is nothing in it, however, to say that these monopoles have
to exist. Whether they exist or not can only be decided by experiment."[49]
In his Lindau lecture given in June 1976, after most physicists had discarded Price's claim, Dirac offered a more optimistic view, stating that
"Price's work points to his having found a magnetic monopole, for it is
difficult to make his observations agree with another explanation."[50]
Although Dirac's attitude toward Price's discovery claim may have
wavered, he felt that in any case Price was justified in publishing his suggestion. In a letter to Alvarez, Price's main antagonist in the monopole
controversy, Dirac wrote in 1978:[51]

I feel you are rather hard on Buford. It sometimes pays to be bold. I remember
Blackett telling me he had evidence for the positron a year before Anderson published his paper about it. Blackett was very cautious and wanted confirmation
before publishing anything, even though he knew about my theory predicting positrons. In the meantime Anderson scooped the discovery on the basis of a single
event, and (apparently) without knowing anything about my supporting theory.
It was very much a leap in the dark for Anderson and it paid off.

As mentioned in Chapter 9, Dirac had then for some years been
engaged in work on a general dynamic theory of streams of matter.
Instead of working with point particles and fields, he based his ideas on
the general dynamic behavior of continuous streams of matter. In 1974,
he investigated the kinematics of a continuous flow of charged matter in
an Einstein gravitational field and set up a comprehensive action principle for its motion.[52] The formulation of dynamic laws in terms of an
action principle always appealed to Dirac as the best and most general
way of formulating physical laws. He used this method in his 1948 theory
of monopoles, as well as in many other cases. In 1974, he argued that[53]

... an action principle is desirable because of the great power that it gives one to
bring in further physical things. One simply has to obtain the action for the further
things, include it in the comprehensive action principle, and then the resulting
equations will describe correctly the interaction of the further things with the
things already present in the theory.

One of these "further things" was the magnetic monopole.
    In 1976, Dirac reconsidered the magnetic monopole in the light of his
recent ideas on stream dynamics.[54] The result was a major change in his

picture of the monopole, which he now considered not to be pointlike but to be endowed with a structure. This change was in agreement with his theory of the electron, proposed fourteen years earlier, in which he had abandoned his lifelong adherence to point electrons (see Chapter 9). In 1976, as in 1962, he believed that what he saw as the unsolved difficulties of quantum electrodynamics justified treating monopoles, electrons, and other elementary particles as extended particles. In his monopole theory of 1976, he built up the monopole as a distribution of magnetic charges, with each element of charge moving according to equation (10.6). In contrast to his electron theory of 1962, he did not introduce any attractive force to keep the charges together and thus had to picture the monopole, (or electron) as an "exploding" particle:[55]

The elements will then be moving apart under the influence of their [magnetic] Coulomb repulsion, so the particle will be exploding. However, it lasts a short time, long enough for one to be able to discuss its equations of motion. The resulting theory is mathematically more satisfactory than any theory involving ill-defined quantities.

Dirac's primary aim was, then as earlier, to set up "satisfactory" equations of motion, not to supply a physical model or picture. He admitted that his new suggestion of exploding particles was "unphysical" but did not regard the obvious conflict with empirical reality a serious objection to the theory. "Now I realize that these equations of motion are departing quite a bit from what the physicist wants, but I think that this blemish in the theory is preferable to the blemish of neglecting infinite quantities," he wrote.[56]

The failure of Price's discovery claim did not stop the monopole hunters, who continued to search for Dirac's elusive particle. In 1981, a conference was held in Trieste on the status of monopole theory and experiments. Dirac's role as initiator of the field was emphasized by reprinting his 1931 paper. He was invited to the conference but, at age seventy-nine, felt unable to travel from Florida to Trieste. In a short letter to Abdus Salam, director of the Trieste institute, he wrote:[57]

I am sorry I cannot come to your conference. It would be too much of a dislocation for me at such short notice. It was very kind of you to invite me. I am inclined now to believe that monopoles do not exist. So many years have gone by without any encouragement from the experimental side. It will be interesting to see if your conference can produce any new angles of attack on the problem.

When Dirac wrote the letter to Salam, writing off the reality of monopoles, experiments designed to detect magnetic poles were under preparation at Stanford University. Blas Cabrera, a high-energy physicist with experience in quark hunting, sought to detect the current induced from monopoles that might pass through a superconducting ring. In February 1982, he recorded a change in magnetic flux that he interpreted as the result of a monopole passing through the ring. Three months later, he reported his discovery, which was, as in Price's case, widely discussed.[58] Although the Cabrera event was not explained in terms of other sources and was not otherwise proved to be a mistake, neither was it confirmed. A single event was not sufficient to change the status of the monopole from being a well-known missing particle to being a real particle.

CHAPTER 11

ADVENTURES IN COSMOLOGY

WHEN Dirac entered the field of cosmology in 1937, it was an unexpected move for a physicist whose entire career had been in quantum theory. Prior to 1937, Dirac had shown very little interest in astrophysics and related areas. Apart from his *Jugendarbeit* on stellar atmospheres, worked out under Milne's supervision, he did not publish on astrophysical subjects. He preferred fundamental problems and considered astrophysics in the same light as solid-state physics, as applied rather than fundamental physics. He was, however, acquainted with the modern development of cosmology and astrophysics. Dirac knew about the cosmological solutions of general relativity from Eddington's textbook on the theory of relativity, and he gained some interest in cosmology from conversations with Howard P. Robertson, the American pioneer in relativistic cosmology.[1] Dirac first met Robertson in 1927 during his stay in Göttingen; Robertson, Dirac's junior by one year, spent the years 1925-7 as a National Research Fellow in Munich and Göttingen and was appointed associate Professor at Princeton in 1929. Later, when Dirac stayed in Princeton, they got to know each other well.

Another motivation for Dirac's growing interest in astrophysical subjects may have been his relationship with the young Indian physicist Subrahmanyan Chandrasekhar, who during the early thirties was a research student under Fowler. Since Fowler was always busy and often away from Cambridge, Dirac worked in practice as Chandrasekhar's supervisor. To Chandresekhar, Dirac commented that if he were to take up astronomical subjects, he would be more attracted by cosmology and general relativity than by astrophysics proper.[2] But Chandrasekhar himself kept to astrophysics, specializing in the physical conditions in the interior of stars. On one occasion, in 1932, Chandrasekhar wrote a paper for the *Proceedings of the Royal Society* that was communicated by Bohr. In this paper he dealt with certain questions of quantum statistics and referred in this context to Dirac, who, however, did not approve of Chandrasekhar's method. Dirac wrote a rather unusual letter to Bohr, requesting that he be allowed to add a critical note to the paper. In the end Chandrasekhar admitted that his argumentation was mistaken and withdrew his paper from publication.[3] But this was a minor incident in Chandrasekhar's career and did not spoil his relationship with Dirac. Later in the thirties, Chandrasekhar became involved in a debate with Eddington concerning relativistic degeneracy in the interior of stars. In discussions in Cambridge in 1938, Peierls, Pryce, and Dirac argued that Eddington had applied relativistic quantum statistics wrongly, and that Chandrasekhar was right.[4]

Also contributing to Dirac's decision to take up a new science may have been the situation in quantum theory at the time. In 1937, he felt that quantum electrodynamics was fundamentally wrong and unattractive; he may have decided to leave the frustrating problems of quantum theory for a period and to devote some of his energy to entirely different problems.

Around 1930, cosmology was in a state of revolution. The static cosmological models of Einstein and de Sitter were recognized to be inadequate, since observational data supplied by Edwin Hubble and others indicated a universe in expansion. In 1927, abbé Georges Lemaître, and some years earlier Alexander Friedmann, had shown that the general theory of relativity contained solutions corresponding to an expanding universe. But the theory of the expanding universe received notice only when Lemaître's paper was translated into English in 1930 and his theory was adopted by Eddington. In the years 1930–4 the expanding universe evolved to become the new paradigm in cosmology.

Lemaître was a postgraduate student in Cambridge from October 1923 to July 1924, studying under Eddington. He must have met Dirac then but did not get acquainted with the shy young student. This happened only later, probably in connection with a talk Lemaître gave to the Kapitza Club in April 1933 on "The Primaeval Hypothesis," that is, his idea of the universe developing from a kind of radioactive super-atom that exploded in a "Big Bang."[5] Later, Dirac met Lemaître on several occasions and got to know him well. Dirac valued Lemaître's scientific contributions highly and wrote in 1968, after Lemaître's death, a biographical survey of his works for the Pontifical Academy of Sciences. From 1960, abbé Lemaître had served as president of the Academy, to which Dirac was elected in 1961. In his survey Dirac recalled:[6]

Once when I was talking with Lemaitre about this question [cosmic evolution] and feeling stimulated by the grandeur of the picture that he has given us, I told him that I thought cosmology was the branch of science that lies closest to religion. However Lemaitre did not agree with me. After thinking it over he suggested psychology as lying closest to religion.

Dirac's cosmology built on the expanding universe and included Lemaître's hypothesis of a beginning of the universe in the distant past. In spirit and content it was indebted to the theories of Eddington and Milne, Dirac's former teachers.

From 1933 until his death in 1950, Edward Milne developed a system of cosmology that was an alternative to the usual cosmological theories based on general relativity. For the purpose of cosmology, Milne considered the general theory of relativity to be philosophically monstrous and wanted to replace it with a system of simple kinematic considerations. Among the results Milne deduced from his new outlook was that the constant of gravitation depended on the epoch.[7] He proposed the relationship

$$G = \frac{c^3}{M_0} t \qquad (11.1)$$

where M_0 is a constant that in usual relativistic cosmology corresponds to the mass of the universe on the assumption of a curved, finite space. According to Milne's conventionalist view, equation (11.1) was valid only in "kinematic time" (t); in usual "dynamic time" (τ), G would reduce to a constant. Milne operated with an infinity of time-scales, from which he considered dynamic and kinematic time (τ-time and t-time) to be of particular physical significance. The two time-scales were related logarithmically by

$$\tau = t_0 \log(t/t_0) + t_0 \qquad (11.2)$$

where t_0 is the present epoch in kinematic time.

In contrast to Milne, Arthur Eddington kept to general relativistic models of the universe. But in opposition to the views of most astronomers and physicists, he was convinced that the cosmological aspects of relativity could be understood only if they were combined with quantum theory. His ideas had their origin in Dirac's 1928 theory of the electron, which impressed Eddington greatly. He elevated Dirac's equation to a status of universal significance and in a number of works applied his own version of the Dirac equation to derive relationships between the macrocosmos and microcosmos, cosmic and atomic constants.[8] Eddington believed that the Dirac equation did not describe an individual electron but instead gave the structural relation of the electron to the entire universe; indeed, in Eddington's philosophy of physics an "individual electron" was a nonsensical notion. Among the relationships derived by Eddington in the mid-thirties were the following:

$$\frac{e^2}{GmM} = \frac{\sqrt{3N}}{\pi}$$

$$\frac{c}{\hbar} \sqrt{\frac{mM}{\wedge}} = \sqrt{\frac{N}{30}} \qquad (11.3)$$

$$\frac{\hbar c}{e^2} = N \frac{\hbar \wedge}{mc}$$

Here m and M denote the mass of the electron and proton, respectively, and N is the so-called cosmical number, the number of particles (protons) in the universe. $\wedge$ is the cosmological constant, a term originally introduced by Einstein in his field equations in 1917 but later abandoned by most cosmologists, including Einstein. Eddington emphasized four interesting properties of the quantities that appear in the relations above: they are dimensionless, they connect the atom with the cosmos, they are independent of the cosmic expansion, and their numerical values are significant.

Dirac was familiar with Eddington's unorthodox theory, not only from his published works but also from talks and private discussions. For example, in March 1933, Dirac hosted a meeting of the $\nabla^2 V$ Club in his room at St. John's, where Eddington gave a talk on "Protons, Electrons and the Cosmical Constant."[9]

Although the physical systems constructed by Milne and Eddington differed in fundamental respects, they had much in common in their general philosophical backgrounds. Both may be seen as manifestations of a particular approach to physical theory, "cosmo-physics," which in the twenties and thirties had considerable appeal in England.[10] The doctrines of cosmo-physics included a peculiar mixture of empiricism and rationalism, which separated this trend from ordinary physical theory. This aprioristic movement in British astronomy and cosmology may have encouraged Dirac to publish his cosmological speculations, for the spirit of cosmo-physics changed, to some extent, the standards for what was acceptable in the field of cosmology. Using this new approach, both Milne and Eddington aimed to perform ambitious reconstructions of the entire physical world-view and believed that the laws of nature could be deduced by rational thought alone. No version of cosmo-physics was received favorably by the majority of physicists, who were uneasy about the rationalism and deductivism of the movement. Born dismissed the

Milne-Eddington approach as "rubbish."[11] Eddington was criticized for his alleged idealism and his claim of being able to bridge cosmology and quantum theory. Most physicists felt that his interpretation of relativity and quantum theory was illegitimate. Dirac was no exception. In 1942, he felt obliged, together with Peierls and Pryce, to protest publicly against Eddington's critique of the standards in quantum mechanics. Referring to Eddington's objections against the customary use of Lorentz transformations, Dirac mildly corrected his senior colleague: "The issue is a little confused because Eddington's system of mechanics is in many important respects completely different from quantum mechanics, and although Eddington's objection is to an alleged illogical practice in quantum mechanics he occasionally makes use of concepts which have no place there."[12]

Although Dirac rejected Eddington's attempt at reconstructing physics, he was to some extent influenced by Eddington's general philosophy of physics and choice of problem areas. Both scientists drifted away from mainstream physics in the thirties, but for different reasons and with different results for their reputations in the physics community. Eddington was perplexed at the almost universal skepticism and indifference that met his theory. He felt that it was more than a bold speculation or imaginative hypothesis, and no more obscure than most of Dirac's contributions, the public success of which he seems to have envied. "I cannot seriously believe that I ever attain the obscurity that Dirac does. But in the case of Einstein and Dirac people have thought it worthwhile to penetrate the obscurity. I believe they will understand me all right when they realize they have got to do so – and when it becomes the fashion 'to explain Eddington,'" he complained in 1944.[13]

Dirac published his ideas on cosmology in a brief note to *Nature* in February 1937, in which he discussed the significance of the following dimensionless combinations of constants of nature:

$$\tau_0 = \frac{t_0}{e^2/mc^3} \simeq 2 \times 10^{39}$$

$$\gamma = \frac{e^2}{GmM} \simeq 7 \times 10^{38} \qquad (11.4)$$

$$\mu = \frac{\rho(c/H)^3}{M} \simeq 10^{78}$$

The symbol t_0 denotes the age of the universe, which Dirac took to be 2×10^9 years, and τ_0 is the corresponding age expressed in atomic units of time (e^2/mc^3 is the time light takes to pass through the diameter of a classical electron). The symbol ρ denotes the mean density of matter in the universe, estimated to be around 5×10^{-31} g/cm^3. H is Hubble's constant, for which Dirac adopted the value 2×10^{-17} sec^{-1}, and c/H is thus the Hubble distance corresponding to the present radius of the universe; these values make μ almost identical with Eddington's cosmical number N. In the spirit of Eddington, Dirac claimed that the regularity exhibited by these large dimensionless numbers was not purely fortuitous. But while Eddington believed the constants to be independent of the cosmic expansion, Dirac regarded them to be contingent quantities, dependent on the history of the universe:[14]

The above-mentioned large numbers are to be regarded, not as constants, but as simple functions of our present epoch, expressed in atomic units. We may take it as a general principle that all large numbers of the order 10^{39}, 10^{78} ... turning up in general physical theory are, apart from simple numerical coefficients, just equal to t, t^2 ... where t is the present epoch expressed in atomic units.

Later, Dirac preferred to call this assumption the Large Number Hypothesis (LNH in what follows). In 1938 he named it the Fundamental Principle, emphasizing that it should be understood as a postulate of correlation between any two of the large dimensionless numbers, whether or not τ_0 was involved. *"Any two of the very large dimensionless numbers occurring in Nature are connected by a simple mathematical relation, in which the coefficients are of the order of magnitude unity,"* he explained.[15] Dirac thus accounted for the vastness of the ratio between electromagnetic and gravitational forces as a consequence of the age of our present universe.

Dirac drew three wide-ranging consequences from the LNH. In his note of February 1937, he focused on the approximate agreement between the present epoch (τ_0), expressed in atomic units, and the ratio of the electric to the gravitational force between two elementary charges (γ). This agreement fascinated Dirac, who, following Eddington, considered it to signify "some deep connexion in Nature between cosmology and atomic theory." The LNH implied that $\gamma(t) = kt$ for all times, k being around unity in magnitude, so that, on the assumption that the atomic constants e, m, and M did not change with time, the "constant" of gravitation would decrease as

$$G \sim t^{-1} \qquad\qquad (11.5)$$

As mentioned, Milne had earlier suggested a time-dependent gravitational constant [see equation (11.1)], but Dirac's result differed from Milne's. It also differed from a suggestion Samuel Sambursky made in 1937; independently of Dirac, Sambursky found an exponential decrease in G.[16]

Another consequence of the LNH that Dirac pointed out in 1937 derived from the fact that

$$\tau_0^2 \simeq (10^{39})^2 \simeq \mu$$

According to Dirac, this implied that the number of nucleons in the universe would increase in proportion to the square of the period:

$$N \sim t^2 \tag{11.6}$$

A third consequence of the LNH, a formula for the recession of the galaxies, came about when Dirac at last worked the cosmic expansion into a more elaborate version of his theory in 1938. Traces of Milne's cosmology showed up distinctly in the 1938 theory, which made wide use of the two time-scales first introduced by Milne and also assumed his so-called cosmological principle, which was first stated in 1933 and held that the large-scale features of the universe should appear the same to any observer, whatever his position in the universe. In Dirac's 1938 paper, he reasoned as follows. Let the cosmic distance function R between receding galaxies be measured in terms of an atomic unit of time, say $e^2 m^{-1} c^{-3}$, so that it becomes a large dimensionless number.[17] Using the model of the expanding universe, the Hubble constant can be expressed as $H = \dot{R} R^{-1}$, where $\dot{R}$ denotes dR/dt. Since H is approximately 10^{-39} in atomic time units, and the mean density of the universe (again in atomic units) was estimated as 10^{-44}, then "allowing for the inaccuracy caused by the uncertainty of which atomic units we ought to use, we see that the average density of matter is of the same order of smallness as Hubble's constant."[18] Using this reasoning and thus applying the LNH to the reciprocal quantities, Dirac found that

$$\rho^{-1} = kH^{-1} = kR(\dot{R})^{-1}$$

Further assuming that matter was conserved in the universe, so that ρ was proportional to R^{-3}, he obtained $R^3 = kR(\dot{R})^{-1}$, from which it followed that

$$R(t) \sim t^{1/3} \tag{11.7}$$

This equation stated Dirac's law of the recession of the galaxies, which implied that the age of the universe was related to the Hubble constant according to

$$t_0 = \frac{1}{3} H^{-1} \tag{11.8}$$

Thus, by 1938, Dirac's cosmological theory yielded three empirically testable consequences, given by equations (11.5), (11.6), and (11.7). They were all problematic. Since the general theory of relativity required a gravitational constant independent of time, equation (11.5) conflicted with Einstein's theory. Dirac was, of course, aware of this disagreement, particularly since he wanted his theory to be expressible in terms of general relativity. In 1938, he hinted that the disagreement could perhaps be resolved by assuming two different metrics, one for atomic phenomena and one for mechanical phenomena, but he did not elaborate the assumption. In any case, there existed no observational evidence whatsoever for the decrease of G over time, and in 1938 Dirac's prediction appeared to be far beyond the possibility of experimental test. According to Dirac's theory, $|\dot{G}G^{-1}|$ was equal to $3H$, which was only about 10^{-11} per year (with the present value of Hubble's constant, the corresponding value is around 10^{-10} per year).

As to equation (11.8), it shared the time-scale difficulty of most other versions of the expanding universe, by implying an absurdly low value for the age of the universe. Dirac's value, $t_0 = 7 \times 10^8$ years, opposed reliable measurements, based on radioactive decay in minerals, that showed the age of the earth to be about 2×10^9 years, and was also in conflict with astronomers' estimate of the age of the galaxies, at least 10^{12} years. Dirac noticed this difficulty but did not regard it as menacing: "This does not cause an inconsistency, since a thorough application of our present ideas would require us to have the rate of radioactive decay varying with the epoch and greater in the distant past than it is now."[19]

Equation (11.6), suggested in the 1937 paper, predicted the spontaneous and accelerating creation of matter in the universe. In his more elaborate theory of 1938, however, Dirac required conservation of matter to derive equations (11.7) and (11.8). He therefore had to change his mind on this crucial point. In February 1937, he had argued:[20]

Present-day physics, both theoretically and experimentally provides no evidence in favour of such an increase [in matter], but is much too imperfect to be able to assert that such an increase cannot occur, as it is so small; so there is no need to condemn our theory on this account.

Ten months later he had this to say:[21]

A spontaneous creation or annihilation of matter is so difficult to fit with our present theoretical ideas in physics as not to be worth considering, unless a definite need for it should appear, which has not happened so far, since we can build up a quite consistent theory of cosmology without it.

Thus, from exactly the same premises, Dirac reached two opposing results, and the only change in argument between his 1937 and 1938 positions was his judgment of the current body of knowledge in physics. Perhaps the change was a reflection of the then current discussion of energy conservation in quantum theory. It seems likely that Dirac's initial willingness to give up matter conservation was related to his interpretation of Shankland's experiment and to his general pessimism over the state of affairs in quantum theory (see Chapter 8). When he recognized that Shankland's result was illusory, his confidence in existing physical theory must have increased and made his physically unfounded suggestion of spontaneous creation of matter unappealing.

The adoption of matter conservation in 1938 gave Dirac a model of the universe that was not very different from Milne's. Dirac concluded that if the universe obeyed matter conservation and the LNH, then it had to be infinite and spatially flat. Discussing his theory in the context of the mechanical metric to which general relativity applied, he further concluded that the cosmological constant $\wedge$ had to be zero; for otherwise, one would be able to construct from $1/\wedge$ a very large dimensionless constant, in conflict with the LNH. In the infinite universe of Dirac's 1938 model, the quantity defined as μ still varied with the square of t because of the LNH, but it no longer signified the total number of particles in the universe, as Dirac, in accordance with Eddington, had assumed in 1937.

Dirac's cosmological theory was not received favorably by people who had observed the rise of rationalistic cosmology with increasing dissatisfaction and annoyance. The astronomer and philosopher Herbert Dingle was among the first to launch a counterattack against cosmophysics. With Milne as his main target, he thundered that "this [Milne's] combination of paralysis of the reason with intoxication of the fancy is shown, if possible, even more strongly in Prof. P. A. M. Dirac's letter in NATURE of February 20 last, in which he, too, appears a victim of the great 'Universe' mania."[22] Exhibiting a rather bad sense of history, Dingle contrasted the alleged inductivism of Galileo and Newton with the Aristotelian methods of Milne, Eddington, and Dirac, and he continued:[23]

Instead of the induction of principles from phenomena we are given a pseudo-science of invertebrate cosmythology, and invited to commit suicide to avoid the need of dying. If anyone is uncertain about the place of imagination in science, let him compare Lord Rayleigh's discovery of argon with Dirac's discovery of the contemporary creation of protons which, according to *The Times*, "alters fundamentally our ideas of the structure of the universe and the nature of time."

Dingle's strongly worded objections caused a heated debate in *Nature* in which many of Britains' most prominent scholars engaged. Among the participants in the debate were the astronomers William McCrea, Gerald Whitrow, and Ralph Sampson, the mathematician and geophysicist Harold Jeffreys, the biologist John B. S. Haldane, the physicist Charles G. Darwin, the philosopher George Hicks, and, of course, Dingle, Milne, Eddington, and Dirac.

Dirac, never much of a debater, kept a low profile in his answer to Dingle. Unlike Milne and Eddington, who clearly enjoyed the controversy and replied to Dingle at length, Dirac was uneasy about the stir. He acknowledged that a proper balance had to be maintained between empirical–inductive and speculative–deductive methods, and believed he had kept such a balance. His "tentative hypothesis" lent itself to experimental verification and built upon the constants of nature as provided by observation. As usual, Dirac avoided philosophical discussion and used most of his reply to restate the main points of his arguments.[24]

Darwin, who had known Dirac since 1924, was one of Britain's most distinguished theoretical physicists. In general, he preferred a more inductive–empirical approach than Dirac, but at the time he felt obliged to defend his famous colleague against Dingle's violent attack and the methodological restrictions he wanted to impose on physical research. He wrote:[25]

It is surely hard enough to make discoveries in science without having to obey arbitrary rules in doing so; in discovering the laws of Nature, foul means are perfectly fair. If Dirac is not to be allowed to conjecture the age of the earth from certain curious numerical coincidences, then Maxwell committed as great a crime in conjecturing that the velocity of light was the same thing as the ratio of the electric and magnetic units. It is absurd to maintain that such guesses are illegitimate.

Darwin remembered well his own attempt to establish a relativistic equation of the spinning electron and how Dirac, by "foul means," had succeeded where he had failed. In a reply to Darwin at the end of this first round of the controversy (which continued during the war years, but without Dirac's participation), Dingle wrote:[26]

I cited Prof. Dirac's letter not as a source of infection but as an example of the bacteria which can flourish in the poisoned atmosphere; in a pure environment it would not have come to birth, and we should still have the old, incomparable Dirac. . . . But my concern is with the general intellectual miasma that threatens to envelop the world of science, and I emphatically disagree with Prof. Darwin's opinion that it does not matter what you think about science as long as you advance it.

It is questionable whether Darwin subscribed to the opinion Dingle attributed to him in the last sentence of the quotation. But Dirac undoubtedly did.

Scientific response to Dirac's cosmological theory was cool, to say the least. His colleagues in quantum theory chose politely to ignore his exotic journey into cosmology. With the one excpetion of Jordan, no quantum physicist of eminence seems to have taken public notice of it. Bohr's reaction to Dirac's cosmology was later recalled by Gamow, who was visiting Copenhagen in the spring of 1937; pointing to Dirac's letter in the new issue of *Nature*, Bohr reportedly quipped, "Look what happens to people when they get married!"[27] Neither did the theory attract much interest among astronomers. For example, on January 27, 1939, a meeting on the subject of the expanding universe was held in London under the joint auspices of the Physical Society and the Royal Astronomical Society. Although a variety of cosmological models was discussed, including the contributions of Eddington, Milne, McCrea, and George McVittie, Dirac's new theory seems not to have been found worthy of mention.[28] About the only exception to this indifference was Chandrasekhar, who at the time of Dirac's cosmological speculations was at Yerkes Observatory. In March 1937, he wrote to Dirac that he was "quite excited about your recent letter in Nature" and enclosed a manuscript in which he deduced, in Dirac's manner, that the number of particles in a star would increase as $t^{3/2}$.[29] One aspect of Dirac's theory did arouse a little interest in Britain, namely, the postulate that the constants and laws of nature were functions of the epoch and, in this sense, "historical." Some Marxist thinkers saw in this a confirmation of Engels's dialectical-materialist view that the laws of nature were historical phenomena and not absolutes laid down by God. Haldane thus referred approvingly to Milne's and Dirac's views as expressions of the fundamental dialectics of nature, the operation of historical process even in exact physics.[30]

For some reason, Dirac seems to have been dissatisfied with his theory. Perhaps he just did not see how to develop it further. At any rate, he left the matter in 1939 and returned to it only after a delay of thirty-three years. In the meantime, it attracted the interest of Jordan, who from 1937 onward engaged in a research program that was strongly indebted to the views of Dirac and Eddington.[31] After the war, when Jordan developed the *Diracsche Gedanke* into a comprehensive mathematical theory, incorporating it into the framework of general relativity, he stated rightly, "I am the only one who has been ready to take Dirac's world model seriously, which even its originator has partly abandoned, and to reconsider its more precise formulation."[32] However, Jordan's theories won no more support than did the earlier theory of Dirac. In the fifties, theories like

Dirac's, Jordan's, and Eddington's were in disagreement with the positivistic climate that permeated much of the science of the period. It was, and is, easy to criticize theories based on numerology. Most scientists and philosophers felt that the LNH and related numerological arguments carried no weight at all and could not be considered serious science.[33]

In the fifties, the Steady-State Theory of the universe, proposed in 1948 by Hermann Bondi and Thomas Gold and further developed by Fred Hoyle, gained considerable popularity. According to this theory, the universe expands but had no origin in the distant past. Steady-state theories rest on the Perfect Cosmological Principle, which holds the universe to be homogeneous not only in space but also in time; that is, according to this principle, the universe will always look the same to any observer, at any location and any time. The theory developed by Hoyle, Bondi, and Gold included continuous creation of matter, and in this respect it had a superficial resemblance to Dirac–Jordan cosmologies. However, an unchanging gravitational constant was essential in the Steady-State Theory. This theory was irreconcilable with Dirac cosmology, since the Perfect Cosmological Principle would obviously leave no room for the LNH. Nonetheless, the LNH was an important inspiration for the Steady-State theorists in 1948, and Dirac was quoted favorably by Hoyle both in his paper and at a seminar Hoyle gave in the Cavendish on March 1, 1948.[34] The seminar was attended by Dirac and also by Heisenberg, who was on a lecture tour to Great Britain. It was the first time in ten years that the two quantum pioneers had met.

Apart from the philosophical objections to Dirac's theory and its incompatibility with steady-state cosmology, it also became unacceptable on empirical grounds. In 1938, the theory was not in obvious conflict with known facts (apart from the time-scale difficulty, which was a problem in most versions of the expanding universe), but after the war an empirical argument against it was raised by Edward Teller.[35] His argument was widely accepted and made Dirac's theory even less appealing than it had been previously. Teller based his argument on the assumption that the temperature of the surface of the earth depends directly on the energy flux received from the sun. From astrophysical assumptions he deduced the temperature to vary as

$$T \sim G^{7/4} M^{5/4} r^{-1/2} \qquad (11.9)$$

where M is the mass of the sun and r the radius of the earth's orbit around the sun. For a circular orbit, classical mechanics yields

$$r^2 v^2 = GMr$$

where v is the velocity of the earth. Teller noticed that this quantity would remain constant even if G varied in time. This follows from the conservation of angular momentum

$$r \sim (GM)^{-1}$$

which, when inserted in equation (11.9), yields

$$T \sim G^{9/4}M^{7/4} \qquad (11.10)$$

If $G \sim 1/t$, as suggested by Dirac, and M is a constant quantity, then the temperature of the earth in the past would be related to its present value T_0 by

$$T = T_0(t_0/t)^{9/4} \qquad (11.11)$$

where t_0 is the present epoch (roughly the Hubble time). With the then accepted value of the Hubble constant, Teller estimated that at a time 200 or 300 million years ago equation (11.11) prescribed a value for T too high to maintain life on earth. Since geological evidence showed that life has in fact existed on earth for at least 500 million years, Teller felt justified in concluding that Dirac's hypothesis was in conflict with sound scientific knowledge. However, he realized that his argument was perhaps oversimplified and that Dirac's theory might in some way escape the difficulty. "Thus our present discussion cannot disprove completely the suggestion of Dirac," he wrote. "This suggestion is, because of the nature of the subject matter, vague and difficult to disprove."[36] But in spite of his cautious conclusion, Teller's paper was often quoted as a disproof of Dirac's theory.

In the fifties and sixties several physicists tried to incorporate cosmologies similar to Dirac's into a relativistic framework by suitable modifications of the field equations of general relativity. Jordan's theory was one of these attempts, and so were theories put forward by C. Gilbert and Markus Fierz.[37] In 1961, Robert Dicke and his student Carl Brans proposed another theory that made use of a variable constant of gravitation.[38] In contrast to earlier theories, the Brans-Dicke theory was much discussed by astronomers and cosmologists. It contained a gravitational constant diminishing with time, but the predicted variation did not accord with Dirac's $G \sim 1/t$. Although Dicke and Brans did not accept Dirac's theory, they felt it was interesting and worthwhile to criticize. In their paper of 1961, they concluded that "although the detailed structure of Dirac's cosmology cannot be justified by the weak empirical evidence on which it is based, the more general conclusion that the number varies with time has a more solid basis."[39] The number referred to was

$$m \left(\frac{G}{hc} \right)^{1/2} \simeq 5 \times 10^{23}$$

which Dicke and Brans wanted to "explain" with Mach's principle rather than Dirac's LNH. Dicke's interest in Dirac cosmology can be seen in an address he gave in 1958. Having reviewed Dirac's theory, he noticed that it contained a logical loophole, namely, the assumption that "now," the epoch of man, is random. Applying an anthropic argument, Dicke claimed that this assumption was unfounded and that the present value of the Hubble time (and other large numbers) should be understood not as a result of the LNH but as a consequence of there being habitable planets with human life.[40] In a letter to *Nature* in 1961, he restated these arguments, which generated a brief reply from Dirac:[41]

On this [Dicke's] assumption habitable planets could exist only for a limited period of time. With my assumption they could exist indefinitely in the future and life need never end. There is no decisive argument for deciding between these assumptions. I prefer the one that allows the possibility of endless life.

This reply was Dirac's first public announcement on cosmology in twenty-two years, and it would be another eleven years before he seriously resumed work in the field. During this period, Big Bang cosmology was established as a new, forceful paradigm, while steady-state theories ceased to be regarded as a serious alternative. A major reason for this change was the discovery in 1965 of the universal 3K background radiation, which was at once interpreted as a remnant of the Big Bang. Also, new determinations of the Hubble constant cleared away the time-scale difficulty associated with Big Bang cosmology, that is, that the predicted age of the universe was less than the age of the stars. In the sixties, the preferred value of $1/H$ was around 10 billion years, a value that made Teller's argument against Dirac's hypothesis less conclusive. In the seventies, time was ripe for a reconsideration of cosmological theories embracing a diminishing gravitational constant. Dirac returned to cosmology in a series of papers in which he altered some of his views but on the whole kept faithful to the LNH program of his original theory.

   Dirac's renewed interest in cosmology was probably indebted to his correspondence with George Gamow, the nuclear physicist turned astrophysicist and cosmologist. On January 10, 1961, Dirac wrote to Gamow about his hope for his old theory:[42]

It was a difficulty with my varying gravitational constant that the time scale appeared too short, but I always believed the idea was essentially correct. Now that the difficulty is removed, of course I believe more than ever. The astronomers now put the age of the universe at about 12×10^9 years, and some even think that it may have to be increased to 20×10^9 years, so that gives us plenty of time.

It is difficult to make any firm theories about the early stages of the universe, because we do not know whether $\hbar c/e^2$ is a constant or varies proportional to log t. If $\hbar c/e^2$ were an integer it would have to be a constant, but the experimenters now say it is not an integer, so it might very well be varying. If it does vary, the chemistry of the early stages would be quite different, and radio-activity would also be affected. When I started work on gravitation I hoped to find some connection between it and neutrinos, but this has failed.

Gamow was fascinated by numerological arguments and strongly felt that the fundamental constants and their combinations in dimensionless ratios were of profound significance in theoretical physics. From 1948 until his death in 1968, he published several papers in which he cultivated "cosmonumerology" in the fashion of Eddington, Dirac, and Jordan.[43] But although he was attracted by the Large Number Hypothesis, he felt that it was contradicted by observational evidence. In 1967, Gamow reviewed Teller's objection and additionally concluded that our sun's being a main-sequence star was incompatible with a decreasing gravitational constant, because with that hypothesis the luminosity of the sun in the past would have been so high that the sun would by now have turned into a red giant star. In spite of this refutation, he thought that "it would be too bad to abandon an idea so attractive as Dirac's proposal."[44]

Upon receiving a reprint of one of Gamow's papers on this matter, Dirac wrote to him:[45]

The total number of nucleons in the universe is roughly 10^{80}, so presumably this number is increasing proportionally to t^2. Thus there must be continuous creation of matter. Continuous creation is required, not only by the steady state theory, but also by the varying γ theory. The continuously created matter is probably protons. The question arises where it is created? The two most natural assumptions are
(1) It is created uniformly throughout space
(2) It is created where matter already exists, and thus in the stars.
With (1) the sun is probably acquiring a good amount of matter by accretion. So in either case the sun is continually getting refuelled, and your argument needs amendment. Can you get a revised estimate of the life of the sun? With assumption (2) we should have the mass of the sun increasing to the law $M = (\text{const})t^2$. If the extra mass is all protons, what result do you get?

Gamow found Dirac's suggestion interesting but in conflict with observational evidence. In a series of letters, he argued that the accretion hypothesis would not help in saving the varying-G hypothesis. Instead, he suggested that the elementary electric charge might vary with time: "Here is an idea which, I am sure, you will like. Since I am completely persuaded that the Newton's constant γ does not decrease inversely porportionally why cannot one assume that e^2 (or, if you want, the coefficient C in the Coulomb's law $F = Cee/r^2$) increases $\simeq t$?"[46]

But Dirac did not share Gamow's enthusiasm over this new idea: "I do not like the idea of $e^2/\hbar c$ varying with time. There is no reason why 137 should be connected with 10^{40}. . . . I am still in favour of γ varying as τ^{-1}. Can we not get over the difficulty of boiling seas in the past by supposing the sun to be increasing in mass?"[47] A week later, Gamow retracted his idea and concluded that "the value of e stands as Rock of Gibraltar for last 6×10^9 years!" But as to the variability of G, he saw no hope:[48]

Of course, one can hypothesize that during its past history, the sun had passed through some very dense interstellar clouds (like Great Orion Nebula), and had accreated just enough material to obscure the effect of changing γ. But, such an assumption would be extremely *un*elegant, so that the *"total ammount of elegance"* in the entire theory would have decreased quite considerably even though the elegant assumption $\gamma \simeq t^{-1}$ would be saved. . . . So, we are thrown back to the hypothesis that 10^{40} is simply the largest number the almighty God could write during the first day of creation.

In spite of Gamow's detailed objections, Dirac was unwilling to give up his theory. He responded to Gamow on November 20:[49]

I do not see your objection to the accretion hypothesis. We may assume that the sun has passed through some dense clouds, sufficiently dense for it to pick up enough matter to keep the earth at a habitable temperature for 10^9 years. You may say that it is improbable that the density should be just right for this purpose. I agree. *It is improbable.* But this kind of improbability does not matter. If we consider all the stars that have planets, only a very small fraction of them will have passed through clouds of the right density to maintain their planets at an equable temperature long enough for advanced life to develop. There will not be so many planets with men on them as we previously thought. However, provided there is *one*, it is sufficient to fit the facts. So there is no objection to assuming our sun has had a very unusual and improbable history. Thanks for your letter. I see that varying $\hbar c/e^2$ is disposed of. I do not have any regrets.

Shortly before Gamow's death, Dirac mentioned a counterargument that he would later elaborate: "I do not think the arguments against a varying G are valid. If G varies, then other "constants" may also be varying, such as the ratio $M_{proton}/M_{electron}$ or the coupling constants of nuclear theory. One then cannot build any reliable models of the universe and arguments depending on the usual models are not valid."[50] The first hint of his new idea appeared in print in a brief article in 1972, dedicated to the memory of Gamow:[51]

Since Gamow's death I have re-examined Teller's arguments and a possible flaw in them has shown up. To calculate the recession of the earth from the sun, Teller assumed conservation of angular momentum of the earth in its orbit around the sun. This is a natural assumption to make, but not really reliable.

Dirac hinted at changes in Newton's laws necessitated by a thorough use of the principle of two time-scales. He hoped that this principle, in connection with a revival of an old theory of Weyl's, would reconcile LNH cosmology with Einstein's theory of gravitation.

Dirac was fascinated by the unified field theory that Hermann Weyl had originally published in 1918.[52] With this theory Weyl gave a purely geometric interpretation of electromagnetism and gravitation, based on a geometry more general than the Riemannian geometry of the ordinary theory of relativity. Weyl pointed out that the integrability of lengths in Riemannian geometry was an arbitrary residual of Euclidean geometry and argued that one should work with a geometry in which only lengths displaced in parallel through infinitely small distances could be compared. He managed to specify the internal metric of space by two quantities, which he interpreted as including, respectively, the Einstein gravitation potentials and the electromagnetic potentials. Weyl's theory at first met widespread interest. It was further developed by the young Pauli and taken over into atomic theory by Schrödinger. However, it was rejected by Einstein, and after the introduction of quantum mechanics it was generally held to be untenable. Weyl himself lost confidence in his ambitious theory, which largely fell into oblivion.[53]

Dirac's interest in Weyl's unified theory stemmed from his student days in Cambridge, when he had studied it thoroughly. In accordance with the consensus at the time, he had concluded that the theory, although mathematically appealing, was physically unsound,[54] but Dirac continued to be fascinated by the mathematical structure of the theory and by Weyl's general approach to physics, which he saw as representative of his own ideal of a "powerful method of advance." The aesthetic basis for Dirac's praise of Weyl's field theory was made clear in 1973 when he returned to cosmology. Dirac characterized the defunct Weyl theory as "a very beautiful synthesis of the electromagnetic field and the gravitational field," a theory that "remains as the outstanding one, unrivalled by its simplicity and beauty."[55] Regarding his new cosmological theory and its foundation in Weyl's geometry Dirac admitted that it lacked empirical support. However, empirical support was not of decisive importance to Dirac, who emphasized methodological reasons for maintaining belief in Weyl's theory. In complete agreement with his general philosophy of physics, as stated most fully in the preface to *Principles of Quantum Mechanics* and in his monopole paper of 1931, Dirac wrote:[56]

It appears as one of the fundamental principles of Nature that the equations expressing basic laws should be invariant under the widest possible group of transformations. . . . The passage to Weyl's geometry is a further step in the direction of widening the group of transformations underlying physical laws. One now has to consider transformations of gauge as well as transformations of curvilinear coordinates and one has to take one's physical laws to be invariant under all these transformations, which impose stringent conditions on them.

Dirac felt that it was imperative to find some way to clear away the objections to Weyl's beautiful theory and to connect it with another idea of great beauty, the Large Number Hypothesis. His means of reconciliation was an elaboration of the idea of the two metrics, which he had already entertained in a vague form in 1938: "[There are] two measures of distance and time that are of importance, one for atomic phenomena and the other for ordinary mechanical phenomena included under general relativity."[57] Although Dirac took over what he sometimes called Milne's hypothesis, he developed it in an original way and obtained results very different from Milne's. In Dirac's version of the hypothesis, one metric (ds_E) applied to Einstein's theory of gravitation and to planetary and related mechanical problems, and the other metric (ds_A) applied to quantum phenomena and all sorts of atomic quantities, including the functioning of usual laboratory equipment. In "Einstein units" the relevant time scale was the dynamical time τ, and on this scale the constant of gravitation was a true constant, but in "atomic units" G depended on the time according to the LHN:

$$G(\tau) = \text{constant} \quad \text{and} \quad G(t) = 1/t$$

Dirac pointed out that the world could be described in two different ways, depending on whether Einstein units or atomic units were used. For example, when atomic units were used, the universe was seen to have expanded from its origin at $t = 0$; but when Einstein units were used, it was a closed static universe in which the redshift had to be explained by some mechanism other than the recession of the galaxies. The same thing had been stressed by Milne in 1935.

Dirac developed his new theory in the years 1973–5 after settling in Florida. He remained faithful to the foundation of his original theory of 1937, the LNH, which he discussed with his senior research assistant,

Leopold Halpern, and other physicists at Florida State University. Considering the implications of the LNH for the expansion of the universe, Dirac concluded that oscillating models of the universe were ruled out; they would involve a maximum size for the universe, amounting to a very large number not dependent on the epoch, which would contradict the LNH. Although in 1938 Dirac had adopted a world model in which R varied as $t^{1/3}$, in the seventies he argued that any decelerating expansion was precluded by the LNH. Suppose, he said, that $R \sim t^n$. If $n < 1$, then there would be some time in the past for which the rate of expansion exceeded the velocity of light; if, for example, $n = \frac{3}{4}$, this epoch would be at $t \cong 10^{27}$ atomic time units, and "this particular epoch would be something which is very characteristic, and it involves a large number, not quite as big as 10^{39}, but still too big to be allowed by our Large Number Hypothesis."[58]

In 1938, Dirac has rejected the idea of continuous creation of matter, but in 1973 he returned to his view of 1937: "I can see no escape from this requirement [of matter creation]. It is just as forced upon us as the variation of the gravitational constant in the first place."[59] Apart from intimating that the postulated continuous creation of matter was perhaps some new kind of radioactivity, Dirac had nothing to say about its mechanism. He distinguished between two types of creation, which he called additive (+) and multiplicative (×). With (+) creation, nucleons would be created uniformly throughout space; with (×) creation, new matter would be created where it already exists, in proportion to the amount existing. In both cases Dirac found that matter would increase as the square of t, which was the same result he had found in 1937. He further found that the relationship between the Einstein and atomic metrics would depend on the sort of creation taking place. On the assumption of (+) creation, the relations were

$$ds_A = t^{-1} \, ds_E \quad \text{and} \quad \tau \sim \tfrac{1}{2}t^2$$

while (×) creation yielded

$$ds_A = t \, ds_E \quad \text{and} \quad \tau \sim \log t$$

According to Dirac this implied that only on the assumption of (×) creation would there have been a Big Bang; in Einstein units, $t = 0$ corresponded to $\tau = -\infty$, which meant no Big Bang. Dirac further drew the conclusion that with (+) creation the earth was approaching the sun, while with (×) creation the earth was receding from the sun. In another paper he also argued that the value of the cosmological constant was related to the sort of creation taking place: additive creation could only occur if $\wedge = 0$, while a nonzero $\wedge$ implied multiplicative creation. Recall that in 1938 Dirac had argued that only a cosmological constant equal to zero would be compatible with the LNH; this argument would thus entail $(+)$ creation, but in the seventies no such conclusion was drawn.

During 1973–5, Dirac wavered as to which of the alternatives he should prefer. Realizing that both were incompatible with standard physics, he thought that the $(\times)$ hypothesis would clash less violently with Einstein's theory of gravitation. Dirac wanted, of course, to keep his theory in essential harmony with general relativity and was ready to apply rather desperate means to do this. He proposed to compensate for matter creation with an additional creation of negative mass.[60] There was no independent justification for this hypothesis, which was clearly ad hoc. Furthermore, to avoid violent disagreements with observational data, Dirac had to assume that the negative-mass atoms had no physical effects at all, that is, were unobservable in principle.

In the Einstein metric the mass of large bodies, like the sun, was constant. But then, Dirac argued, in order to compensate for the $(\times)$ creation, the mass of a nucleon would have to vary as t^{-2} (while in atomic units this quantity was constant). Because of $e^2(GMm)^{-1} \sim t$, the elementary electric charge would therefore vary as $t^{-3/2}$ in Einstein units. Planck's constant would then vary as t^{-3}, since Dirac believed the fine structure constant to be a true constant in both metrics. As a consequence of $(\times)$ creation, assumed to be valid for photons too, Dirac further mentioned that the number of photons in a beam of light would increase as t^2 and then give rise to an increase in the luminosity of distant stars. Although admitting that his new theory was speculative, he was confident that it contained an essential element of truth. He pointed out, as he had done earlier in his correspondence with Gamow, that it would invalidate Teller's objections of 1948 to a changing gravitational constant: if $G \sim t^{-1}$ was supplied with $M \sim t^2$ and $r \sim t$, Teller's formulae yielded a much slower variation in the temperature of the earth, which presented no grave problems regarding the existence of life in the geological past.

In 1973, the cosmic background radiation, isotropic and blackbody distributed with a temperature of 3K, was recognized as a crucial fact with which cosmological models had to comply. In standard Big Bang cosmology the 3K radiation was interpreted as a relic of the primordial decoupling of matter and radiation. The original radiation retains its blackbody structure during the expansion, its temperature decreasing with the radius of the universe. Dirac realized that his cosmological theory was not easily reconcilable with the 3K radiaiton: If photons were continuously created, the present existence of a blackbody radiation would merely be the accidental result of the fact that we just happen to live in an era in which the gradual blackening of radiation has taken the form of a Planck distribution. Although Dirac recognized that the assumption of such a coincidence was not satisfactory, he was unable to provide a better explanation. But he felt his theory was vindicated because it promised to give rebirth to Weyl's old theory. The objection to Weyl's theory was that atomic clocks measure time in an absolute way and hence supply an absolute metric; then there would be no problem in comparing the lengths of a vector under parallel displacement, as assumed by Weyl. But, according to Dirac, Weyl's theory operated with the Einstein metric and could not be criticized with quantum arguments based on the atomic metric. Even if ds_A remained invariant when taken around a closed loop, ds_E would not. "We should reintroduce Weyl's theory. It is such a beautiful theory and it provides such a neat way of unifying the long range forces. And there is really no clash with atomic ideas when we have the two ds's," Dirac wrote.[61]

With the purpose of establishing field equations, in 1973 Dirac worked out an action principle that included gravitational as well as electromagnetic terms. Such an action principle had been included in Weyl's original theory, but Dirac's action principle was much simpler. It led to a scalar–tensor theory of gravitation similar to the one earlier proposed by Dicke and Brans. Dirac further showed that Weyl's theory had consequences with respect to the fundamental symmetries of nature. Einstein's general theory of relativity was invariant with respect to the direction of time, but in Weyl's theory there was no symmetry between future and past, and neither was there symmetry between positive and negative electrical charge. Weyl's theory, as developed by Dirac, preserved parity (P) invariance and the combined charge-time (CT) invariance but neither charge (C) nor time (T) invariance separately. Dirac argued that this kind of symmetry breaking was a result of interacting gravitational and electromagnetic fields and would occur only rarely and not in ordinary elementary particles.[62]

None of Dirac's several versions of his cosmological theory won much support (see Appendix I), which is hardly surprising in view of its unconventional and speculative nature. However, a few researchers of the younger generation did follow the essentials of Dirac's theory, which they developed in various ways.[63] Dirac himself took very little part in these developments. He continued to think about cosmology, and in 1978, at an American Institute of Physics symposium in Tallahassee, reconsidered the question of continuous creation of matter. He then returned to matter conservation, once again: "I have been working with this assumption of continuous creation of matter for a number of years, but find difficulties in reconciling it with various observations, and now believe it should be given up."[64]

Because Dirac still believed in the LNH, he had to maintain equation (11.6). But he interpreted it in a new way, as merely signifiying "a continual increase in the amount of observable matter,"[65] not a genuine creation of matter. With this interpretation, it could easily be reconciled with matter conservation. Galaxies at the limit of the visible universe recede with a velocity of, for example, $\frac{1}{2}c$. Although in Dirac's new cosmology the universe was infinite, one could still talk about a sphere of radius $R_m \simeq ct$ corresponding to the radius of the visible universe. In this part of the universe, the mean density of matter is

$$\frac{N}{R_m^3} \cong \frac{N}{(ct)^3}$$

where N, the number of nucleons in the visible universe, is of the order of magnitude 10^{78}. If N varies as t^2, then ρ must vary as $1/t$. With mass conservation the density would be R^{-3}, where R is the distance between two receding galaxies. Then follow the relations

$$R \sim t^{1/3} \quad \text{and} \quad \frac{dR}{dt} \sim t^{-2/3}$$

This was a return to his 1938 model, the new thing being that $N \sim t^2$ was now seen as an effect of there being more and more galaxies appearing within the visible universe. In Einstein units, Dirac found the result $R \sim \tau^{2/3}$, which agreed with the law of expansion based on the theory of Einstein and de Sitter of 1932. He considered this highly satisfactory and concluded that "the only cosmological model in agreement with the LNH is the ES [Einstein–de Sitter] model."[66] Recall that six years earlier he had maintained that any decelerated expansion, including the Einstein–de Sitter solution, was precluded by the LNH.

Dirac's main reason for proposing his latest version seems to have been a desire to cope with the 3K radiation; by abandoning continuous creation of photons, LNH cosmology was no longer in obvious contradiction with the 3K radiation. However, although Dirac now accepted that the background radiation was a significant fossil of the Big Bang, he could not accept the standard explanation of it. According to standard cosmology, the radiation was a result of an original decoupling between matter and radiation, calculated to have taken place at about 10^{26} atomic time units after the Big Bang. But Dirac objected that "the existence of such a decoupling time, playing a fundamental role in cosmology, would contradict the LNH."[67] His reasoning in dismissing the decoupling was thus of the same type as that applied in 1973 when he dismissed decelerating cosmological models (which he now accepted). The somewhat arbitrary nature of Dirac's theory is further illustrated by the following argument, maintained to be an extra confirmation of the Einstein–de Sitter universe: Consider the energy kT, where T is the present temperature of the background radiation and k is Boltzmann's constant; if expressed in units of the nucleon's rest energy, a large dimensionless number turns up,

$$\left(\frac{mc^2}{kT}\right)^3 \cong 10^{38}.$$

According to the LNH philosophy, this quantity must be proportional to the epoch, and hence the temperature must vary as $t^{-1/3}$, which is the cooling rate to be expected from an Einstein–de Sitter expansion. But why use the nucleon's mass in this argument and not, for example, the electron's? Dirac's answer: "If we had used the mass of the electron instead of the mass of the proton there would have been a small discrepancy, which is not significant in view of the rough nature of the LNH."[68]

Finally, the idea of the two metrics was preserved in Dirac's latest theory; but as a result of retaining matter conservation, he was forced to change the relationships between the metrics. Dirac argued that the time parameters would be connected through $d\tau = t\,dt$, which is the same result as in his earlier (1973) theory on the assumption of $(+)$ creation.

Dirac's interest in cosmology began when he was thirty-five years old but became his major occupation only when he was in his seventies. He published his last research paper on cosmology when he was eighty years old. He was, in cosmology as in quantum electrodynamics, an outsider whose views were not taken very seriously by the astrophysicists. Observational evidence, gathered from studies made by Irwin Shapiro, Thomas Van Vlandern, R. W. Hellings, and others, did not support LNH cosmology (although neither did it unambiguously reject it). Furthermore, by changing his views frequently, Dirac weakened whatever appeal his theory had. In 1982, Hellings and his collaborators concluded from observations from the Viking landers on Mars that if G varies at all, it does so at a rate much smaller than that predicted by the Dirac theory. However, in spite of the negative results, Dirac remained convinced of the correctness of his theory, the beauty of which, he felt, guaranteed its truth.[69]

Dirac's cosmology was not rooted in the general theory of relativity, which he admired so much. Although it was his first love in theoretical physics, he kept general relativity outside his scientific work for a very long time. It was only in the late 1950s that he began to deal seriously with the area, mainly as an offshoot of his interest in establishing a generalized Hamiltonian dynamics for any type of interaction. In 1958,

Dirac succeeded in putting the general theory of relativity into Hamiltonian form by applying the generalized procedures he had developed since the late 1940s.[70] In the following years, he worked extensively on the Hamiltonian formulation of general relativity and the problems of gravitational energy. He presented his results, among other places, at conferences in Paris in 1959, Warsaw in 1962, and Trieste in 1968.[71] In April 1958, the Max Planck centenary was celebrated in Berlin, with Dirac participating as a representative of the Royal Society; from Berlin the guests went to Leipzig, where Dirac gave a report on his ideas of general relativity. At the Warsaw conference in 1962, he presented a model of extended gravitational particles that shared many of the characteristics of his model of the extended electron, developed at the same time (see Chapter 9). Dirac discussed particles with a surface distribution of mass, including a surface pressure to counterbalance the mutual attraction of the mass elements of the shell. He admitted, in response to questions from Wheeler and Bondi, that the model "is rather remote from physical reality" but found that making it more physical would destroy its simplicity.[72] In other works Dirac concluded that accelerating masses emit energy in the form of unidirectional gravitational waves with a well-defined energy density. Gravitational waves had been considered since 1916, but it was only around 1960 that they gained some respectability, partly through the efforts of Joseph Weber at the University of Maryland, who initiated an experimental program in order to detect the waves. Dirac's Hamiltonian formalism for the gravitational field equations made it relatively easy to apply the rules of quantization to gravitational fields, in this way leading to quanta of gravitation. At a meeting of the American Physical Society in New York on January 30, 1959, Dirac proposed that such gravitational quanta be called gravitons.[73] The name went quickly into the physicists' vocabulary. Dirac continued to cultivate general relativity after his retirement in 1969. His lectures on the subject given at Florida State University were published as a textbook in 1975.[74]

CHAPTER 12

THE PUREST SOUL

He was tall, gaunt, awkward, and extremely taciturn. He had succeeded in throwing everything he had into one dominant interest. He was a man, then, of towering magnitude in one field, but with little interest or competence left for other human activities. In conversation he was invariably polite, but it did not follow that he could comprehend his interlocutor. One was never sure that he would say something intelligible. In other words, he was the prototype of the superior mathematical mind; but while in others this had coexisted with a multitude of interests, in Dirac's case everything went into the performance of his great historical mission, the establishment of the new science, quantum mechanics, to which he probably contributed as much as any other man.[1]

I N the preceding excerpt from his memoirs, written in 1978, the German physicist Walther Elsasser, who later in life became a biologist, provided a remarkably apt characterization of his famous British colleague in physics. Paul Dirac was indeed a legendary figure, not only because of his exceptional contributions to physics but also because of his personality. Ever since his childhood, he used as few words as possible and spoke only when asked a direct question or when he felt he had something important to say. And when he said something, he said it in a direct way, without attempting to include hidden significance in his words. He supposed, erroneously, that other people spoke and listened in a similarly direct way. Wigner recalled a luncheon he once attended with Dirac and the scientist and philosopher Michael Polanyi, where they discussed various questions of science and society. During the discussion Dirac did not say a word. Asked to speak up and give his opinion, Dirac said, "There are always more people willing to speak, than willing to listen."[2] This is just one among many anecdotes about Dirac, most of which refer to his introversive nature and his surprising directness in conversation. These stories, which for years have circulated among physicists, constitute an important part of the legend of a man whom very few people would say they knew well. Most often, such anecdotes are not literally true, and some may be purely fictitious; but taken together they make up a picture that has its own life and in an impressionistic way reveals something true about the person in question.

The taciturn nature was deep-rooted in Dirac, who seldom spoke spontaneously. He said (to the talkative Bohr, according to the story) that he was taught that one should not begin a sentence until one knew how to finish it. He behaved in strict accordance with this lesson and clearly gave priority to thinking over talking. The following story, told by Heisenberg, refers to two frequent themes in the Dirac legend: his lack of spontaneity and his timidity with regard to the opposite sex:[3]

Paul always thinks about his formulations very carefully. He does not like to answer spontaneously at once, he first thinks about things. We were on the steamer from America to Japan, and I liked to take part in the social life on the steamer and so, for instance, I took part in the dances in the evening. Paul, somehow, didn't like that too much but he would sit in a chair and look at the dances. Once I came back from a dance and took the chair beside him and he asked me, "Heisenberg, why do you dance?" I said "Well, when there are nice girls it is a pleasure to dance." He thought for quite a long time about it, and after about five minutes he said, "Heisenberg, how do you know *beforehand* that the girls are nice?"

"I still find it very difficult to talk with Dirac," a Cambridge physicist who had known Dirac for many years told Infeld. "If I need his advice I try to formulate my question as briefly as possible. He looks for five minutes at the ceiling, five minutes at the windows, and then says 'Yes' or 'No.' And he is always right."[4] Dirac maintained this attitude even at conferences and scientific meetings. In September 1950, he gave a lecture on field theory at a conference on nuclear physics held at Harwell. The report of the conference, after summarizing the content of Dirac's lecture, ends tersely: "In the following discussion questions were raised but not answered. . . ."[5]

It seems that Dirac's reticence was rooted partly in shyness and partly in an idiosyncratic and exaggerated insistence on logic and intellectual economy. Dirac was famous for his directness and candidness on scientific questions, as well as in daily life. He was difficult to approach and kept a reserved attitude even toward people whom he had known for a long time. But those who knew him well assure us that Dirac was really a very gentle person. Even so, to those who had not penetrated his solitude or were not acquainted with his idiosyncrasies, Dirac's style and propensity for logical conversation must have appeared as mere lack of tact.

His reticence and apparent lack of interest in people did not further his social contact and could not, in some cases, avoid being mistaken for impoliteness or perhaps haughtiness. The following stories illustrate this.

When the young Polish physicist Leopold Infeld came to Cambridge around 1933, Fowler suggested that he work with Dirac on a problem in positron theory. Consequently, Infeld went to see Dirac, and he later wrote of the encounter:[6]

I went along the narrow wooden stairs in St. John's College and knocked at the door of Dirac's room. He opened it silently and with a friendly gesture indicated an armchair. I sat down and waited for Dirac to start the conversation. Complete silence. I began by warning my host that I spoke very little English. A friendly smile but again no answer.

Infeld tried to go further and told Dirac about Fowler's suggestion.

No answer. I waited for some time and tried a direct question: "Do you have any objection to my working on this subject?" – "No." – At least I had got a word out of Dirac. Then I spoke of the problem, took out my pen in order to write a formula. Without saying a word Dirac got up and brought paper. But my pen refused to write. Silently Dirac took out his pencil and handed it to me. Again I asked him a direct question to which I received an answer in five words which took me two days to digest. The conversation was finished. I made an attempt to prolong it. "Do you mind if I bother you sometimes when I come across difficulties?" – "No." – I left Dirac's room, surprised and depressed. He was not forbidding, and I should have had no disagreeable feeling had I known what everyone in Cambridge knew. If he seemed peculiar to Englishmen, how much more so he seemed to a Pole who had polished his smooth tongue in Lwow cafés!

Dirac certainly did not mean to be impolite, although in fact his behavior in a case like the one reported by Infeld can only be described as such. Dirac, the cultivator of logic, just behaved logically, which in some cases is the opposite of behaving in a socially acceptable way. He would answer a direct question, not a comment or other statement that from a logical point of view did not demand an answer. And then he would be candid, not always recognizing that candidness may in some situations signal unkindness. Dennis Sciama, who later became a well known astrophysicist, had Dirac as his supervisor around 1950. Sciama once went enthusiastically to Dirac's office, saying, "Professor Dirac, I've just thought of a way of relating the formation of stars to cosmological questions, shall I tell you about it?" Dirac's answer: "No." Conversation finished.[7]

On another occasion, a French physicist called on Dirac. The Frenchman spoke English with difficulty and struggled very hard to express himself in Dirac's tongue. Dirac listened in silence to the words, which were half in French, half in bad English. "After some time Dirac's sister came into the room and asked Dirac something in French, to which he also replied in fluent French." Recall that the native tongue of Dirac's father was French and that he taught his children the language from an early age. Naturally the visitor was indignant, and he burst out: "Why did you not tell me that you could speak French?" Dirac's terse answer: "You never asked me."[8]

A story to the same effect was reported by Tamm, who in 1931 attended a talk Heisenberg gave in Cambridge on some recent work of Heitler. Heisenberg had forgotten his notes, and consequently his talk, as well as the following discussion, was rather muddled and unsatisfactory. Nobody seemed really to know Heitler's arguments. After the discussion, someone asked Dirac for his opinion. Dirac said that he knew Heitler's ideas well and that Heitler himself had told him about them. "But why didn't you say, Paul?" "Nobody asked me," he replied.[9]

Abraham Pais became acquainted with Dirac during his stay at Princeton's Institute for Advanced Study in 1947–8, and he became used to Dirac's peculiarities in conversation. He recalled a corridor conversation at the Institute: Dirac said, "My wife wants to know if you can come for dinner tonight," to which Pais replied, "I regret, I have another engagement." Then Dirac said "Goodbye." "Nothing else said like 'Some other time perhaps.' The question had been posed and answered, the conversation was finished."[10]

The following anecdote has, I think, also been ascribed to other physicists besides Dirac. After Dirac had delivered a lecture, one of the audience said, "Professor Dirac, I don't understand how you deduced this formula. . . ." Dirac sat quietly, maybe looking out of the window, without any sign of reaction. After some time of silence, the lecture chairman, probably as bewildered as the inquirer, had to request that Dirac answer the question. "It is not a question, it is a statement," Dirac responded.[11] He would not say this to rebuff the inquirer or to be witty, but just to state a fact.

Dirac's predilection for solitude and his reticence resulted from his desire to live as he thought, clearly and logically. Although usually a man of few words, he could on occasion be very articulate if he found it necessary. When Dirac was in Canada in 1950, he once ran into Infeld, who was visiting Banff in the Rocky Mountains. Infeld was not only surprised to meet Dirac but also to notice how talkative he was: "He had spent two weeks away from people and had not spoken a word, thus accumulating a collection of phrases that he suddenly released on me. Despite his natural reluctance to talk – or perhaps because of it – he is intelligent and deep, and never utters a triviality."[12]

The almost manic affection for logic extended to every area of Dirac's life, often with surprising results. Logic and rational thinking do not usually govern practical life, but in Dirac's universe they did, or so he wanted it. "He always seemed to be quite a total rationalist," remarked Mott in 1963.[13] The following story may be characteristic of how Dirac's mind worked: H. R. Hulme, a former research student of Dirac, went by train with Dirac from Cambridge to London. On their way back to Cambridge Dirac noticed that something rattled in Hulme's pocket. Hulme explained that it was some pills he kept in a bottle; on their way to London the bottle had been full and not rattled, but in the meantime he had taken some pills, which was the reason that Dirac only now heard the rattling. After some silence Dirac said, "I suppose it makes a maximum noise when it's just half full."[14]

Dirac lived a modest, almost ascetic life. He did not touch alcoholic drinks and never smoked. As to drinks, he preferred water, which he consumed in great quantities. His lack of concern for personal comfort was an advantage during his many travels. During one of his trips to Russia, he happened to go to the border with a visa valid for a different entry point; he had to spend some days in a miserable little village at the border until he was allowed to pass into Russia. Nice housing facilities, good food, and other kinds of worldly comfort were of no importance to him. When Crowther visited him around 1930 in his room at St. John's, he likened Dirac's situation to a monk living in his cell. Dirac reminded Mott of Gandhi: little flesh and much mind; in a letter of 1931, Mott reported to his parents:[15]

Dirac is rather like one's idea of Gandhi. He is quite indifferent to cold, discomfort, food, etc. We had him to supper here when we got back from the Royal Society in London. It was quite a nice little supper but I am sure he would not have minded if we had only given him porridge. He goes to Copenhagen by the North Sea route because he thinks he ought to cure himself of being seasick. He is quite incapable of pretending to think anything that he did not really think. In the age of Galileo he would have been a very contented martyr.

The austerity and integrity indicated by Mott's description of Dirac showed itself in many ways. Dirac would never compromise on what he thought was right, even if this caused him to become isolated from mainstream developments, as happened in postwar physics. Any idea of joining popular trends, or otherwise bending his ideas in order to adapt to majority views, was totally foreign to him. Niels Bohr once remarked that "of all physicists, Dirac has the purest soul."[16] This intellectual purity was a great scientific and moral strength for Dirac, but socially it was a weakness.

A similar feature in Dirac's psychology was noticed by Harish-Chandra. At a conference in honor of Dirac in 1983, he observed that his former supervisor always preferred to rely on his intuition rather than on established knowledge. Referring to the roles played in Dirac's science by knowledge or experience on the one hand and imagination or intuition on the other, Harish-Chandra said: "I believe that there is a certain fundamental conflict between the two, and knowledge, by advocating caution, tends to inhibit the flight of imagination. Therefore, a certain naïveté, unburdened by conventional wisdom, can sometimes be a positive asset."[17] To describe Dirac as "naïve" may seem surprising, or even offensive, but is not unjustified. Purity and naïveté are closely related. The naïveté of Dirac's thinking is not obvious from his technical contributions to physics, but on the rare occasions when he spoke of matters outside physics, his views were indeed remarkably naïve (a point that will be illustrated shortly).

In September 1972, a symposium, organized by Jagdish Mehra and Abdus Salam, was held in Trieste, where a banquet was given in honor of Dirac on September 21. Wigner, Heisenberg, Peierls, von Weizsäcker, Wheeler, and other prominent physicists celebrated his seventieth birthday. At the banquet Charles P. Snow, the novelist, gave an address on "the classical mind" in which he compared Dirac with Newton: "The minds of Newton and Dirac seem to me to have certain resemblances. . . . A classical mind isn't the only kind of mind, but it's an exceptionally valuable one. It has certain characteristics. Well, Newton had it and he exercised it in his science, but not elsewhere. Paul Dirac has exercised it in all of his human activities." The classical, Diracian mind, according to Snow, included such qualities as "ultimate candour," "rationality," a "strong and prevailing aesthetic sense," and "lucidity, austerity, that is a dislike for unnecessary frills, indeed frills of any kind."[18]

Dirac's mindset made him a very private person who disliked becoming involved in controversies of any kind. He used his reticence as a means of self-protection, to avoid new involvements that would disturb his chosen lifestyle. Because he was a famous Nobel laureate, it was not easy for him to keep out of the public light and to resist becoming involved in extrascientific activities. But with few exceptions he managed to keep a low profile outside physics. He shunned honors and publicity, as indicated in his letter of 1936 to Van Vleck (see Chapter 7, note 11) and in his initial inclination not to accept the Nobel Prize (see Chapter 6). According to a newspaper, on the day it was announced that Dirac had been given the Lucasian professorship, he escaped to the zoo to avoid the many congratulations![19] With respect to the many honorary degrees he was offered, he was firm in refusing them. In 1934, he refused to accept an honorary degree from his alma mater, the University of Bristol, and afterwards he felt that he could not accept honorary degrees from other institutions. All the same, he was more than once awarded honorary doctorates, but always in his absence and apparently without his acquiescence; this took place at the University of Paris in 1946, the University of Torino in 1951, and the University of Moscow in 1966. In spite of his reservation with regard to honors, Dirac actually received most of the honors a scientist could hope for.[20]

In earlier chapters we briefly looked at Dirac's work as a teacher and lecturer. His chief interest in physics lay in fundamental research, not in teaching, and he spent only a relatively small part of his resources on teaching duties and almost none on administration. He created no school, nor did he influence directly the new generation of Cambridge physicists. He preferred to work by himself and only rarely engaged in collaboration. "He is one of the very few scientists who could work even on a lonely island if he had a library and could perhaps even do without books and journals."[21] Of course, Dirac had a great impact on the course of physics, but this was primarily achieved through his research papers, his textbook on quantum mechanics, and his lectures, not through his work as a teacher and supervisor of Ph.D. students. Mott recalled:[22]

I think I have to say his influence was not very great as a teacher. Now and then these extraordinary bombshells came out; the spinning electron, the positron, more or less, and that was it. And he always, of course, has given this lecture based on his book with admirable character. But he never was a man who would advise a student to examine the experimental evidence and see what it means. So his influence would be on the side of the older mathematical development at Cambridge, which results from our educational system. Dirac is a man who would never, between his great discoveries, do any sort of bread and butter problem. He would not interested at all.

As a lecturer, Dirac's most enduring influence was through his course of lectures on quantum mechanics, which he presented for many years at Cambridge University. The course largely followed the material of *Principles,* which itself grew out of Dirac's first courses on quantum mechanics (see Chapter 4). Richard Eden and John Polkinghorne recalled the lectures given by Dirac as follows.[23]

The delivery was always exceptionally clear and one was carried along in the unfolding of an argument which seemed as majestic and inevitable as the development of a Bach fugue. Gestures were kept to a minimum, though there was a celebrated passage near the beginning where he broke a piece of chalk in half and moving one of the bits about the lecture desk said that in quantum theory we must consider states which are a linear superposition of all these different possible locations. There was absolutely no attempt to underline what had been his own contributions, though at times one felt one got a hint of his feelings about what he had done.

But it would hardly be correct to characterize Dirac as a good teacher. Quantum mechanics à la Bach may have been appreciated by a chosen few, but the majority of students would no doubt have benefited from a less majestic, and more pedagogic, presentation. Dirac's lectures were clear and well structured, but for three decades they largely followed his textbook, *the* way of presenting quantum mechanics. Harish-Chandra was accepted by Dirac as a research student in 1945 but had only infrequent contact with his supervisor, about once each term. "I went to his lectures," Harish-Chandra recalled, "but soon dropped out when I discovered that they were almost the same as his book."[24] The same impression was left on Sciama, who remembered that Dirac "didn't particularly help me."[25] When Dirac delivered lectures, he strived to present his text in what he felt was the best way possible, with a maximum of lucidity and directness. He considered it illogical to change his carefully chosen phrases just because they had not been understood. More than once, somebody in the audience asked him to repeat a point the listener had not understood, meaning that he would like a further exposition; in such cases Dirac would repeat exactly what he had said before, using the very same words![26]

It seems that, in general, Dirac was discouraging in his contacts with prospective research students, who often were met with a cold response when they approached him. Several students who initially wanted to do research under Dirac were frightened by his unapproachable character and decided to seek another, more "human" supervisor. He seemed to lack genuine interest in his students, whom he expected to work largely on their own. Subrahmaniyan Shanmugadhasan, who became Dirac's student in 1945, recalled that Dirac did not give any guidance on relevant literature and was unwilling to offer his opinion on anyone's work. He did not even always read the papers he communicated to journals.[27] Although Dirac regularly attended the theoretical physics seminars at Cambridge after his return from Princeton in 1948, he rarely participated actively; sitting in an armchair, he often appeared to be asleep during the seminar.[28] That Dirac was not of much help to students was also the experience of the young Weisskopf, who spent a few months in Cambridge in 1933, mainly to work with Dirac. His verdict: "In Cambridge I was a little disappointed; Dirac is a very great man, but he is absolutely unusable for any student: you can't talk to him, or, if you talk to him, he just listens and says, 'yes.' So that was, from that point of view, a lost experience."[29]

Dirac's indifference toward students was not a result of his position as a distinguished professor. It was just another aspect of his fundamental reticence. When he was on his way to Japan in 1935, Dirac made a stop in Berkeley, where Oppenheimer arranged a meeting with two of his ex-students, Robert Serber and Arnold Nordsieck. The two Americans explained at length their recent work on quantum electrodynamics. "Dirac uttered not a word. At the end of the talk a lengthy silence ensued. Finally, Dirac said, 'Where is the post office?' Exasperated, Serber asked if he and Nordsieck could accompany Dirac to the post office. Along the way, perhaps, Dirac could tell them his reaction. 'I can't do two things at once,' Dirac said."[30]

Schrödinger, a humanist as well as a scientist, once emphasized, "Physics consists not merely of atomic research, science not merely of physics, and life not merely of science."[31] For Dirac, however, life was mostly science and science was physics. Although his close friends report that he did from time to time enjoy activities that were outside physics – reading mystery novels, attending concerts, and visiting art museums, for example – these activities were certainly minor interests. For the most part, he concentrated his resources on theoretical physics, which acted as a substitute for human emotions and a richer social life. Dirac had a reputation for being almost inhuman in his monomaniacal occupation with physics.[32] Compared with other great physicists of the century, his intellectual interests were narrow. He never wrote on anything besides physcis (and cosmology) and felt no temptation to apply his genius to other areas or to deal with other sciences, such as biology or chemistry. These sciences, he implied on some occasions, were not fundamental in the same sense as theoretical physics, from which they could in principle be deduced.

Although Dirac's background and general interests were completely different from those of the intellectually versatile and philosophically oriented Schrödinger, the mentalities of the two physicists had much in common. Neither of them liked the public limelight, and in their scientific careers they both drifted away from mainstream physics. Although they reached different conclusions, their minds operated in much the same way. Dirac recognized this mental similarity and held Schrödinger in great esteem. When Schrödinger died in 1961, his obituary in *Nature* was written by Dirac.[33] Later, in his recollections, Dirac stated:[34]

... of all the physicists that I met, I think Schrödinger was the one that I felt to be most closely similar to myself. I found myself getting into agreement with Schrödinger more readily than with anyone else. I believe the reason for this is that Schrödinger and I both had a very strong appreciation of mathematical beauty, and this appreciation of mathematical beauty dominated all our work. It was a sort of act of faith with us that any equations which describe fundamental laws of Nature must have great mathematical beauty in them. It was like a religion with us.

However, as to interests and insight in areas outside physics, one can hardly imagine two persons more different than Schrödinger and Dirac.

Dirac was largely ignorant of religion, art, literature, philosophy, history, and politics. He was skeptical or even hostile toward these aspects of life, which, he thought, possessed none of the beauty and logic of mathematics and physics. His overall attitude was one of aristocratic rationalism. Indeed, in many ways his general outlook was similar to that which prevailed in the Newtonian "age of reason," when hostile attitudes toward poetry and art were rather in vogue. Dirac might have felt himself more at home with the rationalistic spirit of the early eighteenth century than with the spirit of his own time.

It has already been remarked that Dirac looked at all areas of life, not just physics, from a logical point of view. He occasionally expressed some interest in art, politics, economics, or religion, but then it was in his own peculiar way; after all, these areas lay outside the realm of pure logical analysis. Dirac was not a religious man and had little understanding of the cause of religion. At least in his younger days, he seems to have favored an atheistic or perhaps agnostic view, if he considered religion a worthy subject to think of at all. During the 1927 Solvay Congress, he became involved in an informal discussion of religion with some other physicists, including Heisenberg and Pauli. Dirac, according to Heisenberg's recollections, rejected any religious idea and instead showed himself to be a "fanatic of rationalism."[35] From a modern scientific point of view, the only point of view that Dirac acknowledged, religion is based on irrationalism and silly postulates and hence is without the slightest appeal to the man of science; religion, Dirac maintained, is merely a system of myths, an opium for the people. Pauli commented on Dirac's uncompromising atheism with usual sarcasm: "But yes, our friend Dirac has a religion, and the basic postulate of this religion is: 'There is no God, and Dirac is his prophet.'"[36]

Much later, in 1961, Dirac accepted an invitation to become a member of the Pontifical Academy of Science in Rome. He published several articles in the proceedings of the Academy. However, his membership in the Pontifical Academy does not in itself indicate any particular interest in Roman Catholicism. The Academy does not ask for religious views, only for scientific excellence. Yet late in his life he expressed some interest in religion, a subject that he conceived of in a naïve, rationalistic way. At the 1971 Lindau meeting, he chose to discuss, as one of the fundamental problems of *physics,* the question, does God exist? (The other fundamental problems were: Is there causality? Is there space–time continuity? and Is there an ether?) Dirac wanted to consider the question from the point of view of a physicist, not on the basis of faith or philosophical principles, which "is really just sort of guessing or expressing one's feelings." While classical determinism, according to Dirac, left no place for God, the indeterminism of modern physics makes the existence of a higher being a possibility. He explained that the existence of God could be justified only if the emergence of life required a highly improbable event to have taken place in the past:[37]

It could be that it is extremely difficult to start life. It might be that it is so difficult to start life that it has happened only once among all the planets. . . . Let us consider, just as a conjecture, that the chance life starting when we have got suitable physical conditions is 10^{-100}. I don't have any logical reason for proposing this figure, I just want you to consider it as a possibility. Under those conditions . . . it is almost certain that life would not have started. And I feel that under those conditions it will be necessary to assume the existence of a god to start off life. I would like, therefore, to set up this connexion between the existence of a god and the physical laws: if physical laws are such that to start off life involves an excessively small chance, so that it will not be reasonable to suppose that life would have started just by blind chance, then there must be a god, and such a god would probably be showing his influence in the quantum jumps which are taking place later on. On the other hand, if life can start very easily and does not need any divine influence, then I will say that there is no god.

This was the closest Dirac came to religion in a public address. He did not commit himself to any definite view but just outlined the possibilities for answering the question scientifically.

Dirac's attitude toward politics was not unlike his attitude toward religion. Basically, he was apolitical and felt no commitment to any particular political system or party. Socialism, liberalism, and conservatism were empty labels to him. When he did express an interest in politics, it was in an unusual way, for example, because a certain political system appealed to his sense of logic. As mentioned in Chapter 7, during the thirties Dirac indicated a positive attitude toward the Soviet system, which he, contrary to the common view at the time, considered an interesting social experiment. There is no doubt that he was genuinely interested in the new economic order of Soviet Russia. He found it "completely different" from the system of Western capitalism and urged Tamm to show him a Soviet factory.[38] Whether his interest in the Soviet Union also included an ideological commitment is more doubtful. Although he may have had "vaguely left-wing sympathies," there is no indication that he was ever a communist or a member of any political group.[39]

As to art, music, and literature, Dirac was ignorant and mostly indifferent. He seldom went to the theater and never went by himself. Mott believed that he and his wife Ruth were the first people ever to take Dirac to a theater, which must have been around 1930.[40] Dirac's reactions to art were often peculiar in their lack of appreciation for artistic experience. Once Kapitza gave him an English translation of Dostoyevsky's novel

Crime and Punishment and asked him to read it. When after some time
Kapitza asked if he had enjoyed the book, Dirac's only comment was: "It
is nice, but in one of the chapters the author made a mistake. He
describes the Sun as rising twice on the same day."[41] On another occasion
Dirac listened to Heisenberg playing the piano. An accomplished pianist,
he played several pieces and asked Dirac which he liked best. After thinking for a while, Dirac answered, "The one in which you crossed your
hand."[42]

Dirac did not make a virtue of his scant appreciation for culture and
art, but neither was he the slightest bit embarrassed about it. He would
not think of pretending an interest in order to please people or to adapt
for himself an air of culture. Oppenheimer was very different from Dirac.
He was an open person with many extrascientific interests, including
poetry and Buddist philosophy. This kind of versatility puzzled Dirac.
He once said to Oppenheimer: "How can you do physics and poetry at
the same time? The aim of science is to make difficult things understandable in a simpler way; the aim of poetry is to state simple things in an
incomprehensible way. The two are incompatible."[43] Dirac's reading was
largely restricted to physics. He had no appreciation for literary scholar-
ship and felt that reading might even have an adverse effect on original
thinking. When Oppenheimer met Dirac in Berkeley in 1934, before his
departure for Japan, Oppenheimer offered him two books to read during
the voyage. Dirac politely refused, saying that reading books interfered
with thought.[44]

These anecdotes may easily exaggerate Dirac's single-mindedness. In
fact, he did things other than physics and was, especially during his later
years, interested in other matters. As an illustration of this less well-
known side of Dirac's character, I quote an excerpt of a letter from Dirac
to the author Esther Salaman:[45]

I was especially glad for the talks I had with you on physics and with your husband on his field of work. I often read semi-popular articles on fundamental biological subjects in *Nature, Endeavour* and such magazines, and continually run
into questions which I would like to know the answer to, but there is usually no
one to ask. With talking to your husband I was able to put some of these questions, and get answers – or learn the difficulty of obtaining an answer – which is
a great help to forming a clear picture of the subject. I hope we shall have similar
talks in the future. . . .

When Dirac decided to deal with a matter, he would do it in a systematic
and serious way, as he dealt with a problem of physics. After his marriage,
he became fond of gardening, on which he spent much time. He tried to
solve horticultural problems in the same way that he solved physical

problems, from first principles. The results were not encouraging. He was also a keen chess player and served for many years as president of the chess club of St. John's College.[46] He did not engage in sports as a young man, but in the early thirties, encouraged by Kapitza, he learned to play tennis.

The only major interests, apart from physics, to which Dirac really was devoted, were traveling and walking in the mountains. In earlier chapters I have mentioned some of his many travels, including three tours around the world. During those travels, because he very much appreciated the beauty of nature, he felt inspired to cultivate aesthetic qualities in which he otherwise would express no interest. Not only scenic nature but also museums, palaces, and botanic gardens were frequent targets of his travels. Dirac also was a great walker, and on tours he could show a stamina and physical energy that surprised those who knew him only from conferences or dinner parties. He preferred to visit mountain areas and walked, among other places, in the Rocky Mountains, the Caucasus, the Alps, the Sinkiang, and the Jotunheimen in Norway. Many of these tours were made in company with his friend James Bell and the Russian physicist Igor Tamm. Dirac liked to climb mountains but avoided peaks that required proper mountaineering. Still, he managed to climb some pretty high mountains. With Van Vleck he went to the top of Uncompaghre Peak in the Rockies, altitude 4,360 meters, and with Tamm he later climbed the 5,640-meter Mount Elbruz in the Caucasus, Europe's highest mountain. The latter climb was recalled by K. K. Tikhonov, the leader of the expedition, as follows:[47]

It is summer 1936, the Caucasus, Mount Elbrus. A group of students from the Moscow Transport Engineering Institute led by me was joined by the Soviet physicist Tamm and the British physicist Dirac. After two days of acclimatising at the "Eleven Mountaineers' Shelter" we scaled the eastern side of the Elbrus.... I recall our futile attempts to get "the USSR Alpine Climber" badge for Dirac for having done the ascent....

The Lucasian professor used to practice for the mountains by climbing trees in the Gog-Magog hills outside Cambridge. For this practice he dressed in the formal black suit he always wore.

CHAPTER 13

PHILOSOPHY IN PHYSICS

AMONG the great physicists, Dirac was probably the least philosophical. He was plainly not interested in philosophy, of which he had only the most superficial knowledge, and he was not tempted to draw wider philosophical consequences for his work in physics. As an eighteen-year-old engineering student, he attended Broad's philosophical lectures on relativity and thought for a while that philosophy might be scientifically useful. He did some reading in philosophy, including Mill's *A System of Logic,* but soon decided that philosophy was not important:[1]

My attempts to appreciate philosophy were not very successful. I felt then that all the things that philosophers said were rather indefinite, and came to the conclusion eventually that I did not think philosophy could contribute anything to the advance of physics. I did not immediately have that point of view, but I came to it only after a lot of thought, and studying what philosophers said, in particular Broad.

Dirac retained this view. Forty-three years after his first encounter with philosophy, he said: "I feel that philosophy will never lead to important discoveries. It's just a way of talking about discoveries which have already been made."[2]

Most Nobel Prize-winning theoretical physicists involve themselves with philosophical questions and feel that it is natural to relate their experiences in science with new and old questions of philosophy, or of politics, religion, and morality. Some of the great physicists of the twentieth century, including Einstein, Bohr, Eddington, and Bridgman, were almost as much philosophers as they were physicists. Others, like Heisenberg, Pauli, Schrödinger, Born, Weyl, and Wigner, explicitly concerned themselves with problems of philosophy (especially) during their later years. Dirac was different. With one possible exception, he never wrote papers or gave talks that could reasonably be called philosophical. This does not mean that "philosophy" was absent from his science or that he was not guided by philosophical considerations. But it does imply that whatever philosophy can be found in Dirac's science cannot be analyzed by direct reference to his own philosophical statements. Neither is it useful to seek the roots of Dirac's views in philosophical influences, readings during his youth, his cultural environment, or similar external circumstances.

Dirac's implicit philosophy of science has to be reconstructed piece by piece from his contributions to physics. Obviously, one runs the risk that such a reconstruction will not reflect the actual historical course – that, for example, philosophical elements become artificially extracted from his work in physics. This and the following chapter give an interpretation of certain parts of Dirac's physics. The reconstruction is, of course, based on the historical sources, but as a reconstruction it goes beyond a simple description of the content of those sources. In Dirac's implicit or spontaneous philosophy of science, four themes can be singled out that acted as significant meta-principles guiding his physics. These are:

1. Instrumentalism and the observability doctrine.
2. The unity of nature.
3. The principle of plenitude.
4. The principle of mathematical beauty.

Since these doctrines were, in varying degrees, enduring and partly emotional elements in Dirac's life, they may be viewed as invariant "themata" in Holton's sense.[3] The four themes mentioned are interrelated, or at least they were in Dirac's case, but for him they had unequal status. Thus, the first two issues did not play a crucial role for Dirac in particular, but can be found in many other physicists too. Also, these two principles were expressed rather directly, whereas the third was only applied indirectly by Dirac. The concept of "beauty" in physics was the most influential among the doctrines and was the one most original to Dirac. From the late thirties onwards, it worked as a sort of super-thema to which Dirac was deeply committed and which dominated much of his intellectual life.

Themata, including aesthetic principles, are usually connected to external factors. They may reflect the political or philosophical opinions of a period, and then be concomitant with its *zeitgeist*. No scientist can avoid being to some extent influenced by forces outside his scientific life. Even the greatest scientists are members of a social and cultural world. This was emphasized by Schrödinger:[4]

The scientist cannot shuffle off his mundane coil when he enters his laboratory or ascends the rostrum in his lecture hall. In the morning his leading interest in class or in the laboratory may be his research; but what was he doing the afternoon and evening before? He attends public meetings just as others do or he reads about them in the press. He cannot and does not wish to escape discussion of the mass of ideas that are constantly thronging in the foreground of public interest, especially in our day. Some scientists are lovers of music, some read novels and poetry, some frequent theaters. Some will be interested in painting and sculpture. . . . In short, we are all members of our cultural environment.

However, Dirac was no Schrödinger. The quoted observation has little validity in the case of a physicist who had no particular interest in music, literature, theater, and the like. To a remarkable extent, the philosophical aspects of Dirac's physics were unrelated to external factors. At least, I have not been able to locate relevant external factors that help to "explain" his physics in any reasonable way.[5] My conclusion is that the roots of Dirac's implicit philosophy should be looked for internally, in physics itself. Dirac's philosophical views were a result of the unsophisticated and unphilosophical reflections about his own and his contemporaries' physical theories, that is, of his personal version of the history of modern physics.

Although few philosophers of science will deny the intellectual greatness of Dirac's physics, it has not attracted much philosophical interest. Henry Margenau and Gaston Bachelard were among the few who took up Dirac's works for philosophical analysis. In 1940, Bachelard used Dirac's theory of negative-energy particles to illustrate his "philosophy of no"; Dirac's theory, he claimed, was "de-realized" and an example of "dialectical super-rationalism."[6] The methodology of Dirac's theory of 1931 was a few years later subjected to philosophical analysis by Margenau, who objected that Dirac operated with constructs of explanation to which, by definition, no counterparts in the form of data corresponded. Margenau regarded such constructs, like Dirac's negative-energy electrons and Pauli's neutrinos, as illegitimate because they sinned against requirements of simplicity.[7] In more recent times, Edward MacKinnon and Manfred Stöckler have made philosophical comments on Dirac's physics, but no attempt at a full philosophical analysis exists.[8]

Instrumentalism and the observability doctrine

According to the "observability doctrine," physical theories should build solely on concepts which refer to quantities that can, at least in principle, be observed. Philosophically, the doctrine is closely connected with "operationalism," as developed mainly by Bridgman. In the context of quantum physics the observability doctrine is often referred to as Heisenberg's observability principle, although Heisenberg in fact used it only after Pauli, who stated it clearly in 1919: "However, one would like to insist that only quantities which are in principle observable should be introduced in physics."[9] Pauli's concern was not, at that time, with quantum theory but with the continuum problem associated with Weyl's unified field theory. Quantities that were in principle unobservable would be, according to Pauli, "fictitious and without physical meaning."[10]

Heisenberg was inspired by Pauli's operationalist principle in his construction of quantum mechanics, which was based on a thoroughgoing criticism of the semiclassical atomic models. In the introduction to his pioneering paper of 1925, Heisenberg stated that his program was "to try to establish a theoretical quantum mechanics, analogous to classical mechanics, but in which only relations between observable quantities occur."[11] This doctrine was often reaffirmed during the following years, when Heisenberg's quantum mechanics became generally regarded as the fulfillment of the operationalist program in microphysics. It became commonplace for quantum physicists, especially those affiliated with the Copenhagen school, to highlight in their rhetoric the heuristic validity of the observability doctrine. That the doctrine was in fact never used in a strict way, and hardly played the alleged crucial role in the discovery of quantum mechanics, is in this respect of less importance.

Dirac was impressed by the role of the observability principle in Heisenberg's theory and shared the belief in its general value. As mentioned in Chapter 6, in 1932 he modeled his proposal of an alternative relativistic quantum theory on "[Heisenberg's] principle that one should confine one's attention to observable quantities, and set up an algebraic scheme in which only these observable quantities appear."[12] Similar statements continued to appear often in Dirac's works. Thus in 1967, he wrote, "Only questions about the results of experiments have a real significance and it is only such questions that theoretical physics has to consider."[13] In a less sophisticated, positivistic version, unobservability is taken to imply nonexistence as far as physical reality is concerned, as indicated in Pauli's statement above. But when is an object unobservable and not merely unobserved? In 1925, Heisenberg was aware of the distinction but seems not to have found it important. Concepts like the position and period of revolving electrons are "apparently unobservable in principle," Heisenberg noted; hence they lacked physical foundation "unless one still wants to retain the hope that the hitherto unobservable quantities may later come within the realm of experimental determination."[14] Dirac was not always careful to distinguish unobservable-in-principle from unobserved-so-far and on a few occasions erred in conclusions that rested on his use of the observability principle.

In his short-lived 1936 proposal to abandon energy conservation in atomic processes, Dirac expressed his distrust in the neutrino. He considered it to be unobservable and hence, by virtue of the observability doc-

trine, without physical existence. In this respect he was in agreement with
the methodological criticism put forward by Margenau the year before.
Margenau regarded neutrinos and their accompanying anti-neutrinos as
"monstrosities" and "highly paradoxical entities."[15] Many years later,
and in a completely different context, Dirac was again misled by insisting
on the observability principle. In 1962, he argued briefly that the interior
of black holes should be excluded from the realm of physics.[16] Since sig-
nals cannot be sent from a black hole through the Schwarzschild barrier,
Dirac concluded that the inside region was unobservable and hence without physical meaning. But just as the neutrino was vindicated and eventually detected experimentally in the early fifties, Hawking showed in the
mid-1970s that black holes, if they exist, can emit particles. Hence they
are not unobservable in principle.

   However, these examples are not really representative of Dirac's actual
application of the observability principle, which in general did not affect
his scientific work. In fact, he did not hesitate to propose quantities that
seemed to have only the slightest connection to observables. The negative-energy world of 1941 was such a quantity, and so was the quantum
mechanical ether proposed in the fifties. Further examples include the
particles described by his positive-energy wave equation of 1971 (which
had no electromagnetic characteristics) and, to an even greater extreme,
the negative mass introduced in his 1973 cosmology (which had no
observable effect at all). Dirac realized that, in practice, unobservable
quantities could not be excluded from physical theory and observability
could not be separated from the existing theoretical framework. In 1973,
when looking back at the development of quantum theory, he once again
praised Heisenberg's observability principle as being "a very sound idea
philosophically." But then he added:[17]

It was only possible [to discover quantum mechanics] because Heisenberg didn't
keep too strictly to his idea of working entirely in terms of observable quantities. . . . There must be unobservable quantities coming into the theory and the
hard thing is to find what these unobservable quantities are.

It is well known that Heisenberg believed that the observability doctrine
was at the heart of the Einsteinian theory of relativity and that he was
merely taking Einstein's approach over into atomic physics.[18] However,
Einstein did not accept the doctrine and told Heisenberg in the spring of
1926 that it was "nonsense." On second thought, Heisenberg came to
accept Einstein's dialectical view that although "it may be heuristically
useful to keep in mind what one has actually observed . . . , on principle,
it is quite wrong to try founding a theory on observable magnitudes alone.

In reality the very opposite happens. It is the theory which decides what we can observe."[19] In practice, although not always in rhetoric, this moral was also followed by Dirac. Pauli, too, the other champion of the observability doctrine in early quantum mechanics, came to accept its inadequacy in his later years. In 1955, in a letter to Schrödinger, he turned to the Diracian view that the mathematical structure of a theory, and not its content of observable quantities, was the important thing.[20]

The Copenhagen school of quantum mechanics is often labeled as positivistic or instrumentalistic, and Dirac is mentioned as a chief exponent of Copenhagen quantum positivism. There is some truth in this claim, but Dirac's views were far from being unambiguously positivistic. His entire "mathematics first, then physics" philosophy, his speculations in cosmology, his theories of monopoles, anti-particles, and ether, and his opposition to postwar quantum electrodynamics embodied anything but positivism. The only time that Dirac was involved in a methodological dispute, in connection with his cosmology in the late thirties, he was accused of sinning against positivistic virtues. As Born remarked in his Edinburgh inaugural lecture in 1936, Dirac was less radical than positivists like Jordan: "Whereas he [Dirac] declares himself content with the formulae and uninterested in the question of an objective world, positivism declares the question to be meaningless."[21]

As to instrumentalism, Dirac was indeed committed to the view that physics is basically a formal scheme, or instrument, that allows the calculation of experimental results. This was his view in 1927 (see Chapter 2), and he reiterated it on many later occasions, for example, in his Bakerian Lecture of 1941 in which he argued for what he called "Heisenberg's view about physical theory." This view, according to Dirac, was that "all it [physical theory] does is to provide a consistent means of calculating experimental results."[22] He kept to this instrumentalist view, stressing that the mathematical structure of a physical theory, not its ontological implications, was the important thing. This message was clearly expounded in *Principles* and received as such by its readers. When John Lennard-Jones reviewed the book, he summarized what he called "Dr. Dirac's philosophy of the relation of theoretical and experimental physics" as a case of pure instrumentalism:[23]

A mathematical machine is set up, and without asserting or believing that it is the same as Nature's machine, we put in data at one end and take out results at the other. As long as these results tally with those of Nature, (with the same data or initial conditions) we regard the machine as a satisfactory theory. But as soon as a result is discovered not reproduced by the machine, we proceed to modify the machine until it produces the new result as well.

Dirac's belief in the power of instrumentalism was indebted to Einstein, whose construction of the theory of relativity was seen as the prime paradigm for radical change in physics. I shall elaborate a little on this theme in Chapter 14 and here only note that Heisenberg shared Dirac's methodological indebtedness to Einstein. It is well known, though, that the mature Einstein was hostile to the instrumentalism of established quantum theory and was not at all happy about his work being taken as a methodological model for quantum theory. While Dirac continued to praise instrumentalist virtues, after the war he often stressed that predictive ability is not enough for a physical theory. He increasingly turned to the view that scientific understanding includes criteria like beauty and simplicity, which may well clash with criteria of prediction.

In his later years, Dirac often expressed in a rather vague way a critical attitude toward the Bohr–Heisenberg interpretation of quantum mechanics. He sometimes expressed sympathy with Einstein's view of determinism in quantum theory, which, he suggested, might be a possible way to reformulate relativistic quantum mechanics so as to avoid infinities. In one of his rare comments on the Bohr–Einstein debate, Dirac wrote in 1979: "I think it is very likely, or at any rate quite possible, that in the long run Einstein will turn out to be correct, even though for the time being physicists have to accept the Bohr probability interpretation, especially if they have examinations in front of them."[24] Unlike some exponents of the Copenhagen position, Dirac did not consider quantum mechanics to be a complete or final theory.[25] Such a view was contrary to his idea of the nature of physical theory, especially as it evolved since the mid-1930s under the impact of the troubles of quantum electrodynamics. He believed that any physical theory was always a transient phase of physics that would eventually be superseded by a better theory. The radical departure from the Bohr–Sommerfeld theory that he experienced in 1925–6 served as an exemplary event. To Dirac it illustrated "a general feature in the progress of science – the feature that however good any theory may be, we must always be prepared to have it superseded later on by a still better theory."[26] Quantum mechanics and the theory of relativity were no exceptions. As pointed out in Chapter 4, Dirac was not particularly interested in the philosophy of quantum mechanics and did not care much for the discussions about how to interpret the theory. This might have been interesting to philosophers, but he, as a physicist, was more concerned with the mathematical structure of quantum theory. In one of his last papers, published in 1984, he stated: "I don't want to discuss this question of the interpretation of quantum mechanics. . . . I want to deal with more fundamental things."[27]

With regard to the question of physics as a finished science, it should, however, be mentioned that Dirac did not always adhere to the "recurring revolutions" view expressed above. Around 1930, he viewed the situation in physics optimistically and on some occasions came close to the idea that physics was in principle complete. This view seems for a brief period to have been fashionable among quantum physicists, inspired, no doubt, by the amazing success of the 1928 wave equation of the electron.[28] In an often quoted passage, the architect of this equation wrote in 1929:[29]

The general theory of quantum mechanics is now almost complete.... The underlying physical laws necessary for the mathematical theory of a large part of physics and the whole of chemistry are thus completely known, and the difficulty is only that the exact application of these laws leads to equations much too complicated to be soluble.

The view expressed in this passage is reductionistic and may resemble the view of completeness-in-principle that was frequently expressed in the late nineteenth century. However, Dirac soon changed his mind.

The unity of nature

The principle of nature's unity is a guiding principle for virtually all theoretical physicists. It is closely related to another celebrated meta-principle, the doctrine of "simplicity." The "unity of nature" sometimes refers to the belief that nature is made up of only a few constituents, preferably only one fundamental substance. In this version the principle is an ontological belief that has been of prime importance in theories of matter throughout history. This was the sense in which Dirac applied it in 1930 when he erroneously identified the proton with the anti-electron because he wanted "to have all matter built up from one fundamental kind of particle" (Chapter 5). However, the judgment of when the principle of nature's unity is satisfied is subject to considerable arbitrariness. In 1933, Dirac was ready to accept three particles in the atomic nucleus because, he argued, three was no more ugly than two. In the same year, the American physicist Karl Darrow stated that it would be more "elegant" if there were only two particles in the nucleus.[30]

There is, however, another way to interpret the unity of nature, namely, to postulate that all of nature is amenable to the same kind of theoretical treatment. This version is a methodological principle and should be distinguished from the ontological principle mentioned above. It is, in fact, not a statement of nature's unity but of the unity of physics or, generally, of the ultimate unity of theoretical science. Most physicists agree on the value of this principle: they believe that in the long run it is intolerable to operate with different and perhaps incompatible schemes in different domains of nature.

In 1939, Dirac received the James Scott Prize at Edinburgh University. In his lecture he applied the idea of the unity of nature, in the sense of unity of physics, as an argument against mechanistic determinism in classical physics. His objections were not rooted in the quantum mechanical indeterminacy of observation but in a desire to keep the whole of nature within the realm of mathematical treatment. "I find this position [classical mechanism] very unsatisfactory philosophically, as it goes against all ideas of the *Unity of Nature,*" he said.[31] Dirac's argument was this: Classical mechanics operates with two types of parameters, a complete system of equations of motion and a complete set of initial conditions. With these provided, the development of any dynamical system is completely determined. However, while the equations of motion are amenable to mathematical treatment, the initial conditions are not. They are contingent quantities, determinable only by observation. But then an asymmetry arises in the description of the universe, which methodologically is separated into two spheres: one in which mathematical analysis applies, and another in which it does not. Dirac found this to be an intolerable situation because it ran contrary to the expectation of unity in nature. According to Dirac, all the initial conditions, including the elementary particles, their masses and numbers, and the constants of nature, must be subjected to mathematical theory. He foresaw a mathematical physics of the future in which "the whole of the description of the universe has its mathematical counterpart."[32] The phantom of classical mechanism, Laplace's "supreme intelligence," had to have recourse to the initial conditions in order to predict the development of the universe. In Dirac's philosophical vision, Laplace's daimon reappeared in an even more powerful version, as one able to deduce everything in the universe by pure mathematical reasoning:[33]

We must suppose that a person with a complete knowledge of mathematics could deduce, not only astronomical data, but also all the historical events that take place in the world, even the most trivial ones. . . . The scheme could not be subject to the principle of simplicity since it would have to be extremely complicated, but it may well be subject to the principle of mathematical beauty.

Dirac's mathematical credo may call to mind the ideas of James Jeans more than those of Eddington. Jeans shared Eddington's tendencies toward idealism and conventionalism, but his philosophical views took a more rationalistic turn. Jeans's thinking tended to be a worshiping of mathematics, believed to be the first and last word in science as well as in nature. "From the intrinsic evidence of his creation," Jeans wrote, "the Great Architect of the Universe now begins to appear as a pure mathematician."[34] Since the product designed by the divine mathematician, the universe, is in essence mathematical, all phenomena can be described in mathematical terms. If, Jeans wrote, we are not able to do so, it is not because parts of the world are unamenable to mathematical treatment but because our mathematical knowledge needs to be improved. "*The final truth* about a phenomenon resides in the mathematical description of it; so long as there is no imperfection in this our knowledge of the phenomenon is *complete*."[35] Jeans believed that the ultimate reality of the physical world could be ascribed to mathematics. This ontological aspect was absent in Dirac's thinking, which otherwise remained close to the ideas of Jeans.

In the sixties a powerful research program, known as the bootstrap program, was launched in high-energy physics by the American Geoffrey Chew. It built on the S-matrix theory, originally introduced by Heisenberg in 1943–4 as an attempt to deal with the divergence difficulties of quantum electrodynamics.[36] Heisenberg had constructed his theory out of a refined version of the observability principle and defined a mathematical object, the S-matrix, which in principle comprised everything that could actually be observed in a process, such as scattering cross sections. After the war, the theory was taken up by Christian Møller and Ralph Kronig but soon, in 1947–8, fell into discredit. Most physicists regarded it as an ambitious and mathematically interesting program, but one that was empty as far as physics was concerned. After the breakthrough of quantum electrodynamics, the S-matrix approach was largely forgotten, until it was revived by Chew in 1961. In the new S-matrix theory, also known as the bootstrap theory, all properties of strongly interacting particles were taken to be contained in just one unitary transformation matrix that related the initial to the final asymptotic states of the system.

Dirac briefly discussed Heisenberg's S-matrix theory in 1943 in his correspondence with Pauli, who pointed out the difference between Heisenberg's approach and Dirac's theory of a hypothetical world.[37] Three years later, Dirac attended the Cambridge Conference on Fundamental Particles and Low Temperatures, where a session was devoted to S-matrix theory, which was discussed by Heitler, Møller, and Ernst Stueckelberg.[38] Apparently, Dirac did not at the time take much interest in S-matrix theory. It was only when the theory gained popularity in the sixties that he referred to it, and then to warn against it.[39]

At first glance, Dirac's opposition to the S-matrix or bootstrap program may seem strange, since several things in it agreed with his own ideas. In fact, what is probably the first example of an S-matrix philosophy can be found in Dirac's 1932 paper on relativistic quantum mechanics. The bootstrappers of the 1960s were, like Dirac, opposed to standard quantum electrodynamics founded on Lagrangian field-theoretic methods. According to Chew, quantum electrodynamics should be abandoned because it rested on the concept of the unobservable space–time continuum. In opposition to the standard theory, the bootstrappers endeavored to give an internally consistent and complete mathematical theory in which there were no irreducible fundamentals that had to be accepted as just contingent facts of nature. This endeavor was certainly in agreement with Dirac's view as he stated it in, for example, his James Scott address. Furthermore, the dominant mathematical framework of the bootstrap theorists was the theory of analytic functions, a branch of the general theory of complex variables. Dirac considered this branch of mathematics to be particularly beautiful and promising for application in physics.[40]

However, Dirac did not accept the bootstrap philosophy any more than he accepted the standard ("fundamentalist") theory of elementary particles, because he felt that the bootstrap theory contradicted the principle of the unity of nature. In 1966, he argued as follows:[41]

In physics one should aim at a comprehensive scheme for the description of the whole of Nature. A vast domain in physics can be successfully described in terms of equations of motion. It is necessary that quantum field theory be based on concepts and methods that can be unified with those used in the rest of physics.

Three years later he related his argument explicitly to S-matrix or bootstrap theory:[42]

The reason for this disbelief [in S-matrix theory] comes essentially from my belief in the unity of the whole of physics. High energy physics forms only a very small part of the whole of physics – solid state physics, spectroscopy, the theory of bound states, atoms and molecules interacting with each other, and chemistry. All these subjects form a domain vastly greater than high energy physics, and all of them are based on equations of motion. We have reason to think of physics as a whole and we need to have the same underlying basis for the whole of physical theory.

The principle of plenitude

According to the principle of plenitude, boiled down to its essence, whatever is conceived as possible must also have physical reality. This principle has played an important part in the history of ideas and science, where, by Leibniz and others, it was seen as a manifestation of God's omnipotence. Leibniz related the idea to the principle of continuity, or the "great chain of being," according to which the order of natural quantities formed a single chain, the elements of which exhaust the space of potential being. This belief in nature's fullness and continuity led to several predictions and entire research programs, particularly in the biological sciences. A French philosopher, J. E. Robinet, defined the principle of plentiude in 1766 as follows: "From the fact that a thing can exist I infer readily enough that it does exist."[43] But what is to be meant by "potential existence"? Obviously it should not be identifed with imagined existence; centaurs can exist in the imagination, but they have no real existence. The trouble with interpreting the principle of plenitude in a scientifically useful way was highlighted in the romantic wave at the beginning of the nineteenth century, when authors tended to include even errors, disharmony, and irregularity in the necessary richness of the world. In spite of the vagueness of the principle of plenitude, it worked in the nineteenth century on several occasions as a fruitful scientific meta-principle. The periodic system of the chemical elements is one example among many. Mendeleev and others predicted successfully that new elements would exist from the sole argument that this would be in harmony with the assumed periodicity of the elements. Later, Janne Rydberg and many others less successfully predicted other elements that they thought were prescribed by the periodic law.[44] Because of its ambiguity, the principle of plenitude has been used in very different ways through the course of history, as an argument for the existence of almost everything imaginable. At times plenitude reasoning may seem to amount to little more than asking "why not?" In the eighteenth century, for example, the principle was used to support the claimed observations of mermaids and sea-men. Since the notion of mermaids at the time seemed neither intrinsically contradictory nor in conflict with biological laws, such creatures were assumed to exist in nature – because why shouldn't they?

In modern theoretical physics the principle of plenitude is often implicitly used, in the sense that entities are assumed to exist in nature as far as they are subject to mathematically consistent description and are not ruled out by so-called principles of "impotence," or general statements that assert the impossibility of achieving something.[45] Examples of principles of impotence include the second law of thermodynamics, the uncertainty principle in quantum mechanics, and the impossibility of recognizing absolute velocity in the theory of relativity. It is widely assumed that the basic laws of physics may all be ultimately expressed as similar postulates of impotence. Such postulates are not forced upon us by logical necessity, since universes in which any postulate of impotence is violated are intelligible. On the other hand, neither are they simply generalizations of empirical knowledge. They are assertions of a belief, guided by experience indeed but raised to an a priori status, that all possible attempts to do certain things are bound to fail.

The discussion concerning superluminal particles, so-called tachyons, illustrates how the principle of plenitude works in modern physics. In the sixties, it was shown by Ennackel Sudarshan and others that neither quantum mechanics nor relativity theory precludes the existence of such entities. From the fact that tachyons are not precluded by theory, some physicists drew the conclusion that they exist in nature. This argument, clearly based on the principle of plenitude, was fully recognized by the involved physicists. For example, two tachyon theorists wrote:[46]

There is an unwritten precept in modern physics, often facetiously referred to as Gell-Mann's totalitarian principle, which states that in physics "anything which is not prohibited is compulsory." Guided by this sort of argument we have made a number of remarkable discoveries, from neutrinos to radio galaxies.

Dirac applied similar reasoning on several occasions, perhaps most clearly in his 1931 work on magnetic monopoles. Dirac proved that monopoles could exist according to quantum mechanics. But do they really exist? Dirac thought so, his argument being that "under these circumstances one would be surprised if Nature had made no use of it" (see further Chapter 10). In the introduction to his monopole paper, Dirac referred to what he called "Eddington's principle of identification," by which he apparently meant the realist interpretation of mathematical quantities in terms of physical quantities.[47] In his relativistic quantum theory of the electron, as well as in his theories of monopoles and positrons, he wanted to establish such a realist interpretation, although he admitted that not all mathematical terms in a physical theory could be identified with a physically meaningful term. Dirac's plenitude argument has remained essential for monopole hunters. In a survey article in 1963, the case was stated as follows:[48]

One of the elementary rules of nature is that, in the absence of law prohibiting an event or phenomenon it is bound to occur with some degree or probability. To put it simply and crudely: Anything that *can* happen *does* happen. Hence physicists must assume that the magnetic monopole exists unless they can find a law barring its existence.

This is a clearly formulated principle of plenitude, close to that used by Dirac in 1931. The agreement with Robinet's eighteenth-century formulation, quoted above, is striking, and Dirac's line of reasoning did not differ essentially from the eighteenth-century arguments in favor of mermaids. Although mermaids, tachyons, and monopoles have been claimed to have been observed several times, none of them have found general acceptance among scientists.

At about the same time that Dirac suggested the monopole, he applied similar reasoning in his interpretation of the negative-energy solutions of the relativistic wave equation. He was inclined to think that since these solutions had a mathematical existence, they must also, by virtue of Eddington's principle of identification, represent something physically existing. The methodological similarity between Dirac's monopole theory and his theory of anti-electrons is evident from his *Nature* note of 1930 in which he argued for proton-electron annihilation by means of plenitude reasoning: "There appears to be no reason why such processes should not actually occur somewhere in the world. They would be consistent with all the general laws of Nature. . . ." Peierls seemed to recognize that Dirac's theory relied on the principle of plenitude. In a letter to Pauli in 1933, he wrote: "In quantum electrodynamics one has always succeeded with the principle that the effects, for which one does not obtain diverging results, also correspond to reality. From this I would assume that it is reasonable to proceed in the same way with the hole model."[49] When Pauli and Weisskopf, in 1934, quantized the Klein–Gordon equation and showed that it described hypothetical particles obeying Bose–Einstein statistics, they recalled Dirac's plenitude reasoning and jocularly paraphrased him, asking why "Nature . . . has made no use of" negatively charged spin-zero particles.[50]

While plenitude reasoning was invoked in both the monopole and the hole theory, it functioned differently in the two cases. Anti-electrons were not only allowed in the hole theory; in a sense they were demanded, since the relativistic theory of the electron would be incomplete without them. Monopoles, on the other hand, were predicted solely by a negative argument, since they were not demanded by theory. Quantum electrodynamics is neither more nor less complete or consistent if monopoles are introduced.

Apart from the cases mentioned above, the principle of plenitude is also recognizable in Dirac's introduction of the ether as a physical entity. Dirac concluded that since the new ether was consistent with fundamental physical laws, then "we are rather forced to have an ether" (Chapter 9). As a last example, take the wave equation for particles of higher spin, as introduced by Dirac in 1936 and later developed by Bhabha and others. Referring to these theories, Dirac said in 1946, "Elementary particles of higher spin have not yet been found in nature, but according to present day theory there is no reason why they should not exist."[51]

Dirac always applied plenitude reasoning in the sense of Eddington's principle of identification, by inferring physics from mathematics. As Milne noted in 1936, such procedures, in obvious opposition to the empirical–inductive methods praised by positivists, are common in theoretical physics.[52] To trust the mathematics to look after the physical situation was also Milne's favorite method. But naturally this method depends crucially on which mathematics is accepted as trustworthy. Certainly, not any mathematical consequence of fundamental physics should be accepted as corresponding to a physical reality. Dirac's use of the principle of plenitude relied in practice on what he conceived to be sound or beautiful mathematics. It was closely connected with his idea of mathematical beauty in physics, to which we shall now turn.

CHAPTER 14

THE PRINCIPLE OF MATHEMATICAL BEAUTY

A T the University of Moscow there is a tradition that distinguished visiting physicists are requested to write on a blackboard a self-chosen inscription, which is then preserved for posterity. When Dirac visited Moscow in 1956, he wrote, "A physical law must possess mathematical beauty."[1] This inscription summarizes the philosophy of science that dominated Dirac's thinking from the mid-1930s on. No other modern physicist has been so preoccupied with the concept of beauty as was Dirac. Again and again in his publications, we find terms like beauty, beautiful, or pretty, and ugly or ugliness. The first time he used this vocabulary in an unconventional way was in 1936, when he contrasted the ugly and complex relativistic quantum theory with the general and beautiful non-relativistic quantum theory. His philosophy of beautiful mathematics was no doubt inspired by the difficulties of quantum electrodynamics, which was always Dirac's favorite example of an ugly physical theory.

The allusions to beauty in 1936 and the following years were more than mere casual remarks. This is evident from Dirac's James Scott lecture of 1939, "The Relation between Mathematics and Physics," in which he elaborated on the concept of mathematical beauty in physics. Delivered when Dirac was thirty-six, this address marked an outstanding physicist's reflections on his science, and as such it merits close attention. As usual, Dirac distinguished between the inductive–empirical and the mathematical–deductive method in science. He considered the latter method superior in physics because it "enables one to infer results about experiments that have not been performed."[2] But why is it that the mathematical–deductive method is able to meet with such remarkable success? Dirac answered, "This must be ascribed to some *mathematical quality in Nature,* a quality which the casual observer of Nature would not suspect, but which nevertheless plays an important role in Nature's scheme."[3]

The "mathematical quality" in physics is often identified with a principle of simplicity, according to which the fundamental laws of nature are simple. The principle of least action may be regarded as a mature and quantitative version of the principle of simplicity and shares with it an aesthetic status. Simplicity and minimum principles serve not only as useful heuristic guides but also, at times, as ends to which the entire fabric of theoretical physics becomes subordinated. Thus Max Planck, a great believer in the universality of minimum principles, wrote in 1915 that the principle of least action "occupies the highest position among physical laws . . . [and] appears to govern all reversible processes in Nature."[4] Planck conceived the principle of simplicity to reside in human psychology and not to be an objective feature of nature. This was, and is, a widespread belief. In the same year that Dirac delivered his James Scott lecture, Born commented on the anticipated unified theory of the future in this way:[5]

We may be convinced that it [the universal formula] will have the form of an extremal principle, not because nature has a will or purpose or economy, but because the mechanism of our thinking has no other way of condensing a complicated structure of laws into a short expression.

However, other scientists have conceived of aesthetic principles such as simplicity in a more objective way. According to them, Nature possesses an immanent tendency toward simplicity, and the pragmatic success of the principle is simply ascribed to the fact that the laws of nature are simple. Of course, the two meanings are often mingled together since the mathematical description of nature is usually considered to be a reflection of nature's real constitution. This harmony between methodology and ontology was expressed by Newton in his famous first rule of reasoning in philosophy: "We are to admit no more causes of natural things than such as are both true and sufficient to explain their appearance." For, Newton explained, "Nature does nothing in vain and more is in vain when less will serve; for Nature is pleased with simplicity, and affects not the pomp of superfluous causes."[6]

Dirac's idea of mathematical beauty was in harmony with Newton's, not with Born's. He believed that there are objective laws of nature and that these, because they express nature's mathematical quality, are recognizable by the methods of pure mathematics.[7]

One may describe this situation by saying that the mathematician plays a game in which he himself invents the rules while the physicist plays a game in which the rules are provided by Nature, but as time goes on it becomes increasingly evident that the rules which the mathematician finds interesting are the same as those which Nature has chosen.

However, according to Dirac the relationship between mathematics and physics goes far deeper than is suggested by the principle of simplicity. He claimed that, although this principle is a valuable instrument for research, modern science has demonstrated that it does not apply to natural phenomena in general. Dirac mentioned the laws of gravitation as an example. Newton's theory is much simpler than Einstein's theory of gravitation, which is only expressible by a complicated set of tensor equations. Still, Einstein's theory is a better, deeper, and more general theory. *Mathematical beauty*, not simplicity, is what characterizes the theory of relativity, and this is the key concept in the relationship between mathematics and physics. With this lesson in mind, Dirac gave theoretical physicists the following advice:[8]

The research worker, in his efforts to express the fundamental laws of Nature in mathematical form, should strive mainly for mathematical beauty. He should still take simplicity into consideration in a subordinate way to beauty. . . . It often happens that the requirements of simplicity and beauty are the same, but where they clash the latter must take precedence.

For Dirac this principle of mathematical beauty was partly a methodological moral and partly a postulate about nature's qualities. It was clearly inspired by the theory of relativity, the general theory in particular, and also by the development of quantum mechanics. Classical mechanics, Dirac said, has many "elegant features," which, when carried over into quantum mechanics, "reappear with an enhanced beauty."[9]

In the Scott lecture of 1939 and in many later publications, Dirac asserted that the development of modern physics, when correctly interpreted, shows that there is a perfect marriage between the rules that mathematicians find internally interesting and the rules chosen by nature. This provides the physicist with a "powerful new method of research," namely[10]

. . . to begin by choosing that branch of mathematics which one thinks will form the basis of the new theory. One should be influenced very much in this choice by considerations of mathematical beauty . . . Having decided on the branch of mathematics, one should proceed to develop it along suitable lines, at the same time looking for that way in which it appears to lend itself naturally to physical interpretation

Neither in the Scott lecture nor in later works did Dirac manage to define his concept of mathematical beauty in a satisfactory way. "Mathematical beauty," he wrote, "is a quality which cannot be defined, any more than beauty in art can be defined, but which people who study mathematics usually have no difficulty in appreciating."[11]

This aristocratic view may resemble that earlier advanced by Poincaré, another great champion of mathematical beauty. Poincaré believed that the primary reason scientists study nature is that it is beautiful, and he claimed that the scientific study of nature yields an intellectual pleasure, which is "that more intimate beauty which comes from the harmonious order of its parts, and which a pure intelligence can grasp."[12] Such beautiful combinations of thought, Poincaré continued, are "those that can charm that special sensibility that all mathematicians know, but of which laymen are so ignorant, that they are often tempted to smile at it."[13] In Poincaré's view, beauty in science is quite different from the sensuous beauty connected with, for example, artistic experiences; it is an intellectual, not an emotional, quality. Dirac's notion of beauty was perhaps broader, involving emotions, which are usually associated with non-intellectual aspects of life and which, for Dirac, may have been substitutes for these aspects. In an interview in 1979, he said, "The beauty of the equations provided by nature . . . gives one a strong emotional reaction."[14] On another occasion, at a talk given in 1972, he indicated that his belief in mathematical beauty was rather like a religious belief.[15]

Dirac's idea of beauty in science appealed, not very originally, to concepts like generality, universality, and completeness. For example, the Lorentz group was said to be more interesting and beautiful than the Galilean group because it is more geneal and includes the latter as a special case. But Dirac also regarded non-relativistic quantum mechanics, though less general and universal, as a beautiful theory because it was complete. The statistical basis of quantum mechanics was "an ugly feature," he stated in 1949; with respect to aesthetics, the exact determinism of classical theory would be preferable. But, he said, "In Quantum Mechanics there are certain very beautiful features appearing in the formal mathematical scheme, which I consider are quite adequate to make up for this ugliness, and the net result is that the scheme taken as a whole is not more ugly than Classical Theory."[16]

Mathematical rigor and axiomatic structure, often emphasized by pure mathematicians as beautiful features for a theory to possess, were not elements in Dirac's conception of beauty. In spite of the emphasis he placed on the power of pure mathematics, he did not believe that exact equations and rigorous proofs should be of prime concern for the mathematical physicist. He knew from experience that preoccupation with mathematical rigor might hamper the development of physical ideas. He had experienced success with an intuitive and badly founded mathematics and saw no reason to play the formal game of pure mathematicians. In 1964, he expressed this attitude as follows: "I believe that the correct line of advance for the future lies in the direction of not striving for mathematical rigor but in getting methods that work in practical examples."[17]

In addition to "beautiful," two other aesthetic codewords appeared frequently in Dirac's writings: "convenient" and "complicated." He often contrasted beautiful mathematics with complicated mathematics; if a physical theory, such as the Heisenberg–Pauli quantum electrodynamics, was only expressible with a very complicated mathematical scheme, this was reason enough to distrust the theory. According to Dirac, a fundamental theory should be expressed in a simple and direct way, and for this purpose convenient transformations and notations are essential. If the introduction of such convenient mathematical tools clashed with requirements of rigor, he would not hesitate to abandon mathematical rigor.

In a comment of 1933, Henry Margenau contrasted Dirac's use of mathematics with von Neumann's:[18]

While Dirac presents his reasoning with admirable simplicity and allows himself to be guided at every step by physical intuition – refusing at several places to be burdened by the impediment of mathematical rigor – von Neumann goes at his problems equipped with the nicest of modern mathematical tools and analyses it to the satisfaction of those whose demands for logical completeness are most exacting.

In the same year, the American mathematician Garrett Birkhoff wrote to Edwin Kemble that he disagreed with Dirac's use of mathematics. Birkhoff had attended Dirac's course on quantum mechanics at Cambridge University. His comments on Dirac's mathematical methods supply an interesting perspective on the disagreement between the views of a mathematical physicist and a mathematician interested in physics:[19]

Contrary to my expectations, I have found that while Dirac's method of representation of physical systems is formally convenient, it does not embody any mathematical principles which are not thoroughly familiar. . . . Dirac permits himself a number of mathematical liberties. For instance, that any set of commuting "observables" (i.e., self-adjoint linear operators, and a few others) has a "diagonal representation" – i.e., a complete set of characteristic functions. Again, that the δ-function is susceptible of the conventional methods of analysis; there have been many examples of this in the lecture. More generally he is committed to the dogma that any formula has a meaning, if taken in a proper sense. This position is of course directly opposite to the classical functiontheoretical school, with all its emphasis on criteria of convergence, differentiability, and continuity. All this I think is unjustified by the fact that the abandonment of rigor as far as I can see leads to neither new results nor simplified methods. I think that Dirac's constructive achievements can be traced to brilliant observations concerning principles of symmetry. He has seen that symmetry on the one hand often leads to cancellation of equal and opposite parts, which permits us to greatly simplify relationships, and on the other to a possibility of inference of one kind of thing from another. . . . But while this method has led and probably will lead to a number of very illuminating results, I do not think that it will ever provide a sound theoretical basis upon which to found a new mechanics. I think that Dirac's critical contributions can be traced to a fine appreciation of qualitative principles – such as symmetry, conservation of energy and momentum, relativistic invariance (i.e., symmetry under Lorentz transformations), orders of magnitude, and the like. He impresses me as being at least comparatively deficient in appreciation of quantitative principles, logical consistency and completeness, and possibilities of systematic exposition and extension of a central theory.

Kemble largely agreed with Birkhoff's criticism. In his letter of reply, he wrote:[20]

So far as fallacious reasoning is concerned I have heard of one man who has collected a list of 40 or 50 such mistakes; but nevertheless his [Dirac's] final results seem usually to be in conformity with experiment. He has always seemed to me to be a good deal of a mystic and that is, I suppose, my way of saying that he thinks every formula has meaning if properly understood – a point of view which is completely repugnant to me and is one of the reasons that I have never been able to adopt his methods, as many other physicists have.

Dirac's relaxed attitude toward mathematical rigor may have stemmed from his early engineering training. At least, that is what he said in his recollections:[21]

It seemed to me [around 1918] that if one worked with approximations there was an intolerable ugliness in one's work, and I very much wanted to preserve mathematical beauty. Well, the engineering training which I received did teach me to tolerate approximations, and I was able to see that even theories based on approximations could sometimes have a considerable amount of beauty in them. . . . I think that if I had not had this engineering training, I should not have had any success with the kind of work that I did later on, because it was really necessary to get away from the point of view that one should deal only with results which could be deduced logically from known exact laws which one accepted, in which one had implicit faith.

On many occasions, Dirac worked with mathematics in this antipurist way, relying on his intuition and letting others prove his theorems and present his ideas in rigorous form. One example of this approach was the invention of the Dirac matrices in 1928, which were developed into a spinor theory a few years later. Another example was his introduction of the δ-function in 1927. Dirac recognized that this quantity was not a proper function and introduced it in a rather irregular way, when seen from the mathematicians' point of view. Its practical, unusual appearance did not bother him: "All the same one can use $\delta(x)$ as though it were a proper function for practically all the purposes of quantum mechanics without getting incorrect results."[22] Again, in 1965, he recommended that physicists not aim at complete rigor but follow the more modest path of setting up "a theory with a reasonable practical standard of logic, rather like the way engineers work."[23]

In accordance with this attitude, Dirac was skeptical with regard to grand mathematical syntheses that aimed at producing a unified theory of all physics. During Dirac's career, such Theories Of Everything, to use a later phrase, were suggested by Weyl, Einstein, Klein, Eddington, and Heisenberg, among others. Dirac simply did not believe in any *Weltformel*. Guided by his own experience, he favored a piecemeal strategy for solving the problems in physics. Gradually improving and criticizing existing theories, guided all the way by the beauty of mathematics, was the method he recommended. "One should not try to accomplish too much in one stage. One should separate the difficulties in physics one from another as far as possible, and then dispose of them one by one," he wrote.[24] Dirac felt that his own success in physics was a result of his following this method. The early phase of quantum mechanics, in his reconstruction, proceeded logically, step-by-step, and he took this procedure as a desirable model for the entire development of physical theory.

Heisenberg recalled discussing methodology with Dirac on their trip across the Pacific in 1929 and on other occasions:[25]

Methodologically, his starting points were particular problems, not the wider relationship. When he described his approach, I often had the feeling that he looked upon scientific research much as some mountaineers look upon a tough climb. All that matters is to get over the next three yards. If you do that long enough, you are bound to reach the top.

Heisenberg's own preferred method was more ambitious and revolutionary. In the sixties, Heisenberg engaged in one such ambitious program, which led him and his collaborators to a unified quantum field theory. Dirac was skeptical toward the theory, as he was toward all similar unifying schemes. Characteristically, his objections rested on methodological and aesthetic grounds, as can be seen in this passage from a letter to Heisenberg:[26]

My main objection to your work is that I do not think your basic (non-linear field) equation has sufficient mathematical beauty to be a fundamental equation of physics. The correct equation, when it is discovered, will probably involve some new kind of mathematics and will excite great interest among the pure mathematicians, just like Einstein's theory of the gravitational field did (and still does). The existing mathematical formalism just seems to me inadequate.

Dirac's approach to physics did not always agree with the engineering methods he often praised, and his attitude toward mathematics did not remain just pragmatic. In his characteristically practical vein, he had stressed in the preface to *Principles* that mathematics is a most powerful tool in physics yet "All the same, the mathematics is only a tool."[27] But in the forties and later on, as the principle of mathematical beauty increasingly occupied his thinking, he came close to conceiving the role of mathematics in an absolute, metaphysical way. It would be, after all, difficult to harmonize the gospel of mathematical beauty with a pragmatic, engineering approach in which mathematics was "only a tool."

Although the concept of mathematical beauty did not make its formal entrance into Dirac's physics until the late thirties, it was in his mind at an earlier date. By 1930, he had reached the conclusion that theoretical physics must follow the route determined by beautiful mathematics. The surprising physical consequences of his essentially mathematical approach in establishing the wave equation of the electron in 1928 were instrumental in turning him into an apostle of mathematical beauty. In the introduction to his monopole paper of 1931, he formulated an embryonic form of the principle of mathematical beauty. What he called "the most powerful method of advance" in that paper was in essence the same as his later principle of mathematical beauty. A year earlier, he referred for the first time, I believe, explicitly to beauty in physics. On the first page of *Principles* he wrote that classical electrodynamics "forms a self-consistent and very elegant theory, and one might be inclined to think that no modification of it would be possible which did not introduce arbitrary features and completely spoil its beauty." But this was not so, he continued, since quantum mechanics "has now reached a form in which it can be based on general laws and is, although not yet quite complete, even more elegant and pleasing than the classical theory in those problems with which it deals."[28]

As an interesting mathematical theory that fulfilled his criteria of beauty, Dirac emphasized in 1939 the theory of functions of a complex variable.[29] He found this field to be of "exceptional beauty" and hence likely to lead to deep physical insight. In quantum mechanics the state of a quantum system is usually represented by a function of real variables, the domains of which are the eigenvalues of certain observables. In 1937, Dirac suggested that the condition of realness be dropped and that the variables be considered as complex quantities so that the representatives of dynamical variables could be worked out with the powerful mathematical machinery belonging to the theory of complex functions. If dynamical variables are treated as complex quantities, they can no longer be associated with physical observables in the usual sense. Dirac admitted this loss of physical understanding but did not regard the increased level of abstraction as a disadvantage: "We have, however, some beautiful mathematical features appearing instead, and we gain a considerable amount of mathematical power for the working out of particular examples."[30] With his new method Dirac showed how the hydrogen atom could be treated in an elegant way, but he did not derive new physical results.

Dirac's preference for complex variables seems also to have been connected with his interest in cosmology, although in an indirect way. His cosmological theory of 1937–8 used very little mathematics, and complex variables were not involved at all. The theory appealed not so much to mathematical beauty as to the Pythagorean principle, according to which numerical coincidences and regularities in nature are not fortuitous but are manifestations of the order of the laws of nature. The ancient Pythagorean principle invoked whole numbers. So, too, Dirac imagined that the mysteries of the universe might ultimately find their explanation in terms of whole numbers.[31]

Might it not be that all present events correspond to properties of this large number [10^{39}], and, more generally, that the whole history of the universe corresponds to properties of the whole sequence of natural numbers . . . ? There is thus a possibility that the ancient dream of philosophers to connect all Nature with the properties of whole numbers will some day be realized.

Dirac might have associated whole numbers with cosmology in the following way: Since the laws of the universe are by hypothesis expressible by whole numbers, and the study of such numbers is part of the beautiful theory of functions of a complex variable, then a cosmological theory based upon the large natural numbers is aesthetically satisfying and, assuming beauty to imply truth, also likely to be correct. This interpretation agrees with the following statement of 1939:[32]

One hint for this development seems pretty obvious, namely, the study of whole numbers in modern mathematics is inextricably bound up with the theory of functions of a complex variable, which theory we have already seen has a good chance of forming the basis of the physics of the future. The working out of this idea would lead to a connection between atomic theory and cosmology.

Dirac never gave up his idea of mathematical beauty, to which he referred in numerous publications, technical as well as nontechnical. In 1982, on the occasion of his eightieth birthday, he published a paper entitled "Pretty Mathematics," an unusual title to find in a journal of theoretical physics, but one characteristic of Dirac's inclination.[33] Evidently he was not just aiming at beautiful mathematics for its own sake. Dirac was a physicist, and he believed that the approach would lead to true physics, that is, to results that would eventually be confirmed empirically. This would be so, he thought, because nature happens to be constructed in accordance with the principle of mathematical beauty. In complete agreement with his statement of 1939, he wrote twenty-six years later that "one could perhaps describe the situation by saying that God is a mathematician of a very high order, and he used very advanced mathematics in constructing the universe."[34]

The identification of beauty with truth led Dirac to a one-sided emphasis on the mathematical–aesthetic method at the expense of the empirical–inductive method. To his mind, the latter method was just to "Keep close to the experimental results, hear about all the latest information that the experimenters obtain and then proceed to set up a theory to account for them." This, he said, is an unworthy procedure that leads to a "rat race" (although, he added, with rather intelligent rats taking part in it).[35] No, one should rely on one's basic beliefs and not pay too much attention to experimental results. As far as fundamental physics was concerned, he wanted to subordinate experimental tests to the admittedly vague idea of mathematical beauty. "A theory with mathematical beauty is more likely to be correct than an ugly one that fits some experimental data," he claimed.[36] This is the more controversial interpretation of the principle of mathematical beauty, for this version not only acts as a recommendation in the context of discovery but also intervenes in the very heart of scientific inquiry, the context of justification. I shall examine this aspect a little further.

The problem of the principle of mathematical beauty, apart from its ambiguity, is that it sometimes leads to flat contradictions of experimental evidence. Dirac was well aware of this, but he insisted that such contradictions were problems for the experimentalist, not for the believer in mathematical beauty. In other words, he asserted that mathematical–aesthetic considerations should (sometimes) have priority over experimental facts and in this way act as criteria of truth. "If the equations of physics are not mathematically beautiful that denotes an imperfection, and it means that the theory is at fault and needs improvement. There are occasions when mathematical beauty should take priority over agreement with experiment."[37] I shall refer to this view as the Dirac–Weyl doctrine, although Dirac would probably have preferred to have his name linked with Einstein's, since he was profoundly indebted to Einstein's theory of relativity and to his general philosophy of sicence, which, Dirac believed, agreed well with his own ideas. Of course, ideas similar to the Dirac–Weyl doctrine were held by many scientists and philosophers before Dirac. Considering God, or nature, as a master mathematician is an old theme in the history of ideas. It goes back to Plato and played an important role in the thought of, for example, Galilei, Kepler, Leibniz, and, in

our century, Einstein, Jeans, Milne, Minkowski, and Weyl. Although only
Einstein and Weyl served as inspirations for Dirac, it is probably in Minkowski that we find the most elaborated doctrine of mathematical beauty
prior to Dirac's.

Hermann Minkowski believed that mathematical–aesthetic criteria
should serve as hallmarks of truth. He was convinced of "the idea of a
pre-established harmony between pure mathematics and physics."[38] Conceiving the power of mathematics in an absolute way, he ascribed an
ontological status to mathematical structures. "It would be revealed," he
stated, "to the fame of the mathematician and to the boundless astonish-
ment of the rest of the mankind, that mathematicians, purely in their
imagination, have created a large field to which one day the fullest real
existence should be scribed (though it was never the intention of these
idealistic fellows)."[39] Minkowski's conception of mathematical aesthetics
was thus different from that held by Poincaré and other conventionalists
but agreed with the ideas of Weyl and Dirac. But it may be added that
with respect to elitism there was no major difference: Poincaré (and
Dirac, too) would have agreed with Minkowski in his distinction between
mathematicians and "the rest of the mankind."

Minkowski's most famous pupil, Einstein, served as an important
source of inspiration for Dirac. In particular, Dirac saw in the discovery
of the theory of relativity the prime example of how great physicists could
develop revolutionary ideas by following the mathematical–aesthetic
approach. The worshiping of relativity as a beautiful theory was a recur-
rent theme in Dirac's later writings; for example, in 1980 he wrote:[40]

The Lorentz transformations are beautiful transformations from the mathematical point of view, and Einstein introduced the idea that something which is beautiful is very likely to be valuable in describing fundamental physics. This is really
a more fundamental idea than any previous idea. I think we owe it to Einstein
more than to anyone else, that one needs to have beauty in mathematical equations which describe fundamental physical theories.

Even more than the special theory, the creation of the general theory of
relativity fascinated Dirac. Einstein was guided by principles of simplicity and had such a priori confidence in the equations of his gravitation
theory that he did not care much for attempts to prove the theory obser-
vationally. According to Dirac, this was the proper attitude for a theoretical physicist, an attitude that future generations of physicists should
follow:[41]

Let us now face the question, that a discrepancy has appeared, well confirmed and
substantiated, between the theory and the observations. How should one react to
it? How would Einstein himself have reacted to it? Should one then consider the

theory to be basically wrong? I would say that the answer to the last question is emphatically no. Anyone who appreciates the fundamental harmony connecting the way nature runs and general mathematical principles must feel that a theory with the beauty and elegance of Einstein's theory *has* to be substantially correct. . . . When Einstein was working on building up his theory of gravitation he was not trying to account for some results of observations. Far from it. His entire procedure was to search for a beautiful theory, a theory of a type that nature would choose. He was guided only by the requirement that this theory should have the beauty and elegance which one would expect to be provided by any fundamental description of nature.

In his praise of trans-empiricism and mathematical intuition, Dirac was far from alone. Many outstanding physicists have shared Dirac's belief in Einstein's theory of gravitation as a theory that was created virtually without empirical reasoning, and that has to be true because of its aesthetic merits. Among the quantum pioneers, Bohr and Born were about the only ones who challenged this view.[42]

But was Dirac justified in using Einstein the way he did, as an ally in his advocacy of mathematical beauty? Not really, I think. Dirac reconstructed parts of the history of modern physics so that they fitted his own preferences, but he omitted those parts that contradicted them. As far as Einstein is concerned, his own attitude was not unambiguous. True, Einstein was fascinated by the role played by mathematical simplicity and symmetry. Considerations based on formal or aesthetic reasoning played an important heuristic role in his science, and he often objected to inductive–empirical arguments at the expense of mathematical arguments.[43] In the Herbert Spencer Lecture of 1933, he thus expressed himself in close agreement with Dirac's ideas:[44]

Our experience hitherto justifies us in believing that nature is the realization of the simplest conceivable mathematical ideas. I am convinced that we can discover by means of purely mathematical constructions the concepts and the laws connecting them with each other, which furnish the key to the understanding of natural phenomena. Experience may suggest the appropriate mathematical concepts, but they most certainly cannot be deduced from it. Experience remains, of course, the sole criterion of physical utility of a mathematical construction. But the creative principle resides in mathematics. In a certain sense, therefore, I hold it true that pure thought can grasp reality, as the ancients dreamed.

However, the emphasis on mathematical–aesthetic virtues was primarily a feature in the mature Einstein. In his younger and more creative years, he was less impressed by such arguments and in fact denied that aesthetic considerations could serve even as heuristic aids.[45] He realized that formal beauty does *not* secure physical significance. For example, in 1921 he referred to Eddington's field theory as "beautiful but physically meaningless," and four years later he called his own unified theory of gravitation and electricity "very beautiful but dubious."[46] As late as 1950, Einstein admitted that ultimately "experience alone can decide on truth."[47] Statements like these are hardly in accordance with the Dirac-Weyl doctrine.

More than Einstein, Hermann Weyl may serve as a model for Dirac's philosophy. "My work always tried to unite the true with the beautiful," Weyl once said. "But when I had to choose one or the other, I usually chose the beautiful."[48] He followed this strategy in his unified field theory of 1918, which he defended with aesthetic arguments against Einstein's objection that it had no real physical meaning. Throughout his life Weyl was attracted to the Platonic idea of a mathematical quality residing in nature's scheme, an idea that he used in his attempt to reconcile Christian metaphysics with modern science. "The mathematical lawfulness of nature is the revelation of divine reason," Weyl proclaimed in 1932. "The world is not a chaos, but a cosmos harmonically ordered by inviolable mathematical laws."[49] As mentioned in earlier chapters, Dirac was impressed by Weyl's thoroughly mathematical approach to physics. He ascribed to this approach Weyl's recognition that anti-electrons had the same mass as electrons, and he praised Weyl's unified field theory as "unrivalled by its simplicity and beauty". However, it is worth recalling that this appraisal was a rationalization out of his later years. In the twenties and early thirties, Dirac did not pay much attention to Weyl's theories.

The principle of mathematical beauty, like related aesthetic principles, is problematical. The main problem is that beauty is essentially subjective and hence cannot serve as a commonly defined tool for guiding or evaluating science. It is, to say the least, difficult to justify aesthetic judgment by rational arguments. Within literary and art criticism there is, indeed, a long tradition of analyzing the idea of beauty, including many attempts to give the concept an objective meaning. Objectivist and subjectivist theories of aesthetic judgment have been discussed for centuries without much progress, and today the problem seems as muddled as ever.[50] Apart from the confused state of art in aesthetic theory, it is uncertain to what degree this discussion is relevant to the problem of scientific beauty. I, at any rate, can see no escape from the conclusion that aesthetic judgment in science is rooted in subjective and social factors. The sense of aesthetic standards is part of the socialization that scientists acquire; but scientists, as well as scientific communities, may have widely different ideas of how to judge the aesthetic merit of a particular theory. No wonder that eminent physicists do not agree on which theories are beautiful and which are ugly. Consider the following statement, which contains the core of Dirac's philosophy of physics:[51]

With all the violent changes to which physical theory is subjected in modern times, there is just one rock which weathers every storm, to which one can always hold fast – the assumption that the fundamental laws of nature correspond to a beautiful mathematical theory. This means a theory based on simple mathematical concepts that fit together in an elegant way, so that one has pleasure in working with it. So when a theoretical physicist has found such a theory, people put great confidence in it. If a discrepancy should turn up between the predictions of such a theory and an experimental result, one's first reaction would be to suspect experimental error, and only after exhaustive experimental checks would one accept the view that the theory needs modification, which would mean that one must look for a theory with a still more beautiful mathematical basis.

In this quotation the operational difficulties of the principle of mathematical beauty are clearly, although unwittingly, present. When "*a* theoretical physicist" has found a theory which *he* finds beautiful, then "*people* put great confidence in it," we are assured. But this argument presupposes that the individual physicist's sense of mathematical beauty, and with this his particular psychological constitution, are shared by "people," that is, the community of theoretical physicists. Dirac seems to have been aware of the problematic nature of the idea of beauty, but he claimed that it was not a serious problem within theoretical physics. In a talk given at the University of Miami in 1972, entitled "Basic Belief and Fundamental Research," he said:[52]

It is quite clear that beauty does depend on one's culture and upbringing for certain kinds of beauty, pictures, literature, poetry and so on. . . . But mathematical beauty is of a rather different kind. I should say perhaps it is of a completely different kind and transcends these personal factors. It is the same in all countries and at all periods of time.

However, Dirac's claim lacks justification. There is no evidence, from either psychology, sociology, or history of science, that physicists should have such a common conception concerning the nature of beauty in their science.

On the contrary, the history of modern physics supports the claim that there is no concensus as to which equations and mathematical structures should be termed beautiful and interesting. For example, most physicists probably regard group theory and topology as highly interesting branches, but these did not figure in Dirac's list of beautiful mathematics. Dirac, as well as most other physicists, considered the idea of the Minkowski space as very beautiful; Minkowski himself raised it to the level of divinity.

However, Einstein did not like Minkowski's idea when it first appeared.[53] As another example, one may compare the views of Heisenberg and Einstein on quantum theory. They shared, at least in later periods of their life, the view that all true theories are simple and beautiful, yet they disagreed profoundly in their concrete judgments of particular theories and approaches.[54]

Still another example of the essential subjectivity of aesthetic arguments in science is provided by the discussion between Dirac and Gamow about cosmology (see Chapter 11). Gamow shared Dirac's belief in the value of scientific aesthetics. "[I] agree with Dirac in his conviction that if a theory is elegant, it must be correct," he wrote in 1967.[55] But since Gamow's notion of beauty differed from Dirac's, the two physicists disagreed about how to implement aesthetic reasoning in concrete cases. "Being a theoretician in my heart I have healthy respect for observation and experiment in my brain," Gamow wrote to Dirac in 1967.[56] According to the Dirac-Weyl thesis, such a respect in unwarranted. Dirac wanted the aesthetic qualities of a theory, as *he* conceived them, to decide the truth of experiments.

Dirac's sense of mathematical beauty held that any dynamic system should be able to be formulated in a Hamiltonian scheme satisfying the principle of relativity. Most physicists appreciate the beauty of Hamiltonian theory, but Dirac found it so fascinating that he tended to elevate it to a demand that any fundamental physical theory should satisfy.[57] He was impressed by Hamilton's genius in developing a physical formalism that had no practical importance at the time but could only be appreciated because of its inherent beauty. Delivering the Larmor Lecture in 1963 in Dublin, Dirac paid the following tribute the the Irish theorist: "We may try to make progress by following in Hamilton's footsteps, taking mathematical beauty as our guiding beacon, and setting up theories which are of interest, in the first place, only because of the beauty of their mathematics.[58] But Dirac's commitment to Hamiltonian formalism was not without its problems. According to Wigner, the fact that in 1927 Dirac did not establish the commutation relations for fermions (see Chapter 6) was a direct result of this commitment. "Dirac was a captive and is now a captive of the Hamiltonian formalism and he thinks extremely strongly in terms of the Hamiltonian formalism," Wigner said in 1963.[59]

The worshiping of Hamiltonian schemes in physics further exemplifies how aesthetic judgments may change in time and how two principles, each judged to be beautiful, may be in conflict with each other. Like most other physicists, Dirac valued Lorentz invariance as aesthetically pleasing. For example, it was the main reason he rejected Schrödinger's theory of 1931 in which only positive energies occurred. As he later recalled,

Schrödinger's small change in the wave equation "spoils all the relativistic features of the theory and all the beauty of the theory is gone."[60] In 1958, Dirac managed to express Einstein's equations of gravitation in an exact Hamiltonian form that allowed for a simplified description of gravitational phenomena.[61] His theory thus lived up to the principle of simplicity but, it turned out, only at the expense of giving up four-dimensional symmetry. Hence a conflict arose between two aesthetic criteria: the beauty of four-dimensional symmetry versus the beauty of Hamiltonian formalism. In this case, Dirac chose to give simplicity priority:[62]

This result has led me to doubt how fundamental the four-dimensional requirement in physics is. A few decades ago it seemed quite certain that one had to express the whole of physics in four-dimensional form. But now it seems that four-dimensional symmetry is not of overriding importance, since the description of nature sometimes gets simpler when one departs of it.

Lack of Lorentz invariance implied, in Dirac's view, a certain lack of beauty. More surprisingly, on occasion he was willing to sacrifice even beauty, if this was the price to be paid for logic and clarity. Considering his alternative quantum electrodynamics with a cutoff, he said in 1981: "I think that a theory which is non-relativistic and which is ugly is preferable to one which goes so much against logic as to require discarding infinite terms."[63] A somewhat similar conflict between aesthetic principles occurred in 1930 when Dirac tried to find a candidate for his antielectron. In this case he became caught between two aesthetic principles, the unity of nature and the principle of mathematical reasoning, which unfortunately led to different answers. And when in 1936 he accepted Shankland's result because he saw it as support for beauty in physics, his aesthetic sense simply betrayed him. It would seem that Dirac's view of beauty, when confronted with real cases of physics, was ambiguous, and this ambiguity merely reflected the ambiguity inherent in the concept of beauty and related notions.

An additional problem for the Dirac–Weyl doctrine concerns how strictly it should be obeyed. It is one thing to boldly maintain belief in a theory in spite of some empirical counterevidence, but it is another thing to stick obstinately to the theory and disregard *any* kind of conflicting experimental results. Neither Dirac nor other adherents of mathematical beauty would accept an extreme Cartesianism, divorced from any empirical considerations. Dirac's advice, that one should disregard experimental results which are "ugly," was, wisely but somewhat inconsequently, supplemented with the proviso that "of course one must not be too obstinate over these matters."[64] But this leaves the Dirac–Weyl doctrine rather empty as a methodological guide for research as long as no criterion for defining "too obstinate" is provided (and, of course, no such criterion *can* be provided).

As a final illustration, consider the role played in physics by invariance principles, such as parity and time invariance. These concepts were well known in classical physics, where it was generally believed that any fundamental law must satisfy parity and time invariance. The principles were associated with great aesthetic value, which was only reinforced when Wigner in 1927 and 1932 carried them over into quantum mechanics.[65] As a result of their authority, they were sometimes used to censor theories of physics. For example, when Weyl proposed a two-component wave equation for particles with zero mass and spin one-half in 1929, Pauli rejected the equation as "inapplicable to physical reality."[66] Pauli's objections were not only that no particles of zero mass and spin one-half existed, but, more importantly, that the Weyl equation did not satisfy parity invariance. In his and most other physicists' view, Weyl's equation was aesthetically unsatisfactory for that reason.[67] Pauli was always guided by a strong aesthetic belief in symmetry and conservation properties of the laws of nature. This belief made him reject Weyl's equation and later it made him distrust the evidence in favor of parity nonconservation. Such evidence was produced in 1956–57 by T. D. Lee, C. N. Yang, C. S. Wu, and others.[68] Pauli's absolute confidence in parity invariance turned out to block his scientific imagination.

Contrary to consensus, Dirac did not feel committed to the invariance doctrine. His sense of beauty differed in this respect from that of, for example, Pauli. In his early works Dirac dealt with invariance only once, and then in an unorthodox way. In a little known paper from 1937, he developed Wigner's previously introduced concept of time (or motion) reversal in quantum mechanics.[69] Following Wigner, he made use of a reversal operator that changed the sign of momentum and spin but left position and energy unchanged; but he did not regard this operator as fundamental and introduced another reversal operator that did not correspond to classical time reversal. Dirac's reversal operator was relativistically invariant and reversed the energy as well as the momentum and spin. Applying this operator in his hole theory, Dirac obtained a perfect symmetry between the positive-energy and negative-energy states of a particle. He wrote: "Any occupied positive-energy state will always occur with an unoccupied negative-energy state or hole, the two together representing the same physical reality. Thus we get a theory in which the holes in the negative-energy distribution are physically the same things as the ordinary positive-energy particles."[70]

"It seems fair to assume that every textbook on elementary-particle physics, nuclear physics, and quantum mechanics written prior to 1957 contains a statement of parity conservation," writes Allan Franklin in his detailed study of the history of parity invariance.[71] Remarkably, Dirac's *Principles of Quantum Mechanics* was an exception and was overlooked by Franklin. When Pais, in 1959, asked Dirac why he did not include parity in his textbook, Dirac's direct answer was, "Because I did not believe in it."[72] The only time Dirac referred explicitly to his heretical disbelief in the invariance dogmas was in 1949, at a time when parity and time conservation were taken for granted by almost all physicists. Then he wrote:[73]

A transformation of the type [an inhomogeneous Lorentz transformation] may involve a reflection of the coordinate system in the three spacial dimensions and it may involve a time reflection. . . . I do not believe there is any need for physical laws to be invariant under these reflections, although all the exact laws of nature so far known do have this invariance.

In 1956, it was shown that parity is not conserved in weak interactions, and in 1967, it was established that time reversal does not hold good in decay processes of neutral K mesons. The case here presented cannot be taken as support of the Dirac-Weyl doctrine. Rather, it illustrates the arbitrariness of aesthetic principles in science. The aesthetically founded belief in parity and time conservation was not shared by Dirac, who, without argument, did not find it fundamental. When parity nonconservation was established, the standards of beauty in physics changed too. It should be mentioned here that in his later work Dirac took up the problem of symmetry invariance, now in connection with his Weyl-inspired cosmological theory of 1973 (see Chapter 11). However, this was in the context of cosmology and general relativity and was not directly related to the quantum theory of C, P, and T invariance.

If a conclusion is to be drawn from Dirac's own career with respect to the scientific value of the principle of mathematical beauty, it appears to me to be the following. Many of his most important results were products of his general belief in the power of mathematical reasoning; however, the principle of mathematical beauty, in its more elaborate meaning, proved to be a failure in Dirac's career. He applied it in particular in his persistent attempts to formulate an alternative quantum electrodynamics, and these attempts, as far as we can tell, were failures. In Dirac's scientific life, the mid-1930s marked a major line of division: all of his great discoveries were made before that period, and after 1935 he largely failed to produce physics of lasting value. It is not irrelevant to point out that the principle of mathematical beauty governed his thinking only during the later period.

APPENDIX I

DIRAC BIBLIOMETRICS

Dirac wrote more than 190 publications, the first in 1924 and the last sixty years later. Although he was a prolific author, the length of his publication list is by no means exceptional. The greatness of Dirac's publications lay not in their number, but in their quality and originality. With the exception of a few biographical memoirs, all of his works dealt with theoretical physics and took the form of research papers or of lectures, books, and survey articles. The number of research papers, loosely defined to be contributions intended to advance knowledge in physical theory, is eighty-nine. This number does not include survey articles, lectures, invited talks, or addresses directed at audiences outside the physics community, although the distinction between "research papers" and other works is to some extent arbitrary.

Table I shows the average number of citations Dirac made in his research papers during the three periods 1924–34, 1935–45, and 1946–84. Since citations usually have different functions in research and nonresearch papers, I have included only the first type of publications. In most of his nonresearch works, Dirac did not cite at all.

Citations may be divided into self-references and references to other authors. The ratio of self-references to the total number of references may be taken to indicate how "closed" or self-consistent an author's works in a period are. Scientists who follow an independent research program outside the centers of mainstream physics will tend to cite themselves frequently.

"Citation lag-time" is another measure that may throw light on the independence of research programs or individual scientists. Usually, scientific papers cite only earlier papers, published before the citing paper or perhaps in the same year. That is, the lag time Δt is positive or zero, Δt being the difference between the year of the citing paper and the year of a cited paper. In rare cases one may find $\Delta t < 0$. The average value of Δt within a particular citing paper (or series of papers) reflects how "mod-

Table I. *Statistics on Dirac's publications and citations*

	1924–34	1935–45	1946–84
Total no. of publications	45	19	128
No. of research papers	37	16	38
Total no. of pages	829	251	1,318
No. of pages in research papers	428	239	394
Citations per page (research papers)	0.45	0.23	0.14
Percentage of self-references (research papers)	25	22	44
Average citation time-lag, $\langle \Delta t \rangle$, in years	1.9	7.6	9.5

ern" the paper is or how integrated it is in an evolving research front. In rapidly evolving, "hot" areas $\langle \Delta t \rangle$ will be small and in many cases zero. If $\langle \Delta t \rangle$ is large, say ten years, it usually indicates that the paper (or series of papers) belongs to a stagnating research area or is otherwise out of contact with mainstream research.

Analysis of Dirac's publications results in the following observations:

1. After 1934, Dirac's productivity fell sharply. The decline during the period from 1935 to 1945 was not a result of the war, since almost half of the publications during the period were written between 1940 and 1945; if the classified works on war-related research are included, Dirac was in fact more productive during the war period than in the five previous years.

2. In the first two periods, 1924–34 and 1935–45, most of Dirac's publications were research papers, but in the last period, 1946–84, he increasingly produced nonresearch publications.

3. The self-reference percentage is high for all periods. While self-references by the average theoretical physicist no doubt amount to much less than 20 percent of the citations, in many papers Dirac referred only to himself, and during the entire postwar period almost half of his citations were self-references. For example, in one of his most important papers (1934A; see Appendix II for Dirac references), there are six citations, all of which are to his earlier works.

4. While for the first period $\langle \Delta t \rangle$ was less than two years, which was presumably around the usual citation lag-time for physics papers during the period, in the following years the average lag-time increased remarkably. For example, in one 1936 paper (1936C) the average lag-time for three citations was 21.7 years, and in a 1973 paper (1973A) the figure was 27.6 years. Such exceptionally high citation lag-times are often the result of "historical" citations (e.g., to Maxwell or Boltzmann), in which works

Table II. *Dirac's references to selected physicists (research papers)*

Cited author	1924–34	1935–45	1946–84	Total
Heisenberg	29	3	1	33
Born	24	0	3	27
Jordan	22	2	1	25
Pauli	16	6	2	24
Milne	3	4	6	13
Fock	5	3	4	12
Schrödinger	7	1	1	9
Klein	7	1	0	8
Einstein	6	0	1	7
Compton	6	0	0	6
Weyl	3	0	2	5
Eddington	2	2	1	5
Darwin	4	0	0	4
Oppenheimer	3	1	0	4
Fermi	2	0	2	4
Wigner	2	1	1	4
Tamm	3	0	0	3
Gaunt	3	0	0	3
Neumann	3	0	0	3
Frenkel	1	1	1	3
Robertson	0	2	1	3
Bohr	1	1	0	2
Rosenfeld	1	0	1	2
Tomonaga	0	0	1	1
Schwinger	0	0	1	1

are cited for general historical reasons and not because they are really relevant to the citing work. However, such citations do not figure in the citation structure of Dirac's works. None of the citations made by Dirac were to works earlier than 1910, and he never cited works unless the citation was justifiable for the purpose of presenting facts directly related to his own work at hand. Historical or courtesy citations did not appear in Dirac's research papers.

Table II lists the authors most frequently cited by Dirac. The German quantum pioneers, Heisenberg, Born, Jordan, and Pauli, were by far the most influential sources for Dirac in the first decade; but in the second decade these authors appeared rarely. Also noticeable is the lack of reference to the new generation of quantum physicists that appeared after the war. Dirac quoted Tomonaga once, but leading physicists like Feynman, Yukawa, Dyson, Yang, and Gell-Mann did not figure at all in his citation list. The single reference to Schwinger was to a work from 1962

Table III. *References to Dirac's papers during the period 1924–9*
(see Appendix II for paper titles)

Paper	No. of citations by others	Self-citations by Dirac
1924A	0	0
1924B	0	0
1924C	0	0
1924D	7	1
1925A	1	0
1925B	3	0
1925C	0	0
1925D	18	2
1926A	15	3
1926B	13	3
1926C	33	1
1926D	0	0
1926E	58	2
1926F	0	0
1927A	25	3
1927B	37	2
1927C	16	0
1927D	5	0
1928A	47	2
1928B	28	1
1928C	0	0
1929A	1	0
1929B	1	0

on generalized dynamics and not to his papers on quantum electrodynamics. Again, this indicates Dirac's isolation from mainstream physics.

Table III lists the number of references to Dirac's papers in the period 1924–9, taken from the *Science Citation Index,* 1920–9. If the frequency of citation is taken to be a measure of impact, only one of his early, prequantum mechanics papers (1924D) had some impact. Among the papers on quantum mechanics, "On the Theory of Quantum Mechanics" (1926E) was a real hit, and his work on the Compton effect (1926C) also attracted much interest. In the twenties, "On the Theory of Quantum Mechanics" (1926E) was cited as frequently as the *Dreimännerarbeit* of Heisenberg, Born, and Jordan and was in 1927 the most cited theoretical paper in physics [see Small (1986)]. Not all of Dirac's papers were successes in this sense. Thus, his work on quantum algebra (1926D) received very few, if any, citations. Since the *SCI* bibliography lists citing works only up to 1929, citations of Dirac's papers of 1928 and 1929 are no doubt underrepresented. His epochal work on relativistic quantum mechanics,

"The Quantum Theory of the Electron" (1928A), was widely cited in the early thirties and is probably the most cited of all of Dirac's works. It is likely to be among the most cited theoretical papers in physics ever.

It should be pointed out that the *SCI* 1920–9 bibliography refers only to citations appearing in a restricted number of periodicals. For that reason alone the above figures should be interpreted with some caution. For example, *SCI* does not include journals such as *Proceedings of the Cambridge Philosophical Society, Comptes Rendus, Die Naturwissenschaften,* and *Nature.* There are also other reasons why the *SCI* data should be used with caution.

The thirteen graphs at the end of this appendix show the citations since 1955 of selected Dirac papers, again based on *SCI* data. Some of Dirac's older papers have, of course, ceased to have much scientific impact, if they ever had any; examples are "Relativistic Wave Equations" (1936B) and "Wave Equations in Conformal Space" (1936C). The "Quantum Theory of the Electron" (1928A) is still fairly frequently cited, although it is now only of historical interest. It should be noted that current *SCI* bibliographies (unlike *SCI* 1920–9) include citations from a very comprehensive body of journals and books, including many works in the history of science, biographical notices, *festschriften,* etc. This may account for many of the modern references to "The Quantum Theory of the Electron" (1928A). On the other hand, several of Dirac's works show unusual citation patterns in the sense that interest in them seems to have increased many years after publication. The most remarkable example is the 1931 monopole paper (1931C), but others (1930C and 1950A) also show a pattern that probably signifies a genuine delayed impact. Neither of the latter two papers belongs to Dirac's best-known works, and the delayed success cannot be explained as a result of historical or biographical interest.

The structure of each of the citation graphs can probably be understood in terms of the internal development of various specialties in theoretical physics, but this is not the place for such an analysis. In general, one should be cautious in interpreting citation patterns showing the development of citations over a longer period of time. Although scientific papers have a natural tendency to be less cited after some years, another effect may partly compensate for this decline, namely, the growth in the number of scientific publications. Since the number of scientific publications is much larger today than it was in the late twenties, a comparison of the data in the figures and in Table III will inevitably give a distorted picture. This systematic source of error is often neglected in scientometric papers.

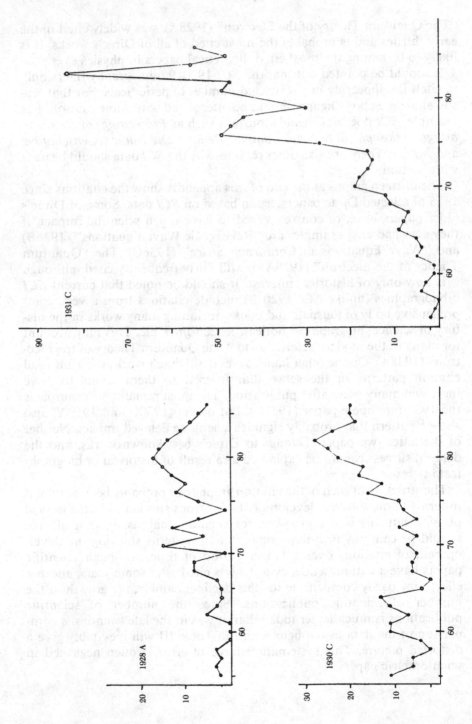

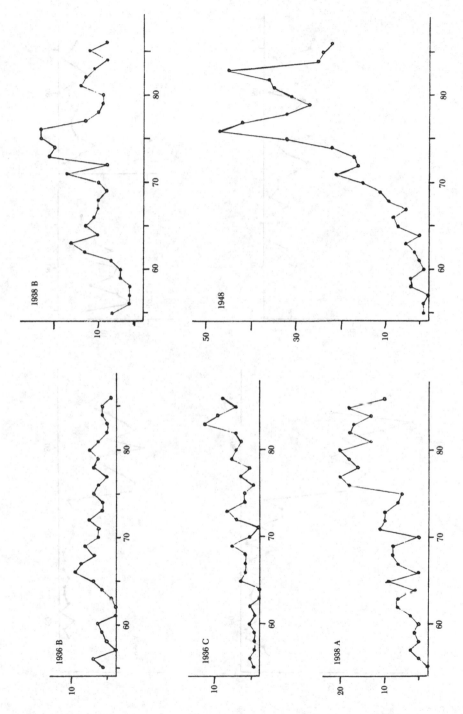

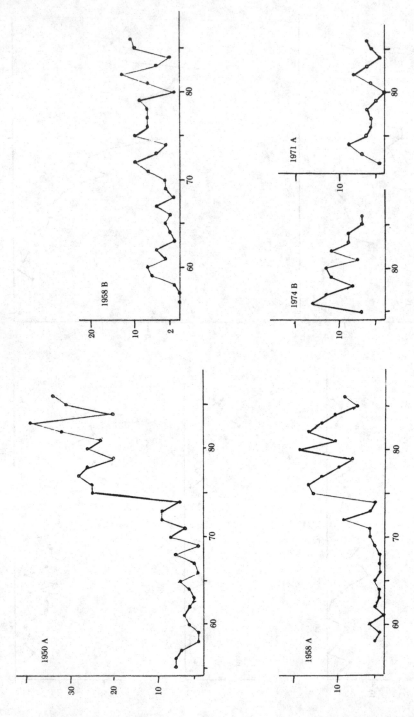

ST JOHN'S COLLEGE,
CAMBRIDGE.
4-11-33

Dear Schrödinger,

Many thanks for your kind letter and congratulations. May I also congratulate you!

I am very glad indeed to be in the company of you and Heisenberg and hope we shall have a good time together in Stockholm.

With best wishes,

Yours sincerely
P. A. M. Dirac

Department of Physics

The Florida State University
Tallahassee, Florida 32306

11 Nov 1981

Dear Abdus,

I am sorry I cannot come to your monopole conference.

It would be too much of a dislocation for me at such short notice.

It is very kind of you to invite me.

I am inclined now to believe that monopoles do not exist.

So many years have gone by without any encouragement from the experimental side. It will be interesting to see if your conference can produce any new developments if further on the problem.

With best wishes,

Yours sincerely
Paul Dirac

Two samples of Dirac's handwriting, from 1933 and 1981. His meticulous handwriting changed remarkably little during his life. Reproduced with the permission of Zentralbibliothek für Physik in Vienna (left) and Abdus Salam (right).

Sketches of Dirac by an unknown artist from "Blegdamsvejens Faust", the parody staged by physicists at Bohr's institute in April 1932. In the upper right, Dirac keeps two oppositely charged monopoles apart, while the figures below refer to Dirac's hole theory. An electron — in the round shape of C. G. Darwin — dives into the Dirac sea and at the bottom Dirac follows thoughtfully the spin of the photon; he is followed by four ghosts, alluding to the troubles of quantum theory: Gauge invariance, the fine structure constant, the negative energies, and the singularities. Reproduced with the permission of the Niels Bohr Archives, Copenhagen.

Two physicists' sketches of Dirac. The artists are George Gamow (left) and Richard Feynman (right). Reproduced from G. Gamow, *The Atom and its Nucleus* (Englewood Cliffs, NJ: Prentice Hall, 1961) and B. N. Kursunoglu and E. P. Wigner, eds., *Paul Adrien Maurice Dirac* (Cambridge: Cambridge University Press, 1987) with the permission of the publishers.

Dirac listening to Richard Feynman during the 1962 Warsaw conference on gravitation. Reproduced with permission of AIP Niels Bohr Library.

APPENDIX II

BIBLIOGRAPHY OF P. A. M. DIRAC

The following abbreviations are used in this bibliography.

AIHP   *Annales de l'Institut Henri Poincaré*
PASC   *Pontificia Academia Scientia, Commentarii*
PCPS   *Proceedings of the Cambridge Philosophical Society*
PR   *Physical Review*
PRSL   *Proceedings of the Royal Society of London*

The present bibliography does not claim to be complete. It includes only Dirac's published works. Earlier bibliographies can be found in Salam and Wigner (1972), pp. xiii–xvi (compiled by J. Mehra) and in Dalitz and Peierls (1986). The latter also includes a number of unpublished manuscripts and reports.

References for works written with others are listed in this bibliography along with other Dirac publications in chronological order, and the names of Dirac's co-authors are given at the end of the reference.

1924A   "Dissociation under a temperature gradient," *PCPS 22*, 132–7 (March 3, 1924).
1924B   "Note on the relativity dynamics of a particle," *Philosophical Magazine 47*, 1158–9.
1924C   "Note on the Doppler principle and Bohr's frequency condition," *PCPS 22*, 432–3 (May 19, 1924).
1924D   "The conditions for statistical equilibrium between atoms, electrons and radiation," *PRSL A106*, 581–96 (July 8, 1924).
1925A   "The adiabatic invariants of the quantum integrals," *PRSL A107*, 725–34 (December 19, 1924).
1925B   "The effect of Compton scattering by free electrons in a stellar atmosphere," *Monthly Notices of the Royal Astronomical Society of London 85*, 825–32.

1925C "The adiabatic hypothesis for magnetic fields," *PCPS 23*, 69–72 (November 5, 1925).

1925D "The fundamental equations of quantum mechanics," *PRSL A109*, 642–53 (November 7, 1925).

1926A "Quantum mechanics and a preliminary investigation of the hydrogen atom," *PRSL A110*, 561–79 (January 22, 1926).

1926B "The elimination of the nodes in quantum mechanics," *PRSL A111*, 281–305 (March 27, 1926).

1926C "Relativity quantum mechanics with an application to Compton scattering," *PRSL A111*, 405–23 (April 29, 1926).

1926D "On quantum algebra," *PCPS 23*, 412–18 (July 17, 1926).

1926E "On the theory of quantum mechanics," *PRSL A112*, 661–77 (August 26, 1926).

1926F "The Compton effect in wave mechanics," *PCPS 23*, 500–7 (November 8, 1926).

1927A "The physical interpretation of the quantum dynamics," *PRSL A113*, 621–41 (December 2, 1926).

1927B "The quantum theory of the emission and absorption of radiation," *PRSL A114*, 243–65 (February 2, 1927).

1927C "The quantum theory of dispersion," *PRSL A114*, 710–28 (April 4, 1927).

1927D "Ueber die Quantenmechanik der Stoßvorgänge," *Zeitschrift für Physik 44*, 585–95 (June 28, 1927).

1927E (Discussion following reports of Bohr and Heisenberg at the Fifth Solvay Congress, October 24–9, 1927), pp. 258–63 in *Electrons et Photons: Rapports et Discussions du cinquième Conseil de Physique*, Paris: Gauthiers-Villars, 1928.

1928A "The quantum theory of the electron," *PRSL A117*, 610–24 (January 2, 1928).

1928B "The quantum theory of the electron, part II," *PRSL A118*, 351–61 (February 2, 1928).

1928C "Zur Quantentheorie des Elektrons," pp. 85–94 in H. Falkenhagen, ed., *Quantentheorie und Chemie*, Leipzig: S. Hirzel, 1928 (lecture given at the Leipziger Vorträge in June 1928; manuscript translated by A. Eucken).

1928D "Ueber die Quantentheorie des Elektrons," *Physikalische Zeitschrift 29*, 561–3 (abridged version of 1928C).

1929A "The basis of statistical quantum mechanics," *PCPS 25*, 62–6 (October 17, 1928).

1929B "Quantum mechanics of many-electron systems," *PRSL A123*, 714–33 (March 12, 1929).

1930A "A theory of electrons and protons," *PRSL A126*, 360–5 (December 6, 1929). Russian translation as "Teoriia elektronov i protonov," *Uspekhi Fizicheskikh Nauk 10* (1930), nos. 5–6.

1930B "On the annihilation of electrons and protons," *PCPS 26*, 361–75 (March 26, 1930).

1930C   "Note on exchange phenomena in the Thomas atom," *PCPS 26*, 376–85 (May 19, 1930).

1930D   "The proton," *Nature 126*, 605–6 (September 8, 1930). Based on paper read before Section A of the British Association for the Advancement of Science on September 8, 1930.

1930E   *The Principles of Quantum Mechanics,* 1st edn., Oxford: Clarendon Press, 265 pp. (preface dated May 29, 1930).

1931A   "Quelques problémes de mécanique quantique," *AIHP 1,* 357–400. From conference held at the Poincaré Institute December 13–20, 1929.

1931B   "Note on the interpretation of the density matrix in the many-electron problem," *PCPS 27,* 240–3 (December 8, 1930).

1931C   "Quantised singularities in the electromagnetic field," *PRSL A133,* 60–72 (May 29, 1931).

1932A   "Photo-electric absorption in hydrogen-like atoms," *PCPS 28,* 209–18 (January 15, 1932). With J. W. Harding.

1932B   "Relativistic quantum mechanics," *PRSL A136,* 453–64 (March 24, 1932).

1932C   "On quantum electrodynamics," *Physikalische Zeitschrift der Sowietunion 2,* 468–79 (October 25, 1932). With V. A. Fock and B. Podolsky.

1932D   *Various Problems in Quantum Theory* (in Japanese), Kei Mekai: Tokyo, 1932 (based on notes of lectures held in Tokyo on September 2–7, 1929; translated by Y. Nishina, M. Kotani, and T. Inui). With W. Heisenberg.

1933A   "The Lagrangian in quantum mechanics," *Physikalische Zeitschrift der Sowietunion 3,* 64–72 (November 19, 1932).

1933B   "The reflection of electrons from standing light waves," *PCPS 29,* 297–300 (March 24, 1933). With P. Kapitza.

1933C   "Homogeneous variables in classical dynamics," *PCPS 29,* 389–401 (May 16, 1933).

1933D   "Statement of a problem in quantum mechanics," *Journal of the London Mathematical Society 8,* 274–7 (May 26, 1933).

1933E   "Théorie du positron," pp. 203–12 in *Structure et Propriétés des Noyaux Atomiques* (7th Solvay Conference Report), Paris: Gauthier–Villars, 1934 (Address given at the Solvay Conference, October 22–9, 1933).

1933F   "Theory of electrons and positrons," pp. 320–5 in *Nobel Lectures, Physics 1922-41,* Elsevier: Amsterdam, 1965 (Nobel lecture, December 12, 1933. First in *Die Moderne Atomtheorie, Die bei der Entgegennahme des Nobelpreises 1933 in Stockholm Gehaltenen Vorträge,* Leipzig: Hirzel; with W. Heisenberg and E. Schrödinger).

1934A   "Discussion of the infinite distribution of electrons in the theory of the positron," *PCPS 30,* 150–63 (February 2, 1934).

1934B   "Teoriya pozitrona," pp. 129–43 in M. P. Bronstein et al., eds., *Atomnoe yadro, sbornik dokladov i vsesoyoznoi yadernoi konferentsii,* Leningrad: Izdatelstvo, Akademii Nauk S.S.R.

1935   "The electron wave equation in de Sitter space," *Annals of Mathematics 36,* 657–63 (April 16, 1935).

1936A   "Does conservation of energy hold in atomic processes?" *Nature 137* (February 22), 298–9.

1936B   "Relativistic wave equations," *PRSL A155*, 447–59 (March 25, 1936).

1936C   "Wave equations in conformal space," *Annals of Mathematics 37*, 429–42 (May 18, 1935).

1937A   "The cosmological constants," *Nature 139* (February 5), 323.

1937B   "Complex variables in quantum mechanics," *PRSL A160*, 48–59 (February 24, 1937).

1937C   "Physical science and philosophy," *Nature* suppl. 139 (June 12), 1001–2.

1937D   "The reversal operator in quantum mechanics," *Bulletin of the USSR Academy of Science*, Series of Mathematical and Natural Sciences, no. 4, 569–82 (in Russian and English).

1938A   "A new basis for cosmology," *PRSL A165*, 199–208 (December 29, 1937).

1938B   "Classical theory of radiating electrons," *PRSL A167*, 148–69 (March 15, 1938).

1939A   "The relation between mathematics and physics," *Proceedings of the Royal Society (Edinburgh) 59* (1938/1939), 122–9 (February 25, 1939). James Scott Prize Lecture, February 6, 1939.

1939B   "A new notation for quantum mechanics," *PCPS 35*, 416–18 (April 29, 1939).

1939C   "La théorie de l'électron et du champ électromagnétique," *AIHP 9*, 13–49.

1940   "Dr. M. Mathisson," *Nature 146* (November 9), 613.

1942A   "The physical interpretation of quantum mechanics," *PRSL A180*, 1–40 (September 23, 1941). Bakerian Lecture, June 19, 1941.

1942B   "On Lorentz invariance in the quantum theory," *PCPS 38*, 193–200 (June 26, 1941). With R. Peierls and M. H. L. Pryce.

1943   "Quantum electrodynamics," *Communications of the Dublin Institute for Advanced Studies, Series A*, no. 1, 1–36.

1945A   "Unitary representations of the Lorentz group," *PRSL A183*, 284–95 (May 31, 1944).

1945B   "On the analogy between classical and quantum mechanics," *Review of Modern Physics 17*, 195–9.

1945c   "Application of quarternions to Lorentz transformations," *Proceedings of the Royal Irish Academy, Section A 50*, 261–70.

1946   "Developments in quantum electrodynamics," *Communications of the Dublin Institute for Advanced Studies, Series A* no. 3, 3–33.

1948A   "The difficulties in quantum electrodynamics," pp. 10–14 in *Report of International Conference on Fundamental Particles and Low Temperatures*, July 1946, vol. 1, London.

1948B   "Quantum theory of localizable dynamical systems," *PR 73*, 1092–1103 (January 12, 1948).

1948C   "On the theory of point electrons," *Philosophical Magazine 39*, 31–4, (June 7, 1948).

1948D   "The theory of magnetic poles," *PR 74*, 817–30 (June 21, 1948).
1948E   "Quelques développements sur la théorie atomique," *Les Conférences du Palais de la Découvertes*, December 6, 1945; Alencon, 1948, 16 pp.
1949A   "La seconde quantification," *AIHP 11*, 15–48.
1949B   "Forms of relativistic dynamics," *Review of Modern Physics 21*, 392–9.
1950A   "Generalized Hamiltonian dynamics," *Canadian Journal of Mathematics 2*, 129–48.
1950B   "A new meaning for gauge transformations in electrodynamics," *II Nuovo Cimento 7*, 925–38 (October 22, 1950).
1950C   "Field theory," pp. 114–15 in E. W. Titterton, ed., *Proceedings of the Harwell Nuclear Physics Conference*, September 1950, Harwell: Ministry of Supply.
1951A   "The relation of classical to quantum mechanics," pp. 10–31 in *Proceedings of the Second Canadian Mathematical Congress*, Vancouver 1949; Toronto: University of Toronto Press, 1951.
1951B   "The Hamiltonian form of field dynamics," *Canadian Journal of Mathematics 3*, 1–23.
1951C   "A new classical theory of electrons," *PRSL A209*, 291–5 (July 3, 1951).
1951D   "Is there an Æther?" *Nature 168* (October 9), 906–7.
1952A   (Reply to H. Bondi and T. Gold), *Nature 169*, 146 (December 15, 1951).
1952B   "A new classical theory of electrons, II," *PRSL A212*, 330–9 (January 17, 1952).
1952C   (Reply to L. Infeld), *Nature 169*, 702 (February 16, 1952).
1952D   "Les transformations de jauge en électrodynamique," *AIHP 13* (1952–53), 1–42.
1953A   "The Lorentz transformation and absolute time," *Physica 19*, 888–96 (April 18, 1953). Also in *Proceedings of the Lorentz-Kammerlingh Onnes Memorial Conference*, Leiden, June 22–26, 1953; Amsterdam, 1953.
1953B   "Méthodes génerales de la mécanique quantique relativiste," pp. 27–45 in *Particules Fondamentales et Noyaux*, Colloques Internationaux du Centre National de la Recherche Scientifique no. 38, Paris: CNRS 1953 (Conference held April 24–9, 1950).
1953C   "Die Stellung des Aethers in der Physik," *Naturwissenschaftliche Rundschau, 6*, 441–6. Lindau Lecture, July 1, 1953.
1954A   "A new classical theory of electrons, III," *PRSL A223*, 438–45 (December 9, 1953).
1954B   Quantum mechanics and the aether," *The Scientific Monthly 78*, 142–6.
1954C   "La mecánica cuántico y el éter," *Ciencia e Investigación* (Argentine) *10*, 399–405.
1955A   "The stress tensor in field dynamics," *Il Nuovo Cimento 1*, 16–36 (August 27, 1954).
1955B   "Gauge-invariant formulation of quantum electrodynamics," *Canadian Journal of Physics 33*, 650–60 (August 23, 1955).
1955C   "Note on the use of non-orthogonal wave functions in perturbation calculations," *Canadian Journal of Physics 33*, 709–12 (August 25, 1955).

1955D "On quantum mechanics and relativistic field theory," *Lectures on Mathematics and Physics,* no. 1, TATA Institute of Fundamental Research (Gen. eds., M. G. K. Menon and R. Navasimhan).

1955E "Symmetry in the atomic world," *Journal of Scientific and Industrial Research* (India) *14A,* 153–5. Address given at the Indian Science Congress, January 1955.

1957 "The vacuum in quantum electrodynamics," *Il Nuovo Cimento,* Supplement, *6,* 322–39.

1958A "Generalized Hamiltonian dynamics," *PRSL A246,* 326–32 (March 18, 1958).

1958B "The theory of gravitation in Hamiltonian form," *PRSL A246,* 333–43 (March 13, 1958).

1958C "The electron wave equation in Riemannian space," pp. 339–44 in B. Kockel, W. Macke, and A, Papatreau, eds., *Max Planck Festschrift,* Berlin: Verlag der Wissenschaft.

1958D "Electrons and vacuum" (in Russian), *Fizika v Skole,* no. 2, 20–32.

1959A "Fixation of coordinates in the Hamiltonian theory of gravitation," *PR 114,* 924–30 (December 10, 1958).

1959B "Energy of the gravitational field," *PR Letters 2,* 368–71 (March 20, 1959).

1959C "Sovremennoe sostojanie reljativistskoj teorii elektrona [The present situation in the relativistic theory of electrons]," *Trudy Instituta Istorii Estestvoznanija i Technii 22,* 32–3.

1959D "Elektrony a Vakuum," *Pokroky Matematiky Fysiky a Astronomie 4,* 309–17.

1960A "A reformulation of the Born-Infeld electrodynamics," *PRSL A247,* 32–43 (March 21, 1960).

1960B "Gravitationswellen," *Physikalische Blätter 16,* 364–6 (Abridged version of Lindau lecture given at the 9th Lindau Meeting, June 29–July 3).

1960C "Gravitationswellen," *Naturwissenschaftliche Rundschau 13,* 165–8 (unabridged version of 1960B).

1961A "Prof. Erwin Schrödinger, For.Mem.R.S." *Nature 189* (February 4), 355–6.

1961B (Reply to R. H. Dicke) *Nature 192* (November 4), 441.

1962A "An extensible model of the electron," *PRSL A268,* 57–67 (February 5, 1962).

1962B "Particles of finite size in the gravitational field," *PRSL A270,* 354–6.

1962C "The conditions for a quantum field theory to be relativistic," *Review of Modern Physics 34,* 592–6 (October 1962).

1962D "Interacting gravitational and spinor fields," pp. 191–200 in *Recent Developments in General Relativity,* Warsaw/Oxford: Polish Scientific Publications.

1962E "The energy of the gravitational field," pp. 385–94 in *Les Théories Relativistes de la Gravitation,* Colloques Internationaux du Centre National Recherche Scientifique, Paris: CNRS (Conference held at Royaumont June 21–7, 1959).

1963A   "A remarkable representation of the $3+2$ de Sitter group," *Journal of Mathematical Physics 4,* 901–9 (February 20, 1963).

1963B   "The evolution of the physicist's picture of nature," *Scientific American 208* (May 1963), 45–53.

1964A   "Hamiltonian methods and quantum mechanics," *Proceedings of the Royal Irish Academy, Section A* A63, 49–59. Larmor Lecture, September 30, 1964.

1964B   "Foundations of quantum mechanics," *Nature 203* (July 11), 115–6.

1964C   *Lectures on Quantum Mechanics,* New York: Belfer Graduate School of Science, 87 pp.

1964D   (Reply to H. S. Perlman), *Nature 204* (November 21), 772.

1964E   "The motion of an extended particle in the gravitational field," pp. 163–75 in L. Infeld, ed., *Proceedings on Theory of Gravitation,* Paris and Warsaw: Gauthier-Villars. Warsaw Conference, July 25–31, 1962.

1965   "Quantum electrodynamics without dead wood," *PR B139,* 684–90 (March 29, 1965).

1966A   *Lectures on Quantum Field Theory,* New York: Belfer Graduate School of Science/Academic Press, 149 pp. (lectures given at Yeshiva University, 1963–4; based on tape recordings transcribed by D. Wisnivesky).

1966B   "Foundations of quantum mechanics," pp. 1–10 in A. Gelbart, ed., *Some Recent Advances in the Basic Sciences,* vol. 1 (Belfer Graduate School of Science Annual Science Conference Proceedings, 1962–4), New York: Belfer Graduate School of Science.

1967   "The versatility of Niels Bohr," pp. 306–9 in *Niels Bohr, His Life and Work,* Amsterdam: North-Holland (first published in Danish in 1964).

1968A   "The scientific work of Georges Lemaitre," *PASC 2* (11), 1–20 (April 25, 1968).

1968B   "The quantization of the gravitational field," pp. 539–43 in *Contemporary Physics: Trieste Symposium 1968,* vol. 1, Vienna: IAEA. International Centre for Theoretical Physics, June 7–28, 1968.

1968C   "The physical interpretation of quantum electrodynamics," *PASC, 2* (13), 1–12.

1969A   "Hopes and fear," *Eureka,* no. 32, 2–4.

1969B   "Methods in theoretical physics," pp. 21–8 in *IAEA Bulletin,* special supplement, Trieste Symposium on Contemporary Physics, Vienna, 1969.

1969C   "Can equations of motion be used?" pp. 1–13 in T. Gudehus, G. Kaiser, and A. Perlmutter, eds., *Fundamental Interactions at High Energy,* New York: Gordon and Breach.

1970   "Can equations of motion be used in high-energy physics?" *Physics Today,* April 1970, 29–31.

1971A   "A positive-energy relativistic wave equation," *PRSL A322,* 435–45 (December 30, 1970).

1971B   *The Development of Quantum Theory,* New York: Gordon and Breach (J. Robert Oppenheimer Memorial Prize acceptance speech).

1971C   "A positive-energy relativistic wave equation," pp. 1–11 in M. Dal Cin, G. J. Iverson, and A. Perlmutter, eds., *Tracts in Mathematics and the Natural Sciences,* vol. 3, New York: Gordon and Breach.

1971D   "Fundamental problems in physics," lecture given at 21st Lindau Meet-
        ing, June 28–July 2. Summarized in German by F. Stricker in *Chemiker-
        Zeitung 95*, 880–1.

1972A   "A positive-energy relativistic wave equation, II," *PRSL A328*, 1–7
        (December 14, 1971).

1972B   "The variability of the gravitational constant," pp. 56–9 in F. Reines,
        ed., *Cosmology, Fusion, and other Matters*, London: Adam Hilger.

1972C   "Evolutionary cosmology," *PASC, 2* (46), 1–15.

1972D   "Discrete subgroups of the Poincaré group," pp. 45–51 in V. I. Ritus,
        ed., *Problemi teoretichestoi fiziki*, Moscow: Publ. House Nauka.

1973A   "Long range forces and broken symmetries," *PRSL A333*, 403–18 (Jan-
        uary 15, 1973).

1973B   "Relativity and quantum mechanics," pp. 747–72 in E. C. G. Sudarshan
        and Y. Ne'eman, eds., *The Past Decade in Particle Theory*, London: Gor-
        don and Breach (First published in *Fields and Quanta, 3* (1972), 139–
        64; presented at symposium in Austin, Texas, April 14–17, 1970).

1973C   "Long range forces and broken symmetries," pp. 1–18 in B. Kursunoglu
        and A. Perlmutter, eds., *Fundamental Interactions in Physics*, New
        York: Plenum Press.

1973D   "Some of the early developments of quantum theory," pp. 1–13 in B.
        Kursunoglu, ed., *Impact of Basic Research on Technology*, New York:
        Plenum Press (abridged version of 1971B).

1973E   "Fundamental constants and their development in time," pp. 45–54 in
        J. Mehra, ed., *The Physicist's Conception of Nature*, Dordrecht: Reidel.
        Abridged version in Hungarian, "Alapvető fizikai állandók és időbeni
        fejlödésük," *Fizikai Szemle 28* (1978), 201–4.

1973F   "Zitterbewegung of the new positive-energy particle," 354–63 in G. Iver-
        son, A. Perlmutter, and S. Mintz, eds., *Fundamental Interactions in
        Physics and Astrophysics*, New York: Plenum Press.

1973G   "New ideas of space and time," *Die Naturwissenschaften 60*, 529–31
        (Lindau Lecture, July 2, 1973).

1973H   "Development of the physicist's conception of nature," pp. 1–15 in J.
        Mehra, ed., *The Physicist's Conception of Nature*, Dordrecht: Reidel.

1974A   "The geometrical nature of space and time," pp. 1–18 in S. Mintz, L.
        Mittag, and S. M. Widmayer, eds., *Fundamental Theories in Physics*,
        New York: Plenum Press.

1974B   "Cosmological models and the Large Number Hypothesis," *PRSL A338*,
        439–46 (December 3, 1973).

1974C   "An action principle for the motion of particles," *General Relativity and
        Gravitation 5*, 741–8 (December 30, 1973).

1974D   *Spinors in Hilbert Space*, New York: Plenum Press, 91 pp.

1974E   "Magnetic monopole," pp. 387–8 in A. M. Prokhorov, gen. ed., *Great
        Soviet Encyclopedia*, vol. 15, New York: MacMillan, 1977 (trans. of the
        3rd edn. of Russian original, Moscow, 1974).

1974F   "The development of quantum mechanics," in *Contributi del Centro
        Linceo Interdisciplinare di Scienze Mathematiche e loro Applicazioni*,
        No. 4, 11 pp., Rome: Academia Nazionale dei Lincei.

1974G   "Dirac, Paul Adrien Maurice," 310–12 in E. Macorini, ed., *Scienziati e Tecnologici Contemporanei,* vol. 1, Milan: Arnoldo Mondadori.

1975A   "Variation of G," *Nature 254* (March 20), 273.

1975B   *General Theory of Relativity,* New York: John Wiley & Sons, 68 pp. Based on lectures given at Florida State University.

1975C   "James Chadwick," *PASC 3* (6), 1–5.

1975D   "Does the gravitational constant vary?" *PASC 3* (7), 1–7.

1975E   "An historical perspective on spin," pp. 1–11 in J. B. Roberts, ed., *Proceedings of Summer Studies of High-Energy Physics with Polarized Beams* (July 1974), Argonne National Laboratory.

1976A   "Theory of magnetic monopoles," pp. 1–14 in A. Perlmutter, ed., *New Pathways in High-Energy Physics, I,* New York: Plenum Press.

1976B   "Heisenberg's influence on physics," *PASC 3* (14), 1–15.

1977A   "Ehrenhaft, the subelectron and the Quark," pp. 290–3 in C. Weiner, ed., *History of Twentieth Century Physics,* New York: Academic Press.

1977B   "Recollections of an exciting era," pp. 109–46 in Weiner, ed., op. cit.

1977C   "The dynamics of streams of matter," pp. 1–11 in A. Perlmutter and L. Scott, eds., *Deeper Pathways in High Energy Physics,* New York: Plenum Press.

1977D   "The relativistic electron wave electron," pp. 15–34 in L. Jenik and I. Montvay, eds., *Proceedings of the 1977 European Conference on Particle Physics* held in Budapest July 4–9, 1977, vol. 1, Budapest: Central Research Institute for Physics.

1977E   "The relativistic electron wave equation," *Europhysics News 8.* Slightly abridged version of (1977D).

1977F   "Annahmen und Voreingenommenheit in der Physik," *Naturwissenschaftliche Rundschau 30,* 429–32 (Lindau Meeting, 1976).

1978A   *Directions in Physics,* New York: John Wiley & Sons, 95 pp. Lectures delivered during a visit to Australia and New Zealand in August and September 1975.

1978B   "Consequences of varying G," pp. 169–74 in J. E. Lannutti and P. K. Williams, eds., *Current Trends in the Theory of Fields,* New York: AIP Conference Proceedings.

1978C   "The mathematical foundations of quantum theory," pp. 1–8 in A. R. Marlow, ed., *Mathematical Foundations of Quantum Theory,* New York: Academic Press.

1978D   "The large number hypothesis and the cosmological variation of the gravitational constant," pp. 3–20 in L. Halpern, ed., *On the Measurement of Cosmological Variations of the Gravitational Constant,* Miami: University of Florida Press.

1978E   "The monopole concept," *International Journal of Theoretical Physics 17,* 235–47.

1978F   "New approach to cosmological theory," pp. 1–16 in A. Perlmutter and L. F. Scott, eds., *New Frontiers in High Energy Physics,* New York: Plenum Press.

1978G   "The prediction of antimatter," Ann Arbor, University of Michigan. The 1st H. R Crane Lecture, April 17, 1978.

1978H "New Ideas about gravitation and cosmology," *PASC 3* (24), 1–10.

1979A "The relativistic electron wave equation," *Soviet Physics: Uspeki 22,* 648–53. Presented at the European Conference on Particle Physics, Budapest, July 4–9, 1977. Hungarian translation in *Fizikai Szemle 27* (1979), 443–50.

1979B "The excellence of Einstein's theory of gravitation," *Impact of Science on Society 29,* 11–14. Also in M. Goldsmith, A. Mackay, and J. Woudhuysen, eds., *Einstein: the First Hundred Years,* Oxford: Pergamon, 1980. Article adapted from talk at the UNESCO Symposium at Ulm, September 19, 1978.

1979C "The Large Number Hypothesis and the Einstein theory of gravitation," *PRSL A365,* 19–30 (June 26,1978).

1979D "Developments of Einstein's theory of gravitation," pp. 1–13 in A. Perlmutter and L. Scott, eds., *On the Path of Albert Einstein,* New York: Plenum Press.

1979E "The test of time," *The Unesco Courier 32,* 17–23. Abridged version of lecture given at the UNESCO Symposium in Ulm, September 1978. German translation, "Der Zeittest," *Naturwissenschaftliche Rundschau, 33* (1980), 353–6.

1979F Interview with P. A. M. Dirac conducted by P. Buckley and F. D. Peat, pp. 34–40 in P. Buckley and F. D. Peat, *A Question of Physics: Conversations in Physics and Biology,* London: Routledge and Kegan Paul.

1979G "Einstein's influence on physics," pp. 19–23 in *Einstein Galileo. Commemoration of Albert Einstein,* Citta del Vaticano, Pontifical Academy of Science: Liberia Editrice Vaticana.

1980A "The variation of G and the problem of the moon," pp. 1–7 in A. Perlmutter and L. Scott, eds., *Recent Developments in High Energy Physics,* New York: Plenum Press.

1980B "Why we believe in the Einstein theory," pp. 1–11 in B. Gruber and R. S. Millmann, eds., *Symmetries in Science,* New York: Plenum Press.

1980C "A little prehistory," *The Old Cothamian,* 9.

1980D ("Dirac recalls Kapitza"), *Physics Today, 33* (May), 15.

1981A "The arrival of quantum mechanics," pp. 185–190 in I. A. Dorobantu, ed., *Trends in Physics 1981,* Bucharest: European Physical Society.

1981B "Models of the universe," pp. 1–9 in B. Kursunoglu and A. Perlmutter, eds., *Gauge Theories, Massive Neutrinos, and Proton Decay,* New York: Plenum Press.

1981C "Einstein and the development of physics," pp. 13–23 in C. M. Kinnon, A. N. Kholodilin, and J. G. Richardson, eds., *The Impact of Modern Scientific Ideas on Society,* Dordrecht: Reidel.

1981D "Does renormalization make sense?" pp. 129–30 in D. W. Duke and J. F. Owens, eds., *Perturbative Quantum Chromodynamics,* New York: AIP Conference Proceedings no. 74. Conference held at Florida State University March 25–8, 1981.

1982A "The variation of G and the quantum theory," pp. 1–6 in R. Ruffini, ed., *Proceedings of the Second Marcel Grossman Meeting on General Relativity,* Amsterdam: North Holland.

1982B "Pretty mathematics," *International Journal of Theoretical Physics 21,* 603–5. Contribution to symposium at Loyola University, New Orleans, May 1981 in honor of Dirac's 80th birthday.

1982C "The early years of relativity," pp. 79–90 in G. Holton and Y. Elkana, eds., *Albert Einstein, Historical and Cultural Perspectives* (Jerusalem Centennial Symposium, March 14–23, 1979), Princeton: Princeton University Press.

1983A "The present state of gravitational theory," pp. 1–10 in *Field Theory in Elementary Particles,* New York: Plenum Press. Conference at Coral Gables, January 18–21, 1982.

1983B "The origin of quantum field theory," pp. 39–55 in L. Brown and L. Hoddeson, eds., *The Birth of Particle Physics,* Cambridge: Cambridge University Press.

1983C "My life as a physicist," pp. 733–49 in A. Zichichi, ed., *The Unity of the Fundamental Interactions,* New York: Plenum Press (Proceedings of the 19th course of the International School of Subnuclear Physics, July 31–August 11, 1981, in Erice, Sicily).

1984A "Blackett and the positron," pp. 61–2 in J. Hendry, ed., *Cambridge Physics in the Thirties,* Bristol: Adam Hilger Ltd.

1984B "The requirements of fundamental physical theory," *European Journal of Physics, 5,* 65–7 (Lindau Meeting, 1982). Abridged German translation in *Naturwissenschaftliche Rundschau 30* (1983), 429–31.

1984C "The future of atomic physics," *International Journal of Theoretical Physics 23,* 677–81.

1987 "The inadequacies of quantum field theory," pp. 194–8 in B. M. Kursunoglu and E. P. Wigner, eds., *Paul Adrien Maurice Dirac,* Cambridge: Cambridge University Press.

NOTES AND REFERENCES

The following abbreviations are used throughout the Notes and References.

AHQP   Archive for History of Quantum Physics, Niels Bohr Institute, Copenhagen

AIP   Centre of History of Physics, American Institute of Physics, New York

BSC   Bohr Scientific Correspondence, Niels Bohr Institute, Copenhagen

CC   Dirac Papers, Churchill College, Cambridge (now moved to Florida State University, Tallahassee)

EA   Ehrenfest Archive, Museum Boerhaave, Leiden

LC   Manuscript Division, Library of Congress, Washington, D.C.

NA   Nobel Archive, Royal Swedish Academy of Science, Stockholm

PB1   Volume 1 of Pauli's correspondence [Pauli (1979)]

PB2   Volume 2 of Pauli's correspondence [Pauli (1985)]

PB3   Volume 3 of Pauli's correspondence [in press]

PQM   *The Principles of Quantum Mechanics;* the first edition [Dirac (1930E)] was published in 1930, the second in 1935, the third in 1947, and the fourth in 1958

SUL   Sussex University Library

TDC   Tamm–Dirac Correspondence, USSR Academy of Science, Moscow

Complete references to works by Dirac are given in Appendix II, which precedes this section; references to works by others are in the General Bibliography, which follows this section.

315

Chapter 1

1 This story is told in Van Vleck (1972), 10.
2 Much of the information in this and the following chapter is based on the works of Jagdish Mehra: Mehra (1972) and Mehra and Rechenberg (1982+), vol. 4. Additional biographical information on Dirac, including details of his family background, can be found in Dalitz and Peierls (1986). Dalitz (1987B) provides a useful chronological sketch, and Crease and Mann (1986), 75–91, a vivid but rather uncritical portrait of the young Dirac.
3 Mehra and Rechenberg (1982+), vol. 4, p. 7.
4 Interview with P. A. M. Dirac, April 1962, conducted by T. S. Kuhn and E. P. Wigner (AHQP); hereafter referred to as Dirac interview, 1962 (AHQP). Dirac was interviewed by Kuhn again in May 1963. Although no doubt a bright and thoughtful boy, Paul hardly lived up to C. P. Snow's claim that "before he [Dirac] was three . . . he was playing with mathematical ideas, such as the singular properties of infinite numbers" [Snow (1981), 75].
5 Margit Dirac (1987), 5.
6 Dirac interview, 1962 (AHQP).
7 Ibid. Emphasis added.
8 L. R. Phillips, as quoted in Dalitz and Peierls (1986), 142.
9 Mehra and Rechenberg (1982+), vol. 4, p. 11.
10 Salaman and Salaman (1986), 69. Dirac's confessions to Esther Salaman are reported to have taken place in 1973. There are several inaccuracies in the memoirs of the Salamans, so they should be read with some caution. For example, Dirac did not visit Russia in 1927, but in 1928.
11 Dirac (1980C), 9.
12 Dirac (1982C), 80.
13 Mehra and Rechenberg (1982+), vol. 4, p. 9.
14 A detailed account of this episode is given in Earman and Glymour (1981).
15 Dirac (1977B), 110. Also Dirac (1982C), 80–2.
16 Dirac (1977B), 113.
17 Details about the Cambridge physicists and their works are given in Mehra and Rechenberg (1982+), vol. 4, pp. 24 ff.
18 For example, Cunningham (1915).
19 Interview with E. Cunningham, June 1963, conducted by J. L. Heilbron (AHQP).
20 Hendry (1984B), 104.
21 Dirac (1977B), 115. Dirac kept lecture notes on Fowler's course on quantum theory. These include "The Quantum Theory" (probably winter 1924/25; 45 pp.), "Recent Developments" (probably spring 1925; 33 pp.), and "Spinning Electrons" (probably spring 1926; 8 pp.) (AHQP).
22 Dirac (1977B), 115.
23 Dirac, Notebook on the Theory of Relativity, 1923, 34 pp. (CC). Cunningham's lectures followed closely his textbook [Cunningham (1915)], which Dirac used.
24 Cunningham interview, 1963 (AHQP).
25 Slater (1975), 131.

26 Interview with L. H. Thomas, November 1962, conducted by T. S. Kuhn and G. E. Uhlenbeck (AHQP).
27 Dirac (1977B), 116.
28 Minute books of the Kapitza Club and $\nabla^2 V$ Club (AHQP).
29 Interview with N. Mott, May 1963, conducted by T. S. Kuhn (AHQP). See also Mott (1986), 22.
30 Dirac (1977B), 116.
31 Dirac (1924A). The technical details of this and other early papers of Dirac are examined in Mehra and Rechenberg (1982+), vol. 4.
32 Darwin to Dirac, undated but probably December 1924 (AHQP); Darwin to Dirac, January 19, 1925 (AHQP).
33 Dirac (1977B), 120.
34 Dirac, Notebook on the Physics of Stellar Atmospheres, 70 pp.; undated but probably 1924 (CC).
35 Dirac (1924B).
36 Eddington to Dirac, 1924, as quoted in Douglas (1956), 111.
37 Dirac (1977B), 118.
38 Heisenberg (1971A), 87. The quotation is Heisenberg's free reconstruction of what Bohr said.
39 Feinberg (1987), 169.

Chapter 2

1 Sources on this episode do not agree; witness the following: (I) In an interview of June 26, 1961, Dirac told Van der Waerden that "I was not present at the [Heisenberg] lecture and did not know anything about it" [Van der Waerden (1967), 41]. (II) In the AHQP interview of January 1962 Dirac said: "He [Heisenberg] gave a talk about a new theory to the Kapitza Club, but I wasn't a member of the Club so I did not go to the talk. I did not know about it at the time." However, Dirac was in fact a member of the Kapitza Club in the summer of 1925. (III) In 1972, Dirac recalled the events in a very different light; namely, he said that he attended the lecture and followed most of it. He even recalled that Heisenberg "was talking about the origins of his ideas of the new mechanics" (which he did not do). However, Dirac did not actually remember that the new theory was part of the lecture: "By that time I was just too exhausted to be able to follow what he said, and I just did not take it in. . . . Later on I completely forgot what he had said concerning his new theory. I even felt rather convinced that he had not spoken about it at all, but other people who were present at this meeting of the Kapitza Club assured me that he had spoken about it" [Dirac (1977B), 119]. If Dirac was in Cambridge on July 28, he almost certainly would have attended Heisenberg's lecture. We know that Dirac was in Cambridge a week later, for on August 4 he personally gave a talk to the Kapitza Club, entitled "Bose's and L. de Broglie's Deductions of Planck's Law" (manuscript draft in AHQP, 11 sheets). The lack of agreement among Dirac's various recollections, as illustrated in this case, is also to be found in other cases. In general, one should

not rate Dirac's recollections as very reliable – at least not when it comes to details.

2   Heisenberg (1925). The paper was received by *Zeitschrift für Physik* on July 29. The proofs, read by Fowler and Dirac, are dated August 13, 1925. On the top of the first page, Fowler wrote, "What do you think of this? I shall be glad to hear" (CC). For analysis of and background to Heisenberg's paper, see Van der Waerden (1967) and Mehra and Rechenberg (1982+), vol. 3.

3   Dirac interview, 1962 (AHQP).

4   Dirac, "Heisenberg's Quantum Mechanics and the Principle of Relativity," 4 pp., dated October 1925 (AHQP). The manuscript, apparently intended to be published as a paper, contains no trace of Poisson brackets and was therefore supposedly written shortly before Dirac got the idea of connecting the quantum mechanical commutator with the Poisson bracket. See also Dirac (1983C), 738.

5   Dirac (1977B), 121.

6   Dirac (1924C). At about the same time, in another unpublished manuscript, Dirac examined how to relate the frequencies of the radiation emitted by a moving harmonic oscillator to the frequencies emitted by the same oscillator at rest [Dirac, "Radiation from a Moving Planck Oscillator," 2 pp., undated (CC)]. This manuscript is analyzed in Mehra and Rechenberg (1982+), vol. 4, p. 198.

7   Dirac (1977B), 121–2. The Poisson bracket expression appears in Whittaker (1917) on p. 299, where it is written as (x,y). This was the usual notation, while the notation $[x,y]$ was reserved for the so-called Lagrange symbol, defined

$$\sum_k \left(\frac{\partial p_k}{\partial x} \frac{\partial q_k}{\partial y} - \frac{\partial q_k}{\partial x} \frac{\partial p_k}{\partial y} \right)$$

Dirac introduced the notation of square brackets for Poisson symbols, which were also taken over to signify the quantum mechanical commutator. Lagrange brackets play no role in quantum mechanics. Dirac, always fond of inventing new notation and terminology, was proud that use of square brackets became standard practice. See Dirac (1983C), 739.

8   Dirac (1977B), 124.

9   Ibid. However, Dirac reported differently to Van der Waerden: "I did get the general formula [equation (2.5)] on the lines indicated in the paper, by working with the case of large quantum numbers. At that time I was expecting some kind of connection between the new mechanics and Hamiltonian dynamics . . . and it seemed to me that this connection should show up best with large quantum numbers" [Van der Waerden (1967), 42].

10   Dirac (1925D), 649.

11   See Van der Waerden (1967), 52–3.

12   Heisenberg to Dirac, November 20, 1925 (CC). Here quoted from Brown and Rechenberg (1987), 150, which also includes the German original. Dirac's translation of the letter, and of excerpts of other Heisenberg letters from the same period, appear in Dirac (1977B), 124 ff. The same day, November 20, Heisenberg reported to Bohr about Dirac's work: "In its style in writing, some of it pleases me better than that of Born and Jordan" [Heisenberg to Bohr, November 20, 1925 (BSC); original in German].

13 Born and Jordan (1925).

14 Slater to Bohr, May 27, 1926, as quoted in Bohr (1972+), 501. See also Slater (1967).

15 J. C. Slater, "A Theorem in the Correspondence Principle," unpublished manuscript contained in Slater's "Excerpts from Personal Letters and Records Bearing on Visit to Copenhagen," a copy of which is at the Niels Bohr Archive, Copenhagen. Slater's work included most of Dirac's results, including equations (2.5), (2.6), and (2.8), the only difference being a change of sign; specifically, Slater wrote equation (2.6) as $xy - yx = -ih/2\pi\{x,y\}$, (etc.), using the symbol $\{x,y\}$ instead of Dirac's $[x,y]$. Slater was not only "on the track of making use of Poisson brackets," as stated in Mehra and Rechenberg (1982+), vol. 4, p. 266, but actually had the entire formalism with several applications.

16 Kramers (1925). See Mehra and Rechenberg (1982+), vol. 3, p. 143, and Dresden (1987), 315, 450.

17 Born (1978), 226.

18 Heisenberg to Pauli, November 24, 1925 (PB1, 266). Bohr had also received news of Dirac's work before publication and wanted to have a copy of it. Bohr to Fowler, November 26, 1925 (BSC).

19 From October 1925 to April 1926, Dirac gave the following lectures to the Kapitza Club: "Direction of Ejection of Photo-electrons" (October 13); "Debye's Note on the Scattering of X-rays by Liquids" (October 23); "The Light-Quantum Theory of Diffraction" (December 15); "Fundamental Equations of Quantum Mechanics" (January 19); "Kramers' Derivation of the Quantum Mechanics" (April 27). On March 2, he presented to the $\nabla^2 V$ Club a lecture on "Quantum Mechanics and a Preliminary Investigation of the Hydrogen Atom."

20 Van Vleck (1968), 1234.

21 Dirac (1926A).

22 Dirac (1926A), 561, and *PQM*, 1st edn., p. 96.

23 For a treatment of the hydrogen spectrum in early quantum mechanics and the reasons for the delay of Pauli's paper, see Kragh (1985), 113–22. Pauli's paper was received by *Zeitschrift für Physik* on January 17 and was published March 27; Dirac's paper was received by *Proceedings* on January 22 and was published March 1.

24 Dirac (1977B), 130.

25 Heisenberg to Dirac, April 9, 1926, as quoted in Dirac (1977B), 130. Parts of the original German text are reproduced in Brown and Rechenberg (1987), 150. The reference to Thomas's "factor 2" is to the spin.

26 Pauli to Kramers, March 8, 1926 (PB1, 307).

27 Wentzel (1926), received by *Zeitschrift für Physik* on March 27.

28 Schrödinger to Lorentz, June 6, 1926, as quoted in Przibram (1963), 64. Schrödinger's theory of the hydrogen atom was received by *Annalen der Physik* on January 27 and appeared in print about two months later; see Schrödinger (1926A).

29 Van Vleck (1968). The Van Vleck–Dirac method was eventually published in 1934, when Van Vleck gave calculations of the mean values of $1/r^n$ for $n = 1$ to $n = 6$ [Van Vleck (1934A)]. Dirac, who communicated Van Vleck's paper to the *Proceedings*, wrote to him: "I sent your paper in to the Royal Society a little while ago. It seemed to be quite interesting. There is a rule that every paper sent to the Royal Society must be accompanied by a summary, but in your case the paper was so short and the title was so long that I thought the title would do as a summary. I suggested this to them and it seems to be alright." Dirac to Van Vleck, November 8, 1933 (AHQP).

30 Whittaker (1917) and Baker (1922). As Mehra and Rechenberg have shown, Dirac's formulation of q-number algebra, including much of the notation, was largely taken over from Baker's book. Mehra and Rechenberg (1982+), vol. 4, p. 163–5.

31 Dirac (1926A), 563.

32 Dirac (1977B), 130. In a letter to Dirac in March 1926, Léon Brillouin pointed out that Dirac had made some mistakes because he did not realize the equivalence between q-numbers and matrices. I have not been able to find Brillouin's letter, which was probably lost, with most of his correspondence and papers, during World War II.

33 Dirac (1926D).

34 Jordan to Dirac, undated but postmarked April 14, 1927 (AHQP); in German.

35 Dirac (1926A), 565.

36 Dirac (1926B).

37 Ibid.

38 Ibid., 299.

39 Ibid., 300.

40 The g-factor determines the splitting of doublets and triplets in a magnetic field. A formula for g was derived in 1923 by Alfred Landé, who used the vector core model of the old quantum theory. Landé's g-formula was first derived in matrix mechanics by Heisenberg and Jordan in March 1926.

41 Dirac (1926C).

42 Ibid., 421. As Dirac mentioned in a note, the same result, as far as unpolarized light is concerned, was found by Gregory Breit three months earlier without using quantum mechanics proper. See Breit (1926).

43 Interview with Klein, February 1963, conducted by T. S. Kuhn and J. L. Heilbron (AHQP).

44 Eddington to Dirac, March 10, 1926, quoted in Douglas (1956), 111; also Heisenberg to Dirac, May 26, 1926, as referred to in Dirac (1977B), 132.

45 Dirac's entire production of technical publications, more than 120 papers, includes only five figures and no tables at all; the figure from 1926 was the only one in which Dirac graphically compared theory with experiment.

46 Dirac (1926C), 421.

47 Compton to Dirac, August 21, 1926 (AHQP).

48 Dirac (1926F).

49 Dirac, "Quantum Mechanics," Dissertation, Cambridge University, May 1926 (unpublished, copy in AHQP). In a brief abstract of his dissertation [Dirac (1927F)], Dirac summarized his work as follows: "The dissertation consists of a development of Heisenberg's new quantum mechanics (introduced in the Zeits.f.Phys. XXXIII, 1925, 879) from the point of view that the dynamical variables are simply 'numbers' that do not obey the commutative law of multiplication, but satisfy instead certain quantum conditions. The laws of the ordinary quantum theory can then be deduced."

50 Mott (1986), 11–13.

51 Dirac (1977B), 132. For the effect of the strike on Cambridge University, see Howarth (1978), 148–50.

52 Reported by Lady Jeffreys, then Bertha Swirles, who became a research student under Fowler in 1925 and attended Dirac's lectures in the spring of 1926. Quoted from Eden and Polkinghorne (1972), 2. In 1940, Miss Swirles married Sir Harold Jeffreys, a well-known astronomer, mathematician, and geophysicist.

53 Hendry (1984B), 104. For a discussion of Dirac as a teacher, see Chapter 12.

54 N. Mott to his parents, November 27, 1927, as quoted in Mott (1936), 17.

55 Schrödinger (1926A). It was only in the second communication that Schrödinger referred to the Göttingen matrix mechanics and also to Dirac's quantum mechanics.

56 Unpublished manuscript (AHQP). See Mehra and Rechenberg (1982+), vol. 4, p. 109. A year earlier, Dirac had dismissed the wave–particle duality advocated by de Broglie [Dirac (1924D)]. Dirac knew de Broglie's theory primarily through a paper by de Broglie [de Broglie (1924)], the publication of which had been arranged by Fowler.

57 Heisenberg to Dirac, April 9, 1926. German original in Brown and Rechenberg (1987), 151.

58 Dirac (1977B), 131.

59 Pauli to Jordan, April 12, 1926 (PB1, 315–20).

60 Heisenberg to Dirac, May 26, 1926 (AHQP).

61 Dirac interview, 1962 (AHQP).

62 Schrödinger, unpublished notebook entitled "Dirac," probably from June 1926 (AHQP). In spite of its title, the main part of the notebook does not deal with Dirac's works but with wave mechanical calculations of the relativistic Kepler problem.

63 See Kerber et al. (1987), 65–6.

64 The term eigenfunction, a hybrid of English and German, was probably Dirac's own invention. It soon became generally used in English physics literature, though "proper function" was also used. In the same paper Dirac introduced the verb "commute," defined as "the statement a commutes with b means $ab = ba$ identically" [Dirac (1926E), 674].

65 Schrödinger (1926B). This paper was received by the *Annalen* on June 21 but did not appear in print until September 5.

66 Dirac (1926E), 667.

67 Ibid.

68 Heisenberg (1926A). Dirac mentioned in a note that "Prof. Born has informed me that Heisenberg had independently obtained results equivalent to these" and added in proof a reference to Heisenberg (1926A); see Dirac (1926E), 670. Born probably informed Dirac in late July, while he (Born) was in Cambridge to give a talk to the Kapitza Club. For the relationship between Heisenberg's and Dirac's contributions, see Mehra and Rechenberg (1982+), vol. 5, pp. 757–71.

69 Heisenberg to Dirac, April 9, 1926 (CC).

70 Fermi (1926B). An earlier version in Italian was Fermi (1926A). For historical details, see Belloni (1978) and Mehra and Rechenberg (1982+), vol. 5, pp. 746–57.

71 Fermi to Dirac, October 26, 1926 (CC). Neither did Heisenberg refer to Fermi's work, which had appeared at the time when Heisenberg submitted his paper.
72 Dirac (1977B), 133.
73 Dirac seems first to have suggested the names in a lecture given in Paris on December 6, 1945; see Dirac (1948E), 8. He reiterated the proposal in 1947, in the third edition of *PQM* (p. 210). See also Walker and Slack (1970).
74 Fowler (1926).
75 See Hoddeson and Baym (1980).
76 Schrödinger to Bohr, October 23, 1926. Schrödinger expressed his puzzlement over a point in Dirac's paper: "According to the proof on p. 662 below, p_n always 'means' $\partial/\partial q_n$. Now look at the following equation arising from eq. (4) on the following page: $dx/ds = [x,F]$, in which x can represent one of the p's, for example, and the quantity s can represent the time. Now what does $d/dt(\partial/\partial q)$ mean? What is it supposed to mean, differentiating the operator $\partial/\partial q$ totally with respect to the time?" [Bohr (1972+), vol. 3, p. 461].
77 In reply to Fowler's request of April 9, Bohr wrote: "He shall certainly be welcome to work here next term. We are all full of admiration for his work. Dirac's coming will also be a great pleasure to Heisenberg. . . . Already last autumn . . . he suggested a visit of Dirac as most desirable" [Bohr to Fowler, April 14 1926 (AHQP)]. "I think he [Dirac] has done enough already to need no further introduction," Fowler wrote (AHQP).
78 Dirac (1977B), 134.
79 Møller (1963), 59.
80 Dirac (1977B), 134.
81 The development of transformation theory during 1926 is considered in Jammer (1966), 293–323, and in Hendry (1984A), 102–10.
82 Pauli to Heisenberg, October 19, 1926 (PB1, 340–9). Pauli's letters were much discussed in Copenhagen and thus also known to Dirac. Pauli expressed his vague feelings of a complementarity between position and momentum measurements as follows: "One can look at the world with p-eyes, or one can look at the world with q-eyes, but with both eyes open at once one makes mistakes" (Ibid., 343; trans. from German).
83 Heisenberg to Pauli, October 28, 1926 (PB1, 349–51).
84 Heisenberg (1926B) and Lanczos (1926).
85 Dirac (1977B), 137.
86 Klein interview, September 1963 (AHQP).
87 Heisenberg to Pauli, November 23, 1926 (PB1, 357–60).
88 Dirac (1927A), 622.
89 Ibid., 635
90 See Lützen (1982).
91 Dirac interview, 1962 (AHQP).
92 The lectures of Dirac and Schwartz were published in *Proceedings of the Second Canadian Mathematical Congress,* Toronto: University of Toronto Press, 1951.
93 Dirac (1927A), 623 and 641.
94 Ibid., 641.

95 Heisenberg to Pauli, February 23, 1927 (PB1, 377); see also Heisenberg to Jordan, March 7, 1927 (AHQP). For background, see Beller (1985).
96 Pauli to Bohr, November 17, 1925 (PB1, 260).
97 Jordan (1927A).
98 Dirac to Jordan, December 24, 1926 (AHQP). Jordan received notice of Dirac's forthcoming paper only after he had submitted his own work [Jordan to Dirac, December 22, 1926 (CC)]. In April, Jordan reported to Dirac in detail about the relationship between the two transformation theories [Jordan to Dirac, April 24, 1927 (AHQP)].
99 Kudar to Dirac, January 29, 1927 (CC).
100 Ehrenfest to Pauli, September 22, 1928 (PB1, 474).
101 Dirac (1929B), 716.
102 Birge to Van Vleck, March 10, 1927, quoted from Schweber (1986B), 55.
103 Heisenberg (1927). See MacKinnon (1982), 252–9.
104 Heisenberg to Dirac, April 27, 1927 (AHQP). Dirac's letter, to which Heisenberg replied, is missing, but the content of it can be inferred from Heisenberg's reply.
105 Lorentz to Dirac, June 9, 1927 (CC). Lorentz referred to Dirac (1927A), which he had recently read. Dirac answered that his obligations at Cambridge prevented him from accepting the invitation [Dirac to Lorentz, June 11, 1927 (AHQP)]. According to the memoirs of Adriaan Fokker, Lorentz had great trouble in assimilating Dirac's ideas [interview with A. Fokker, April 1962, conducted by J. L. Heilbron (AHQP)].
106 Ehrenfest to Dirac, October 1, 1926 (AHQP); Ehrenfest to Dirac, June 3, 1927 (CC); Dirac to Ehrenfest, June 20, 1927 (EA).
107 Ehrenfest to Dirac, June 16, 1927 (CC).
108 Ehrenfest to Uhlenbeck, July 23, 1927 (AHQP).
109 Ehrenfest to Planck, June 1, 1927; Planck to Ehrenfest, July 24, 1927 (EA).
110 Oppenheimer to Dirac, November 28, 1927, as reproduced in Smith and Weiner (1980), 108. Dirac almost kept his promise. At the end of June, he completed a paper on the quantum mechanics of collision processes [Dirac (1927D)], which was his last publication for the year. He lectured twice on the same subject at the Kapitza Club, on October 4 and October 11.

Chapter 3

1 Dirac (1926C), 407.
2 Pauli to Heisenberg, January 31, 1926 (PB1, 284).
3 For references and further details, see Kragh (1984).
4 Kragh (1982B) and Mehra and Rechenberg (1982+), vol. 5, pp. 430 ff.
5 Dirac (1971B), 38–9. Similar accounts are given in Dirac (1961A, 1963B, 1969A, 1969B, 1975E, and 1977B). As another illustration of the hope-and-fear moral, Dirac used to mention Heisenberg's first, unintended introduction of noncommuting variables. Heisenberg was "really scared" and "afraid this was a fundamental blemish in his theory and that probably the whole beautiful idea would have to be given up" [Dirac (1971B), 23]. Dirac, on the other hand, had an advan-

tage over Heisenberg in 1925, because "I was not afraid of Heisenberg's theory collapsing" (ibid.).

Dirac's narrative concerning the birth of wave mechanics is not completely reliable. I have criticized it in Kragh (1982B). Dirac's other reminiscences and semi-historical works should also be viewed with caution. For example, in Dirac (1975E, 1977D, and 1977E) he stated that Schrödinger worked from "de Broglie's equation" – which Dirac claimed to be identical to the Klein–Gordon equation for a free particle – and that he (Dirac) read the paper in which de Broglie proposed his wave equation. The facts are that de Broglie did propose a wave equation in February 1925, but this equation was not exactly the same equation as the one which Dirac referred to in 1977, and it had no impact at all on Schrödinger's work. It is unlikely that Dirac read de Broglie's little-known, brief paper in *Comptes Rendus* [de Broglie (1925)]; Dirac knew of de Broglie's ideas from a review article that appeared in *Philosophical Magazine* in 1924, but no wave equation appeared in this article [De Broglie (1924)]. The conclusion is that Dirac sometimes mixed things up in his recollections.

6 Dirac (1971B), 24.

7 Dirac (1926E). Dirac probably came to the KG equation independently of Klein, whose work he did not quote. But, as proved by the letter from Heisenberg of May 26, 1926, mentioned in Chapter 2, Dirac already knew about the relativistic equation in May. The substitutions given by equation (3.13) were due to Schrödinger, who stated them in a slightly different form in Schrödinger (1926B).

8 Dirac (1926F).

9 Ibid., 507.

10 "Dirac has given some thoughts to the theory of relativity, which in effect leads back to Klein's fifth dimension" [Heisenberg to Pauli, November 4, 1926 (PB1, 352)].

11 Ehrenfest to Dirac, October 1, 1926 (AHQP). Most of the letter is in German.

12 *PQM*, 1st edn., p. 240.

13 Kudar to Dirac, December 21, 1926 (AHQP). In a letter of Janaury 20, 1927, Kudar and Gordon referred to a letter in which Dirac probably replied to Kudar's December letter: "We have studied your letter with great interest. We believe that through your accomplishments, the matrix problem in the theory of relativity has now been solved" [Gordon and Kudar to Dirac (AHQP)]. I have not found Dirac's letter.

14 Pauli to Schrödinger, November 22, 1926 (PB1, 356).

15 Heisenberg (1927), 184.

16 Heisenberg to Pauli, February 5, 1927 (PB1, 374). Pauli apparently made a similar bet with Kramers about the same time, arguing that it would not be possible to construct a relativistic spin quantum theory; see Dresden (1987), 64. In the AHQP interview of May 10, 1963, Dirac had forgotten all about his own bet, which he found "unlikely" ever to have taken place.

17 A couple of years later, Heisenberg told Carl-Friedrich von Weiszäcker, who was then one of his students, about "a young Englishman by the name of Dirac ... [who] is so clever [that] ... it's of no use to work in competition with him" [Weiszäcker interview, June 9, 1963 (AHQP)].

18 Dirac (1977B), 138. No independent sources have been found to support Dirac's claim.

19 Pauli (1927).

20 Darwin (1927).

21 Pauli (1927), 619.

22 Dirac, "Lecture Notes on Modern Quantum Mechanics," unpublished manuscript, 115 pp. (AHQP); undated but probably from early fall 1927. The notes contain a detailed and original treatment of spin quantum theory, including applications to the anomalous Zeeman effect and the fine structure of hydrogen.

23 Dirac (1978A), 15. See also Dirac (1977B), 141, and Dirac (1975E), 8. Dirac's accounts of the event are not entirely consistent. In the AHQP interview and also in a conversation with Mehra, he recalled his interrupted talk with Bohr as having taken place in Copenhagen [Mehra (1972), 44].

24 Slater reports that the stumbling block to Dirac's invention of his theory was the negative energies, which he did not know how to interpret; Dirac, says Slater, had the necessary time to think over the problem while traveling on the Trans-Siberian Railway [Slater (1975), 148]. However, there is nothing to Slater's story. Not only did the negative energies play no role in the creative phase of the theory, but the longest journey Dirac made during the period was from Brussels to Cambridge.

25 Darwin to Bohr, December 26, 1927 (AHQP).

26 Mott interview, 1963 (AHQP). See also Mott (1987), 231.

27 Dirac (1928A), 610. The following account relies on Kragh (1981A). See also Moyer (1981A) and Dirac (1977D).

28 Dirac (1928A), 612.

29 Dirac matrices are closely related to the so-called Clifford algebra, used as early as 1884 [Lipschitz (1884)]. For this reference I am indebted to B. L. Van der Waerden.

30 Dirac (1977B), 142. In his later years, Dirac often stressed his peculiar way of producing theories in physics. An example: "A great deal of my work is just playing with equations and seeing what they give. . . . I don't suppose that applies so much to other physicists; I think it's a peculiarity of myself that I like to play about with equations, just looking for beautiful mathematical relations which maybe can't have any physical meaning at all. Sometimes they do" [Dirac interview, 1963 (AHQP)].

31 Equation (3.21) is the so-called Weyl equation, first published in Weyl (1929) and later shown to be applicable to the neutrino. In 1982, Dirac said that he constructed this equation in 1927, before he got the idea of introducing 4×4 matrices [Dirac (1982B), 603].

32 Dirac (1977B), 142.

33 See Wigner (1969).

34 Darwin (1928), 664. The empirical method of Darwin, and also of Pauli, was "to proceed by empirically constructing a pair of equations to represent the fine-structure of the hydrogen spectrum" [Darwin (1927), 230].

35 Dirac (1928A), 619. A second term, corresponding to an electric moment, also came out, but Dirac dismissed it as unphysical because it was imaginary. In

1935, Andrew Lees, a student of Dirac, showed that in a more rigorous treatment the electric moment does not appear [Lees (1935)].

36 Dirac (1977B), 139.

37 Dirac (1971B), 42.

38 According to an interview with Rosenfeld, July 1963, conducted by T. S. Kuhn and J. L. Heilbron (AHQP).

39 According to Wigner (1973), 320.

40 For Kramers's equation, see Kramers (1934) and Kramers (1957), paragraph 64. According to his biographer, Kramers completed his derivation only a few weeks after Dirac submitted his paper [Dresden (1987), 316]. For the unpublished Jordan–Wigner attempt, see Wigner (1973) and also Birtwistle (1928), 213: "Pauli has recently done some work on the application of quantum mechanics to the spinning electron, which is now being extended by Jordan so as to include relativity." The preface to Birtwistle's book was dated October 1, 1927. The results of the Russian physicists are Frenkel (1928) and Iwanenko and Landau (1928).

41 The remark was made in a conversation with E. J. Eliezer in 1947; see Eliezer (1987), 60.

42 Fowler to Bohr, January 5, 1928 (BSC); Darwin to Pauli, January 11, 1928 (PB1, 424); Bohr to Dirac, February 27, 1928 (BSC). For the impact of Dirac's letter to Born, see Wigner (1973), 320.

43 Rosenfeld interview, 1963 (AHQP). See also V. A. Fock in an interview of October 11, 1967, conducted by T. S. Kuhn (AIP): "It [the Dirac equation] was quite new for me . . . I found that it was quite, in the German sense *genial*."

44 Heisenberg to Dirac, February 13, 1928 (CC).

45 Ehrenfest to Joffe, April 13, 1928, as quoted in Feinberg (1987), 168.

46 Gordon to Dirac, January 15, 1928 (CC). Gordon first reported his results to Dirac in a letter of January 13 (CC).

47 Darwin to Bohr, March 31, 1923 (BSC).

48 The agreement between Sommerfeld's and Dirac's results, separated by thirteen years and a revolution in physics, has usually been ascribed to coincidence. For example, commenting in 1932 on what they assumed was a chance canceling of two errors in Sommerfeld's theory, Otto Halpern and Hans Thirring called this a "highly remarkable result, which reminds us of the dramatic climaxes of the stories by Jules Verne" [Halpern and Thirring (1932), 82]. Only recently has it been argued that this is not the case and that the puzzle can be explained through the underlying symmetry that the two theories share; see Biedenharn (1983).

49 Dirac to Jordan, January 23, 1928 (AHQP).

50 Dirac (1928C). See also Dirac (1928D).

51 Dirac (1928D), 561.

52 Darwin (1928).

53 At the 1930 Solvay Congress, Pauli lectured on the magnetic electron and concluded that spin effects would not be observable for free electrons. In the discussion that followed the lecture, Bohr argued that "One can see in a very general way that it is impossible experimentally to detect the intrinsic magnetic moment of the electron." Dirac's only comment in the discussion was, "I agree with Bohr in a general way . . ." (*Le Magnetisme,* 6th Solvay Congress, Paris: Gauthier-Villars, 1932, p. 239). For details, see Darrigol (1984B) and Franklin (1986), 39–72.

54 Dymond (1934), 657.

55 After more than a decade's discussion, it turned out that the experiments, not the theory, were wrong. During the World War II, improved experiments vindicated the Mott effect and then supplied Dirac's theory with yet another success.

56 Houston and Hsieh (1934), 271. They suggested that the discrepancy might be due to neglect of the interaction between the electron and the radiation field. Other researchers tried to account for the discrepancy by suggesting a breakdown of the Coulomb law for very small distances, and some experiments were interpreted to be in complete support of Dirac's predictions. It was only in 1947 that the discrepancies between the observed values and those predicted by the Dirac theory were conclusively verified by W. Lamb and R. Retherford, and it was shown that the Dirac equation was in need of correction. The case of fine structure measurements in the thirties is discussed in Morrison (1986) and in Aramaki (1987).

57 "Pauli told me that Stern has measured the moment of the proton and found c. $3(h/4\pi)$ (e/Mc). That would mean, then, that the Dirac theory is not applicable to the proton, a result Pauli is very happy about" [F. Bloch to Bohr, December 30, 1932 (BSC), trans. from German].

58 "The Electron and the Proton," unpublished manuscript, undated but from 1933 or 1934; reproduced in Bohr (1972+), vol. 9, pp. 123–7.

59 For details of the Meitner–Hupfeld effect, see Brown and Moyer (1984).

60 Dirac (1928C), 94. Exactly what Dirac meant by referring to the asymmetry of the time is unknown. The remark may have been inspired by Bohr's contemporary speculations about relating time asymmetry with the irreversibility of the quantum mechanical measurement process. In a letter of March 24, 1928, Bohr mentioned to Dirac that "I have been very interested lately in the question of the uni-direction of time" (AHQP). One may speculate that Dirac played with the idea that negative-energy particles formally correspond to positive-energy particles running backwards in time, an idea which was developed much later by Stueckelberg and Feynman. For this suggestion, which unfortunately lacks documentation, I am indebted to Kurt Gottfried. Alternatively, Dirac might have had in mind unified-field theories, such as Weyl's, which he had studied. In this theory the asymmetry between positive and negative charges was sometimes believed to have its roots in the asymmetry between past and future; see Weyl to Pauli, May 10, 1919 (PB1, 3).

61 Heisenberg to Jordan, June 25, 1928 (AHQP); the German original is in Brown and Rechenberg (1987), 152.

62 Heisenberg to Bohr, July 23, 1928 (AHQP). See also Heisenberg to Bohr, March 31, 1928 (AHQP).

Chapter 4

1 Bohr to Dirac, March 24, 1928 (BSC); Dirac to Ehrenfest, April 24, 1928 (EA).

2 Casimir (1983), 73. See also Casimir (1985).

3 Tamm to relative, March 4, 1928, quoted in Feinberg (1987), 300. Dirac did not arrive in Leiden until April 28.

4 Tamm to relative, undated [Feinberg (1987), 301].

5 Dirac to Tamm, January 3, 1929 (TDC). In early 1930, when Gamow stayed in Cambridge, Dirac also learned to ride a motorcycle, which Gamow had bought.
6 Tamm to V. I. Yakovleva, October 4, 1928, quoted in Feinberg (1987), 302.
7 L. J. Mordell to Dirac, July 4, 1928 (CC).
8 Draft for an answer to Manchester University, written in pencil on the back of Mordell's letter of July 4 (CC).
9 A. Compton to Dirac, September 14, 1928 (CC). Compton had earlier invited Dirac to lecture at the University of Chicago during the fall and winter of 1928–9, but Dirac did not accept the invitation [Compton to Dirac (telegram), February 16, 1928 (CC)].
10 Dirac to Klein, July 24, 1928 (AHQP).
11 For an evaluation of Soviet physics in the period, see Josephson (1988).
12 See Joffe (1967), which includes, on p. 114, a photograph from the congress, showing Dirac in conversation with Lewis. See also Born (1928).
13 Dirac to Tamm, November 4, 1928 (TDC).
14 Birge to Van Vleck, March 10, 1927, quoted in Schweber (1986B), 55.
15 Note 9 and K. T. Compton to Dirac, 2 January 1929; cf. Schweber (1986B), 56.
16 Dirac to Veblen, March 21, 1929; Veblen to Dirac, March 22, 1929 (LC).
17 Dirac to Van Vleck, December 6, 1928 (AHQP).
18 See Van Vleck (1972).
19 *Wisconsin State Journal,* April 31, 1929, quoted from typewritten copy kept at the Churchill College Archive, Cambridge.
20 Heisenberg to Dirac, undated, quoted in Brown and Rechenberg (1987), 134.
21 Dirac to Van Vleck, June 9, 1929 (AHQP), sent from Glacier Point Hotel in Yosemite National Park.
22 Heisenberg to Dirac, February 13, 1928 (CC). Dirac mentioned his travel plans in a letter to Tamm of January 3, 1929 (TDC).
23 See excerpts of letters from Heisenberg to Dirac (May 24, 1929, June 17, 1929, July 5, 1929, and July 11, 1929), as quoted in Brown and Rechenberg (1987).
24 Van Vleck (1972), 11.
25 Dirac in an unpublished talk of 1969, quoted in Mehra (1972), 57.
26 Dirac to A. Pais, October 21, 1982, quoted in Pais (1987), 99.
27 Nishina to Bohr, November 25, 1929 (BSC).
28 Dirac and Heisenberg (1932D).
29 Dirac to Tamm, September 12, 1929 (TDC).
30 Nishina to Dirac, July 16, 1929, and July 30, 1929 (CC). Tamm helped Dirac by arranging his travel from Japan to Moscow and also arranging a visa for him [Tamm to Dirac, June 28, 1929, and July 19, 1929 (TDC)].
31 Dirac, notes for seminar talks on the London-Heitler theory, May 1931 and February 1932 (CC). The Dirac collection at Churchill College (now at Florida State University) contains many sheets of notes, intended either for seminars or for personal use, on current developments in solid-state physics, nuclear theory, and other subjects outside Dirac's expertise. His lectures to the Kapitza Club during the period included "String and Scissors Problem to Illustrate that Two Rotations Are Not Equivalent to One (July 7, 1931), "Solvay Congress. The Curie

Joliot Measurement of the Neutron. J. D. Cockcroft. Lewis and Adam's Theory" (November 14, 1933), and "Nascent Electricity" (July 17, 1934).

32 Dirac (1929A).

33 Dirac (1929B and 1930C).

34 Van Vleck (1934B).

35 Slater (1929).

36 Dirac, notes for "Lectures on Modern Quantum Mechanics" (AHQP).

37 Crowther (1970B), 39.

38 Dirac to Tamm, January 3, 1929 (TDC). A year later Dirac had completed the book and was dealing with the galley proofs [Dirac to Tamm, February 21, 1930 (TDC)]. The "Frenkel" as a unit of writing speed is an allusion to the Russian physicist Yakov Frenkel, who wrote and published books very quickly.

39 Van Vleck to Dirac, April 13, 1930 (CC). At the time, Van Vleck was preparing another monograph in the series edited by Fowler and Kapitza [Van Vleck (1932)] and was given the proofs of Dirac's work as an example of the typography of Oxford University Press.

40 First edition, 1930; second edition, 1935; third edition, 1947; fourth edition, 1958 (revised in 1967 and reprinted in 1984). A German translation appeared in 1930, French in 1931, Russian in 1932, and Japanese in 1936.

41 Quoted in Weiner (1972), 131.

42 Heisenberg to Dirac, February 27, 1958 (CC); the German original is given in Brown and Rechenberg (1987), 157.

43 *PQM*, 1st edn., p. vi.

44 *PQM*, 1st edn., p. 18.

45 *PQM*, Russian translation, Moscow, 1932, publisher's foreword.

46 Einstein (1931), 73.

47 Eddington (1928), 210. Eddington emphasized the same virtues in a review of *PQM* in which he hailed Dirac's approach as a victory of "pure thought." See *Nature*, 1931, p. 699; the review was anonymous, but its style reveals Eddington as its author.

48 *PQM*, 4th edn., p. 10.

49 Pauli (1931).

50 Fokker interview, 1963 (AHQP); interview with Uhlenbeck, March 1962, conducted by T. S. Kuhn (AHQP).

51 Bloch (1931) and Oppenheimer (1931). "It is clear," wrote Oppenheimer, "with a clarity dangerous for a beginner, deductive, and in its foundation abstract; its argument is predominantly analytical . . ., physical ideas are seldom used to advance the argument, and occur chiefly as an aid to exposition. The book remains a difficult book, and one suited only to those who come to it with some familiarity with the theory. It should not be the sole text, nor the first text, in quantum theory, just as that of Gibbs' should not be the first in statistical mechanics." Other reviews of *PQM* include F. Rasetti in *Scientia 51*, 1932, p. 371, and P. Frank in *Zeitschrift für angewandte Mathematik und Mechanik 13*, 1933, p. 63.

52 Darwin (1935), 411.

53 Dirac, notes for "Lectures on Modern Quantum Mechanics" (AHQP).

54 *PQM*, 2nd edn., p. 4.

55 Ibid.
56 *PQM*, 4th edn. p. 4.
57 Born (1936/37), 12.
58 Einstein to Ehrenfest, August 23, 1926 and August 28, 1926, quoted in Pais (1982), 441.
59 Dirac (1982C), 84–5.
60 Dirac (1927E), 261 (Dirac's emphasis).
61 Ibid., 262.
62 Ibid.
63 See De Maria and La Teana (1982).
64 *PQM*, 1st edn., pp. iv–v.
65 Heisenberg 1969, 98.
66 The reference is to Bohr's favorite verse, at the end of Schiller's poem "Sprüchen des Konfuzius," which reads: "Nur Beharrung führt zum Ziel / nur die Fülle führt zur Klarheit / und im Abgrund wohnt die Wahrheit."
67 Dirac interview, 1963 (AHQP).
68 See Balibar (1985).
69 Bohr (1928).
70 Bohr to Dirac, March 24, 1928 (BSC).
71 Dirac (1951A), 17.
72 Dirac to Bohr, November 30, 1930 (BSC).

Chapter 5

1 Dirac (1928A), 612. Before Dirac addressed it, the problem of the negative energies was, as far as I know, mentioned only once, and then only casually. Klein mentioned in a footnote that the Klein–Gordon equation yields "a class of solutions for which the energies turn out to be negative and which have no direct connection to the motion of the electrons. . . . *Naturally we will leave these out of consideration.*" [Klein (1927), 411, emphasis added].
2 Dirac (1977E), 4.
3 Heisenberg to Pauli, July 31, 1928 (PB1, 468).
4 Jordan (1928), 206.
5 Heisenberg to Pauli, July 31, 1928 (PB1, 467). See also Heisenberg to Bohr, July 23, 1928 (AHQP). In his Leipzig lecture Dirac had stated without proof that the transition probability would be of the order $(v/c)^4$, where v is the speed of the electron.
6 Dirac to Klein, July 24, 1928 (AHQP).
7 Klein (1929).
8 Klein and Nishina (1929).
9 Waller (1930).
10 Bohr to Nishina, January 26, 1934 (BSC).
11 Weyl (1929B), 332. The same suggestion was made a little earlier in Weyl (1929A) and was made independently in January 1930 by the Leningrad physicist H. Mandel, who based his idea on five-dimensional Kaluza–Klein theory [see Mandel (1930)].

12 Heisenberg to Dirac, undated but late March 1929, as quoted in Brown and Rechenberg (1987), 134. Heisenberg also mentioned that he had recently read Eddington (1929), in which the Cambridge astronomer attempted to deduce the value of the fine structure constant from Dirac's wave equation. "I cannot find anything reasonable in it," Heisenberg wrote, "but perhaps you understand the deeper reasons and can tell me about them." Heisenberg's skepticism toward Eddington's work was shared by most physicists. Pauli considered it to be "complete madness . . . , for romantic poetry and not for physics" [Pauli to Klein, February 18, 1929 (PB1, 491)].

13 Bohr to Dirac, November 24, 1929 (AHQP). Bohr felt at the time that the mystery of the continuous β-spectrum necessitated that energy conservation be abandoned in nuclear processes. See Bohr (1932), Proceedings of Volta Congress (Rome), October 11–18, 1931.

14 Dirac to Bohr, November 26, 1929 (AHQP).

15 Bohr to Dirac, December 5, 1929 (AHQP). The letter is reproduced in full in Moyer (1981B), 1057.

16 Ibid.

17 Dirac to Bohr, December 9, 1929 (AHQP). The papers by Pokrowski and Waller, mentioned by Dirac, are Pokrowski (1929) and Waller (1930). As remarked by Pais, Dirac's letter to Bohr announced a monumental novelty in physical theory since it described for the first time the idea of virtual electron-positron pair formation (although the positron was still not introduced). Dirac was on the verge of realizing that the simple problem of the scattering of a photon on an electron involves infinitely many particles. See Pais (1986), 350.

18 Casimir (1983), 325.

19 Dirac (1931A).

20 Heisenberg to Dirac, December 7, 1929 (CC); Jordan to Dirac, December 4, 1929 (CC); and Pauli to Jordan, November 30, 1929 (PB1, 525). According to Gamow's recollections, he sat in the Maxwell Library of the Cavendish when Dirac approached him and, uninvited, told him about his new theory – a most remarkable thing for Dirac to do [interview with G. Gamow, April 25, 1968, conducted by C. Weiner (AIP)].

21 The typewritten copy is in the Ehrenfest Archive.

22 Iwanenko to Dirac, December 11, 1929; Tamm to Dirac, February 5, 1930; and Fock to Dirac, February 12, 1930, and March 1, 1930 (CC).

23 Heisenberg to Dirac, December 7, 1929 (CC).

24 Heisenberg to Bohr, December 20, 1929 (AHQP); original in German.

25 Heisenberg to Dirac, January 16, 1930 (CC).

26 Dirac (1930A), 361.

27 Ibid.

28 Ibid., 362.

29 Ibid.

30 For a description of the problem and Dirac's alleged solution, see Singh (1970), 207.

31 Dirac (1930A), 363. See also Dirac (1977B), 144.

32 Dirac (1977B), 144.

33 Dirac (1927B), 260. The analogy between the hole theory and the radiation theory is pointed out in Bromberg (1976), 187. Dirac's radiation theory is outlined in Chapter 6 of the present work.

34 Dirac (1930A), 363.

35 Dirac interview, 1962 (AHQP), and Dirac (1977B), 144.

36 Dirac interview, 1962 (AHQP).

37 See the letter to Bohr of November 26, quoted in the text (note 14).

38 Dirac said on several occasions that his disinclination to propose a new particle was due to a "lack of boldness" [e.g., Dirac (1969A) and Dirac (1978A), 17].

39 Dirac (1930D), 605.

40 "New Atomic Theory Given," *New York Times,* September 11, 1930.

41 Thomson (1907). In Dirac's theory the electron was not really a more fundamental particle than the proton. Because of the symmetry, the two particles were equally fundamental, both being a manifestation of the same basic substratum.

42 Dirac (1930B).

43 Dirac (1930D), 606. See also *PQM,* 1st edn., p. 257: "Such processes probably actually occur in Nature."

44 See Bromberg (1976). The notion of electron–proton annihilation was also a part of Oliver Lodge's attempt to revive the old ether theory, at that time far from mainstream physics [Lodge (1925), 95]. The early role of annihilation processes in cosmic ray research is considered in Galison (1983), 264–9.

45 Tolman (1930).

46 Jeans (1931), 110.

47 Dirac did not use the term "anti-particle," which entered the vocabulary of physics only in the mid-1930s. But he invented the term "anti-electron," which first appeared in Dirac (1931C), 61.

48 Dirac (1930A), 364. Eddington sought to generalize the Dirac equation and interpret it in an alternative way [Eddington (1929)]. He claimed to have deduced the numerical value of the fine structure constant and believed that it was also possible to deduce the value of other physical constants from his version of Dirac's theory. Dirac studied Eddington intensively but did not accept his point of view. His few references to Eddington merely reflect that at the time he was willing to consider almost anything in order to keep his theory alive.

49 Dirac (1930B), 361.

50 Tamm (1930); Tamm to Dirac, March 3, 1930 (CC); and Tamm to Ehrenfest, February 24, 1930 (EA). Tamm at first believed that he had derived from Dirac's theory a scattering formula different from the Klein–Nishina formula, but he was informed by Dirac that he was wrong.

51 Weiszäcker interview, 1963 (AHQP). Walter Heitler recalled a heated discussion in Copenhagen in which Bohr described the hole theory as madness to Dirac's face [interview with Heitler, March 1963, conducted by J. L. Heilbron (AHQP)].

52 Delbrück (1972), 281. According to Delbrück, Bohr was not informed of the code and hence was puzzled to receive the telegram until it was deciphered by Gamow.

53 Oppenheimer (1930A and 1930B). Oppenheimer's result for the mean life was incorrect by a factor $(2\pi)^4$. He was known as a rather careless physicist when it came to arithmetic.

54 Oppenheimer (1930A), 562. See also Oppenheimer (1930C).

55 Dirac (1930D), 606.

56 Iwanenko and Ambarzumian (1930). Gamow also tried to use Dirac's hypothesis in explaining β-decay, but his attempt never reached the stage of publication [Gamow to Bohr, February 25, 1930 (AHQP)].

57 Tamm to Dirac, September 13, 1930 (TDC).

58 Weyl (1931), 263. The preface of the German edition, which Dirac read, is dated November 1930. Weyl found Dirac's theory "attractive" but noted as a flaw that "according to it the mass of a proton should be the same as the mass of an electron" (p. 263). Referring to Dirac's theory of the proton, Weyl stated in the preface, "I fear that the clouds hanging over this part of the subject will roll together to form a new crisis in quantum physics."

59 Dirac (1931C), 61.

60 Dirac (1971B), 55.

61 Dirac (1931C), 61.

62 Ibid., 62.

63 Dirac (1933F), 325.

64 Placintéanu (1933), Fürth (1933), and Gamow (1934). Not all speculations about negative protons identified them with anti-protons in Dirac's sense. For example, Gamow introduced negative protons as nuclear constituents, but he believed that neither the proton nor its negative counterpart would be described by Dirac's equation [Gamow (1935B)]. On June 11, 1934, Gamow gave a talk on "Negative Protons" to the Kapitza Club.

65 Mohorovicic (1934) suggested that positron–electron atoms existed in stellar atmospheres, as the constituents of a new element which he named "electrum." The name "positronium" was first suggested by Arthur Ruark in 1945.

66 The question of a possible relationship between the two concepts of the hole is considered in Hoddeson, Baym, and Eckert (1987), 294, which also contains references to the solid-state literature of the period.

67 Dirac to Tamm, July 9, 1930 (TDC).

68 Mott (1986), 42.

69 Tamm to Ehrenfest, February 23, 1931 (EA), and Feinberg (1987), 278.

70 Dirac to Van Vleck, April 24, 1931 (AHQP).

71 Veblen to Dirac, May 12, 1930; Dirac to Veblen, May 22, 1930; and Veblen to Dirac, June 3, 1930 (LC).

72 Veblen to Dirac, December 2, 1930 (LC).

73 Van Vleck (1972), 12.

74 Dirac to Van Vleck, October 2, 1931 (AHQP). For the remarks on "neutrons and magnetic poles," see Chapter 10. Wigner's result, referred to by Dirac, was published the following year; see Wigner (1932). The book by Van Vleck, also mentioned in the letter, is Van Vleck (1932).

75 Fellows (1985), 247.

76 "Lectures on Quantum Mechanics," Princeton University, 1931, 134 pp., mimeographed (CC); based on notes taken by Banesh Hoffmann.

77 Dirac to Tamm, January 21, 1932 (TDC).
78 E. F. Burton to Dirac, February 20, 1932 (CC).
79 See Anderson (1961), Hanson (1963), Brown and Hoddeson (1983), and De Maria and Russo (1985). The role of Dirac's theory is discussed in Moyer (1981C) and Dorfman (1981).
80 See Bromberg (1975). See also Nernst (1907), who suggested the existence of "neutrons," massless composites of positive and negative electrons. Nernst's neutrons filled the entire space, making it "weightless and electrically non-conducting, but electrically polarizable; that is, [they] must possess such properties which physics normally only ascribes to the light aether" (p. 392). Eight years earlier, William Sutherland had proposed a similar idea, including the name "neutron," which appeared first in Sutherland (1898). The similarity is neither deep-rooted nor historically important. Dirac was not inspired by Reynolds, Nernst, or other ether theorists.
81 Pearson (1891) and Schuster (1898). I cannot refrain from quoting Schuster, who wrote more than thirty years before Dirac's crowded quantum vacuum: "When the atom and anti-atom unite, is it gravity only that is neutralized, or inertia also? May there not be, in fact, potential matter as well as potential energy? And if that is the case, can we imagine a vast expanse, without motion or mass, filled with this primordial mixture, which we cannot call a substance because it possesses none of the attributes which characterise matter ready to be called into life by the creative spark? Was this the beginning of the world?" (p. 367).
82 See, e.g., Lodge (1906), 148; Larmor (1894); and Lenard (1909). During the Nazi reign, when Lenard became a leader of the alternative "Aryan Physics" movement, his early writings were sometimes quoted as a prediction of the positron!
83 Nernst (1907), 393; Becquerel (1908); and Wood (1908).
84 But not completely. Thus the imaginative Oliver Lodge suggested in 1922 that positive electrons did indeed exist but were confined in the proton [Lodge (1922)].
85 Pauli (1919B) and Einstein (1925). Pauli did not explicitly mention positive electrons in 1919, but he showed that in Weyl's unified theory negative and positive charges appeared completely symmetrically.
86 Anderson (1932).
87 Anderson (1933).
88 Anderson (1961), 825.
89 Blackett and Occhialini (1933).
90 Blackett was at the Cavendish Laboratory and often met Dirac. The climate among the Cavendish experimentalists was more receptive to theory than that in Millikan's group [interview with P. M. S. Blackett, December 1962, conducted by T. S. Kuhn (AHQP)]. In his recollections, such as Dirac (1978G) and Dirac (1984A), Dirac recalled that Blackett had photographs of positrons earlier than Anderson but that he was too cautious to publish his evidence until after Anderson's announcement. See also Dirac to Alvarez, February 3, 1978, as quoted in Chapter 10. However, as noted in De Maria and Russo (1985), Dirac's recollection is almost certainly incorrect on this point.

91 Blackett and Occhialini (1933), 716. In his Solvay address in October of the same year, Dirac reported that calculations of the lifetime of positrons in air yielded a mean lifetime of 3×10^{-7} seconds.
92 Dirac interview, 1962 (AHQP).
93 Birkhoff to Kemble, March 9, 1933 (AHQP). Other parts of the letter are reproduced in Chapter 14.
94 Kemble to Birkhoff, March 27, 1933 (AHQP). See also Chapter 14.
95 For references and further information, see Pais (1986).
96 Schrödinger (1931A and 1931B).
97 Dirac, Notes on Schrödinger's Theory, undated but 1931 (CC).
98 Examples of misunderstandings of Dirac's theory, including the claim that it postulated the real existence of negative-energy particles, may be found in Halpern and Thirring (1932), 101, Langer (1932), and Placintéanu (1933).
99 Tamm to Dirac, June 5, 1933 (TDC).
100 Landau and Peierls (1931), 68.
101 Ibid., 57. Bohr found Landau and Peierls's interpretation of his views on quantum mechanics to be much too extreme, in particular because their interpretation implied that electromagnetic field strengths or potentials were not measurable; he had no objection to their criticism of current relativistic quantum theory.
102 Condon and Mack (1930), 582.
103 "Blegdamsvejen Faust," 1932 (AHQP). The prime mover behind the project seems to have been Delbrück. The play is reproduced in English translation in Gamow (1966), and in its original language, mostly German, in the Danish translation of Gamow's book.
104 Max Delbrück to T. S. Kuhn, March 13, 1962 (AHQP). As late as the Solvay Conference, October 22–29, Bohr warned against identifying the positive electron with the Dirac anti-electron. In a discussion following Frederic Joliot and Irène Joliot-Curie's report on artificial radioactivity, Bohr stressed that the best thing to do was "to draw as many conclusions as possible from the positive–electron experiments without relying on the Dirac theory" [Bohr (1972T), vol. 9, p. 136].
105 Pauli to Dirac, May 1, 1933 (PB2, 159).
106 Pauli to Blackett, April 19, 1933 (PB2, 154). The idea of the neutrino as a positron–electron pair would seem to contradict conservation of spin. But Pauli believed at that time that the positron might be a boson, in which case there would be no violation of conservation laws. The Dirac positron was, of course, a fermion. The idea of "Bose positrons" was suggested by several physicists in 1933 [Darrigol (1988A), 256–9].
107 Pauli and Solomon (1932).
108 Pauli (1933), 246. The text has "γ-*Strahl-Protonen*," which is obviously a misprint.
109 Pauli to Heisenberg, June 14, 1934 (PB2, 329).
110 Quoted from Moyer (1981C), 1123.
111 Heisenberg (1973), 271.
112 Pauli to Heisenberg, June 14, 1934 (PB2, 329).
113 Ibid. The Pauli–Weisskopf "anti-Dirac theory" was presented in Pauli and Weisskopf (1934). Pauli's feelings against Dirac's physics were exhibited even

more strongly by Kramers. For example, Dresden (1987) mentions that Kramers always referred to the "Dirac–Heisenberg hole theory" and the "Dirac–Pauli matrices," rather than attributing them to Dirac alone (p. 481).

114 Oppenheimer and Furry (1934). Oppenheimer sent preprints of the work to Pauli, Bohr, and Dirac; while Pauli and Bohr responded in a friendly way to the ideas, Oppenheimer remarked that "from Dirac we have not had a murmur" [Oppenheimer to Uhlenbeck, March 1934, quoted in Smith and Weiner (1980), 175].

115 Fermi (1934), quoted in English translation in Strachan (1969), 108.

116 Beck and Sitte (1933) proposed that an electron–positron pair was created just outside the nucleus, the positron being absorbed by the nucleus and the electron leaving as a β-particle. The Beck–Sitte theory did not involve neutrinos. Fermi's renunciation of the hole theory may have been rooted in his wish to stress the difference between his theory and the Beck–Sitte theory.

117 Wick (1934).

118 Peierls to Bethe, July 24, 1933, trans. from German (Bethe papers, Cornell University Archive).

119 Peierls to Bethe, August 14, 1933, trans. from German (Bethe papers, Cornell University Archive).

120 Thirring to Nobel Committee, January 29, 1929 (NA). In 1928, Einstein argued that the founders of quantum mechanics should be rewarded with a Nobel Prize and mentioned de Broglie, Schrödinger, Heisenberg, Born, and Jordan as possible candidates. In 1931, he proposed Schrödinger and Heisenberg, and the following year again Schrödinger. Notably, Einstein never mentioned Dirac to the Nobel Committee. For the rules of Nobel nominations and a complete census of nominators and nominees, see Crawford, Heilbron, and Ullrich (1987).

121 Nobel Committee 1929, trans. from Swedish (NA).

122 Bragg to Nobel Committee, December 9, 1932 (NA). Dirac was also nominated as a second choice by the Polish physicist Czeslaw Bialobrzeski, who listed Paul Langevin and Robert Wood as his first choices.

123 Oseen, Memorandum on Dirac, March 1933, trans. from Swedish (NA).

124 Ibid.

125 Peierls (1987), 37.

126 *Sunday Dispatch,* November 19, 1933.

127 Bohr to Dirac, November 22, 1933 (AHQP).

128 Dirac to Bohr November 28, 1933 (AHQP).

129 Dirac to Bohr, December 20, 1933 (AHQP).

Chapter 6

1 For the historical development of quantum electrodynamics, see in particular Darrigol (1984A) and Pais (1986), 328–96; and for the early development, Darrigol (1986). Other useful works include Wentzel (1960), Jost (1972), Schweber (1986A), and Weinberg (1977). Schwinger (1958) provides a collection of some of the classic papers of quantum electrodynamics, starting with Dirac's paper of 1927.

2 Einstein (1909).

3 Einstein (1917). See also Einstein (1916).

4 Einstein (1917), 124.

5 Jordan's dissertation dealt with the quantum theory of radiation; see Jordan (1924). In the Born–Jordan and Born–Heisenberg–Jordan papers, it was Jordan who was responsible for the parts dealing with quantization of radiation. See van der Waerden (1967), 40, and Heisenberg to Pauli, October 23, 1925 (PB1, 102).

6 Born and Jordan (1925) and Born, Heisenberg, and Jordan (1925).

7 Born, Jordan, and Heisenberg (1925), 615. See Darrigol (1986).

8 Dirac (1927B).

9 Dirac (1926E), 677.

10 Slater (1927).

11 Dirac (1977B), 151.

12 Dirac interview, 1963 (AHQP).

13 The term second quantization was first used in 1932 by Jordan and Fock; see Jordan (1932) and Fock (1932).

14 Dirac (1927B), 260–1.

15 The subscript s denotes the state to which the electron is scattered from state r. A couple of months later, Dirac pointed out that n should not be the number of photons in the initial state, but the number of photons existing after the process. However, the error was of no consequence for the results he obtained [Dirac (1927), 711].

16 Pauli (1923).

18 Dirac (1927B), 245.

19 Slater (1975), 135; and the theory in Slater (1927).

20 Kramers and Heisenberg (1925); Heisenberg (1925); Schrödinger (1926B); Gordon (1927); Bursian (1927); Klein (1927).

21 Dirac (1927C), 710.

22 Dirac to Bohr, February 19, 1927 (BSC).

23 Weisskopf and Wigner (1930).

24 Klein (1927).

25 Klein to Dirac, March 24, 1927 (AHQP). Klein interview, 1963 (AHQP). Dirac's "curious letter" seems to be lost.

26 Dirac (1927C), 712.

27 Wentzel (1960), 49.

28 Dirac (1927C), 720.

29 Dirac (1983B), 48.

30 Fermi (1932).

31 Slater (1975), 135.

32 Heisenberg to Pauli, February 23, 1927 (PB1, 376). Pauli's letter is lost. Heisenberg, who was at the time concentrating on the measurement problems of quantum mechanics and the still not fully hatched uncertainty principle, was not yet ready to join forces with Pauli. In May, he wrote: "Now I really want to start familiarizing myself with Dirac's work. . . . I still have no definite opinion about the Hamiltonian interaction function between radiation and matter. I need to study Dirac first" [Heisenberg to Pauli, May 16, 1927 (PB1, 395)].

33 Jordan and Klein (1927).

34 Jordan (1927B) and Jordan and Wigner (1928).

35 Pauli to Dirac, February 17, 1928 (PB1, 435).
36 Dirac (1927E), 270.
37 Dirac (1983B), 49–50. See also Dirac (1977B), 140, where he attributed his dislike of the Jordan–Wigner theory to his mind "being essentially a geometrical one and not an algebraic one."
38 Jordan and Pauli (1928).
39 Jordan (1927B), 473. See also Cini (1982).
40 Cini (1982), 232, and also Darrigol (1986). Dirac always kept a relaxed attitude to the wave–particle problem. In 1982, he expressed his opinion in terms much like those he used fifty-five years earlier: "One can treat light as composed of electromagnetic waves, each wave to be treated like an oscillator; alternatively, one can treat light as composed of photons, the photons being bosons and each photon state corresponding to one of the oscillators of the electromagnetic field. One then has the reconciliation of the wave and corpuscular theories of light. They are just two mathematical descriptions of the same physical reality" [Dirac (1983B), 49].
41 Pauli to Dirac, February 17, 1928 (PB1, 437). See also Heisenberg to Dirac, February 13, 1928 (CC).
42 Ibid.
43 Heisenberg and Pauli (1929 and 1930).
44 Jordan (1929), 711. The infinities of the Heisenberg–Pauli theory also caused Oppenheimer to worry, and he pessimistically concluded: "It appears improbable that the difficulties . . . will be soluble without an adequate theory of the masses of the electron and proton; nor is it certain that such a theory will be possible on the basis of the special theory of relativity" [Oppenheimer (1930A), 461].
45 Pauli to Dirac, February 17, 1928 (PB1, 436).
46 Ibid.
47 Dirac, notes on "Critical Exposition of Heisenberg–Pauli Paper", undated but late 1929, 8 pp. (CC). In *PQM* he referred to the Heisenberg–Pauli approach as follows: "This appears to be the most straightforward way of dealing with general dynamical systems on relativity lines, but it involves complicated mathematics and appears to be too difficult for practical application" (*PQM*, 1st edn., p. 238).
48 Dirac (1932B).
49 Ibid., 454.
50 Ibid., 457.
51 Stöckler (1984), 116, points out that Dirac's arguments were untenable, because it would be unwarranted to conclude, from the fact that fields are necessary for the observation of particles, that fields have ontological priority. This is true but – since Dirac did not draw this ontological conclusion – historically irrelevant.
52 In his Nobel Prize memorandum, Oseen evaluated Dirac's theory as follows: "If the approach which Dirac has pointed to turns out to be current, a new great revolution in our worldview has occurred. A new and great displacement of the central point of physics from the world of matter to the world of radiation has then taken place" (NA, trans. from Swedish).
53 Dirac (1932B), 459.

54 Bohr to Dirac, November 14, 1932 (BSC). Bohr's letter was a much delayed reply to a letter from Dirac of May 6 in which he told Bohr, "I am continuing with my work on relativistic quantum mechanics and hope to get a satisfactory theory soon" (BSC). For the Landau–Peierls paper, see Chapter 5, note 100.

55 Klein interview, 1963 (AHQP).

56 Pauli to Meitner, May 29, 1932 (PB2, 114).

57 Pauli to Dirac, September 11, 1932 (PB2, 115). The opinion of Dirac's work was similarly low in Berlin: "There is nothing particularly new here in Berlin. People here are not enthusiastic about Dirac's work" [Wigner to Ortvay, May 30, 1932 (PB2, 115)].

58 Fock and Podolsky (1932A and 1932B) and Nikolsky (1932).

59 Rosenfeld (1932).

60 Dirac to Bohr, March 23, 1932 (BSC).

61 Dirac to Rosenfeld, May 6, 1932 (AHQP). Dirac's use of the phrase "Nur um zu lernen" referred to one of Bohr's favorite expressions, *nicht um zu kritisieren, nur um zu lernen* (not to criticize, just to learn), known to any Copenhagen visitor.

62 Fock and Podolsky (1932B).

63 Dirac, Fock, and Podolsky (1932C). The paper was mainly authored by Podolsky, who communicated with Dirac by letter. It was originally entitled "On Dirac's Form of Relativistic Quantum Mechanics," but Dirac objected that an author's name should not appear in the title of one of his papers [Dirac to Podolsky, November 2, 1932 (CC)]. See also Fock to Dirac, October 16, 1932 (CC), and Podolsky to Dirac, November 16, 1932 (CC).

64 Pauli to Dirac, April 13, 1933 (PB2, 161). Pauli wrote "approvement," which is obviously a linguistic error. He was informed of the content of the Dirac–Fock–Podolsky paper prior to its publication and discussed it in several letters with Dirac [Pauli to Dirac, November 10, 1932, March 10, 1933, and April 1, 1933 (PB2)].

65 Bloch (1934).

66 Tomonaga (1973). As a physics student, Tomonaga attended the lectures of Dirac and Heisenberg when they visited Tokyo in 1929. He published his covariant quantum field theory, based on a generalization of Dirac's many-time theory, in 1943, but it became known only in 1948, two years after it had been translated into English. For Dirac's influence on Tomonaga's theory, see Darrigol (1988B). Dirac recognized the importance of Tomonaga's formalism, which he discussed in a paper published in early 1948 [Dirac (1948B)].

67 Dirac (1933A).

68 Feynman (1966). Feynman encountered Dirac's old paper in 1941. Dirac had written that the transformation function $(q_{t+dt}|q_t)$ "corresponds to" the quantity $\exp(iLdt/h)$, where L is the Lagrangian, and had concluded that this "analogy . . . suggests that we ought to consider the classical Lagrangian, not as a function of the coordinates and velocities, but rather as a function of the coordinates at time t and the coordinates at time $t + dt$" [Dirac (1933A), 68]. Feynman was puzzled over what Dirac had meant by "correspond to" and "analogy," and he eventually discovered that the two quantities were proportional. When he met Dirac in 1946, he asked him if he had known about the proportionality. Apparently Dirac had not. His only comment was "Are they? Oh, that's interesting." For a detailed account of Feynman's early career, see Schweber (1986C), from which the above information is taken.

69 Dirac (1933C).
70 Dirac (1933D), 274.
71 Casimir (1985), 74.
72 Dirac to Bohr, August 20, 1933 (BSC); Dirac described his trip to Russia in a letter to J. Crowther of October 16, 1933 (SUL).
73 Dirac to Van Vleck, November 8, 1933 (AHQP). The mention of the King's drinking habits was a reference to an observation of Van Vleck when, three years earlier, he had attended the royal dinner: "You will probably be interested to know that the King didn't drink *any* hootch at all. Instead he drank one individual pitcher of H_2O, and then sent for a second pitcher. So we have a high opinion of him" [Van Vleck to Dirac, October 1930 (CC)].
74 *Structure et Propriétés des Noyaux Atomiques,* 7th Solvay Report (1933), Paris: Gauthier-Villars, 1934, p. 218.
75 At the Solvay Conference Chadwick wavered between the two views, presenting arguments in favor of both the composite and the elementary neutron. For details, see Stuewer (1983).
76 Gamow to Bohr, June 13, 1934 (BSC). Published references to Dirac's suggestion include Gamow (1935A) and Gapon and Iwanenko (1934).
77 Dirac's notes, intended for seminars or personal use, demonstrate that he did not completely ignore the applications of quantum theory. In addition to the subjects mentioned in Chapter 5, Dirac was occupied with Heisenberg's theory of ferromagnetism, group theory, London's theory of superfluidity, Møller's collision theory, Rosenfeld's nuclear theory, Landau and Weisskopf's statistical theory of nuclei, and a few other subjects (CC). In early 1932, together with the research student J. W. Harding, Dirac also investigated photo-electric absorption in atoms; see Dirac (1932A). However, he seems not to have shown any interest at all in solid-state physics. Perhaps he shared Pauli's attitude that solid-state physics was "dirty physics" (see PB2, xiv).
78 Dirac and Kapitza (1933B).
79 Dirac (1980D).
80 Gould et al. (1986).
81 Rutherford to Fermi, April 23, 1934 (AHQP). For Dirac's work with isotope separation during the war, see the following chapter.
82 Tamm to Dirac, May 13, 1934 (TDC).
83 For details about the electromagnetic mass, see, e.g., Hirosige (1966). For the development of electron theory up to 1949, see Weisskopf (1972), 96–128.
84 Frenkel (1925), 527.
85 Dirac (1928A), 610.
86 Oppenheimer (1930A).
87 Weisskopf (1934). By mistake, Weisskopf first got a quadratic divergence. The mistake was pointed out by Furry, and Weisskopf published the correct result as a correction in *Zeitschrift für Physik* 90 (1934), 53–4. See Weisskopf (1983).
88 Oppenheimer and Furry (1934).
89 Pais (1986), 377.

90 Pauli to P. Langevin, August 26, 1933, in French; translated from A. Langevin (1966), 92.

91 Dirac to Bohr, August 10, 1933 (BSC).

92 Dirac (1934A), 150.

93 Dirac (1931B).

94 Dirac (1933E), 211.

95 The expression given by Dirac contains a numerical error: the factor $4/15\pi$ should be $1/15\pi$.

96 Dirac (1933E), 212.

97 Ibid.

98 Dirac to Bohr, November 10, 1933 (BSC). Dirac wrote to Pauli the following day, but the letter seems to have been lost. Pauli responded in letters of November 14 and December 2 (PB2, 229 and 234). In the letter of November 14, Pauli congratulated Dirac for the Nobel Prize and wrote, "I am very happy that the difficulties originating from the infinities have now been overcomed."

99 Pauli to Heisenberg, February 6, 1934 (PB2, 275–6).

100 Heisenberg to Pauli, February 8, 1934 (PB2, 279).

101 Heisenberg to Pauli, April 19, 1934 (PB2, 318); and Heisenberg to Born, May 16, 1934 (AHQP).

102 Pauli to Heisenberg, June 14, 1934 (PB2, 327–9).

103 Heisenberg (1934).

Chapter 7

1 Joffe to Bohr, March 23, 1934 (BSC); and Dirac to Tamm, June 7, 1934 (TDC).

2 Dirac to Van Vleck, July 22, 1934 (AHQP).

3 Van Vleck (1972), 12 ff.

4 The newspaper was the *Lake City Silver World*. Unfortunately, the archival files of the paper do not include the year 1934, in which Dirac was interviewed. See Van Vleck (1972), 16.

5 Dirac to Van Vleck, October 30, 1934 (CC). The "Pauli effect," which was known to every physicist at the time, refers to Pauli's legendary clumsiness with experimental apparatus; at Bohr's institute it was usual to blame the Pauli effect whenever a device broke down or would not work.

6 Wigner (1939), 156. In 1983, Wigner did not remember these conversations; see the interview conducted by M. G. Doncel, L. Michel, and J. Six in Wigner (1984), 188.

7 Wigner (1984), 188.

8 Dirac to Veblen, July 4, 1935 (LC).

9 Dirac to Tamm, November 7, 1935 (TDC).

10 The entire Kapitza case, including the correspondence between Peter and Anna Kapitza, is the subject of Badash (1985). In a letter [reproduced in Badash (1985), 92] of July 26, 1935, to Anna Kapitza, her husband referred to Dirac's arrival.

11 Dirac to Van Vleck, November 14, 1935 (AHQP).

12 Margit Dirac (1987), 6.

13 Dirac to Tamm, June 17, 1936 (TDC).

14 Dirac (1937D).

15 Margit Dirac to Veblen, June 17, 1937 (LC). The hardened ideological climate in the USSR was indicated in one survey article on Soviet theoretical physics. The author, Dimitrii Blochinzew, concluded that the most urgent task for Soviet physicists was to "put an end to the light hearted attitude toward the idealistic concepts which soak into our physics from bourgeois science" [Blochinzew (1937), 549].

16 Gamow (1966), 121.

17 Gabriel Andrew Dirac (March 13, 1925–July 20, 1984) studied mathematics at Cambridge University, from which he received a Ph.D. in 1951. He became a British citizen in 1949. Gabriel Dirac's mathematical career included positions at the universities of London, Vienna, Hamburg, and Dublin. From 1970 until his death, he lived in Aarhus, Denmark, where he was an associate professor at the University of Aarhus, specializing in graph theory.

18 "Dirac is much preoccupied with his marriage," wrote Peierls, who missed the discussions he used to have with his colleague [Peierls to Bethe, February 13, 1937, trans. from German (Bethe papers, Cornell University Archive)].

19 According to his wife and biographer, Heisenberg "saved" the Teszler family; see E. Heisenberg (1984), 95. However, when Josef Teszler asked for help in March 1943, Heisenberg was not able to assist. In February 1944, Teszler, then in Budapest, requested a statement confirming Beatrice's Aryan descent and got one from Heisenberg.

20 Pauli to Dirac, November 11, 1938, November 17, 1938, and November 26, 1938 (PB2, 607–10). I do not know whether Dirac's request for a petition was granted.

21 Wigner to Dirac, March 30, 1939. The entire letter is reproduced in Weart and Szilard (1978), 71–2.

22 The Frisch–Peierls memorandum, dating from February 1940, is reproduced in Gowing (1964), 389–93. Otto Frisch, a German Jew, worked with Bohr in Copenhagen and went to Birmingham in the summer of 1939.

23 The initials M.A.U.D. had a curious origin. They appear to form an acronym but in fact they referred to a British governess, Maud Ray, who had taught Bohr's sons English and at the time lived in Kent. A cable from Lise Meitner contained a mysterious reference to "Maud Ray Kent," and the British physicists, who believed it was an anagram for "radium taken," decided to use Maud as a suitable name for their committee. Only later did they learn about their mistake.

24 Dirac to Peierls, August 25, 1942, quoted from Dalitz (1987A), 80. For further details and references to Dirac's war work, Dalitz's careful study is the best source. See also Dalitz and Peierls (1986), who list the following reports on the bomb project authored or co-authored by Dirac (the reports are located at the Harwell Archives): "Expansion of U-Sphere Enclosed in a Container" (Report MSD2, 1942); "Neutron Multiplication in a Sphere of Uniform Density Surrounded by a Shell of Non-uniform Density" (Report MSD3, 1942); "Estimate of the Efficiency of Energy Release with a Non-scattering Container" (Report MSD4, 1942); "Approximate Rate of Neutron Multiplication for a Solid of Arbitrary Shape and Uniform Density" (Report MSD5, Part I, 1943); "Application to the

Oblate Spheroide Hemisphere and Oblate Hemispheroid" (Report MSD5, Part II, 1943; with K. Fuchs, R. Peierls, and P. Preston).

25 Dirac to Peierls, October 9, 1942, quoted from Dalitz (1987A), 82.

26 Hoyle (1986), 18. Hoyle did not witness the episode himself but reports it "on good authority." See also Gowing (1964), 261.

27 Quoted from Eden and Polkinghorne (1972), 7. See also Dalitz and Peierls (1986). Dirac wrote two reports on the theory of isotope separation during the war, "The Theory of the Separation of Isotopes by Statistical Methods" (1941) and "The Motion in a Self-Fractionating Centrifuge" (1942).

28 Whitley (1984), 59, in which the importance of Dirac's work is emphasized and his contributions are placed in their historical context.

29 Much later, in the 1970s, Dirac's method was rediscovered and actually used for large-scale uranium enrichment. At first, it was not recognized that the process had been worked out by Dirac thirty years earlier. See Dalitz (1987A), 87.

30 Born to Einstein, April 10, 1940, quoted from Born (1972), 144.

31 Crowther (1970B), 216.

32 Dirac to Crowther, February 27, 1945 (SUL).

33 Dirac (1948E), 12.

34 Crowther to Dirac, December 18, 1942 (CC), marked "Confidential." Dirac replied that he would like to go provided that the British authorities did not have any objections. Dirac to Crowther, December 21, 1942 (SUL). I do not know the purpose of Dirac's planned travel.

35 Werskey (1979), 21. See also Howarth (1978).

36 Kemenov to Dirac, December 28, 1941 (CC).

37 Dirac to J. R. Oppenheimer, November 11, 1949; J. R. Oppenheimer to Dirac, November 18, 1949; Dirac to J. R. Oppenheimer, November 24, 1949 (LC). Frank Oppenheimer was forced to leave physics for a period of eleven years, until he was hired by the University of Colorado in 1959.

38 *New York Times,* May 27, 1954.

39 *The Times* (London), May 28, 1954.

40 *New York Times,* June 3, 1954. "Dirac Denied Visa," *Physics Today,* July 1954, 7. According to the physicist Lew Kowarski, Dirac was also excluded from a nuclear physics conference in Chicago in 1951. Kowarski believed that the reason was Margit Dirac's being from a country behind the Iron Curtain [interview with Kowarski, May 3, 1970, conducted by C. Weiner (AIP)].

41 Dirac to Oppenheimer, September 25, 1955 (LC).

42 Dirac (1955D and 1955E).

43 For details about Dirac's years in Florida, see contributions in Kursunoglu and Wigner (1987).

Chapter 8

1 Heitler to Bohr, November 16, 1933 (BSC). Details of the development of quantum electrodynamics in the 1930s can be found in Pais (1986), Cassidy (1981), Galison (1983), and Brown and Hoddeson (1983).

2 Oppenheimer to Uhlenbeck, March 1934, quoted from Smith and Weiner

(1980), 175. The work referred to by Oppenheimer was published the same year; see Oppenheimer and Furry (1934).
3 J. Robert Oppenheimer to Frank Oppenheimer, June 4, 1934, quoted from Smith and Weiner (1980), 181.
4 Born and Infeld (1934).
5 Heisenberg (1938).
6 Van Vleck to Dirac, October 23, 1934 (CC); Dirac to Van Vleck, October 30, 1934 (CC). Van Vleck, who had recently moved into an apartment in Cambridge, Mass., wrote to Dirac: "Unfortunately we haven't any oil burner for you to experiment on." This reference is to the incident mentioned in Chapter 5.
7 Dirac, unpublished lecture notes for Harvard lecture on quantum electrodynamics, February 19, 1935 (CC).
8 Dirac, *Lectures on Quantum Electrodynamics 1934–35,* 71 pp., unpublished lecture notes based on notes taken by B. Podolsky and N. Rosen. See also Dirac's Kapitza Club lecture on "Nascent Electricity," as mentioned in note 27, Chapter 10.
9 Schrödinger (1940).
10 Dirac (1935, 1936C). "In Princeton I have been working more on mathematics than on physics." [Dirac to Tamm, April 2, 1935 (TDC)]. The content of these works was the subject of a lecture Dirac gave to the $\nabla^2 V$ Club on December 12, 1935, entitled "Tensor Representation of Physical Quantities."
11 Dirac (1936C), 429.
12 *PQM,* 1st edn., p. v.
13 Dirac (1936B).
14 Petiau (1936) and Proca (1936). Petiau was a student of de Broglie, who himself investigated generalizations of the Dirac equation in the hope that it would describe photons of nonzero mass. Proca's equation, a generalization of the second-order Klein–Gordon equation earlier revived by Pauli and Weisskopf, was believed to describe particles with spin one.
15 Dirac (1936C), 448.
16 See Mituo (1974).
17 Yukawa and Sakata (1937); Fierz and Pauli (1939).
18 For references, see Krajcik and Nieto (1977). Dirac gave a talk on "Kemmer's Work on Nuclear Forces" to the Kapitza Club on May 31, 1938.
19 Shankland (1936).
20 Dirac, "Does Detailed Conservation of Energy Hold in Atomic Processes?" Kapitza Club lecture by Dirac on December 10, 1935. At the same meeting, P. I. Dee and C. W. Gilbert discussed the experiments of 1925 performed by W. Bothe and H. Geiger, and A. H. Compton and A. Simon, which proved the incorrectness of the BKS theory. At another Kapitza Club meeting, on March 10, 1936, Dirac and Peierls discussed "Bohr–Kramers–Slater and the Uncertainty Principle."
21 Dirac to Tamm, December 6, 1935 (TDC). Tamm found Dirac's ideas interesting but not convincing [Tamm to Dirac, January 2, 1936 (TDC)].
22 Bohr, Kramers, and Slater (1924). The BKS theory was rejected soon after its appearance when experiments by Bothe and Geiger on scattering of radiation by electrons showed disagreement with the theory. For details of the history of the BKS theory, see Hendry (1981) and Dresden (1987), 159–215.

23 Dirac (1936A), 298. See also "Atomic Physics Must Discard Conservation of Energy Law," *Science News Letter 29* (1936), 213.
24 Ibid., 299.
25 Ibid., 298. Notice that Dirac referred to neutrinos as "unobservable" rather than "unobserved" particles. Although in 1936 it was generally assumed that the neutrino was beyond experimental detection, it was certainly never meant to be unobservable in principle. Furthermore, the neutrino was not introduced in an ad hoc way, merely to balance the missing energy in β-decay. It accounted for other anomalies too, especially the quantum statistics of nuclei, and was able to be incorporated in a larger theoretical framework. This was well known in 1936, but was not mentioned by Dirac.
26 Kramers to Bohr, March 20, 1936, trans. from Danish [Bohr (1972T), vol. 9, p. 604].
27 Oppenheimer and Carlson (1932), 763. Pauli's wave equation for what he called the "magnetic neutron" was presented at a seminar held in Ann Arbor, Michigan, in the summer of 1931. The neutrino equation was identical to the Dirac equation with no fields. Pauli discussed the equation in his *Handbuch* article of 1933; see Pauli (1933), 233.
28 Dirac to Tamm, May 14, 1936 (TDC).
29 Heisenberg to Pauli, May 23, 1936 (PB2, 442).
30 Williams (1936).
31 Williams to Dirac, January 22, 1936 (CC).
32 Bohr (1936); Jordan (1936), vii. Originally, Bohr had rejected the neutrino hypothesis, but now he supported it. As reported by Gamow, as late as 1934 Bohr wanted to avoid Pauli's hypothesis: "Bohr, . . . does not like this chargeless massless little thing, thinks that continuous β structure is compensated by the emission of gravitational waves which play the role of neutrino but are much more physical things" [Gamow to Goudsmit, March 8, 1934, quoted in Weiner (1972), 180]. Bohr also wrote to F. Bloch: "I don't yet feel fully convinced of the physical existence of the neutrino" [Bohr to Bloch, February 17, 1934, in Bohr (1972+), vol. 9, p. 541].
33 Bohr to Kramers, March 14, 1936, trans. from Danish [Bohr (1972+), vol. 9, p. 599].
34 Jacobsen (1936). Experiments in Germany further undermined the conclusion drawn by Shankland; see Bothe and Maier-Leibnitz (1936).
35 Peierls (1936).
36 Einstein to Schrödinger, March 23, 1936 (AHQP).
37 Schrödinger to Dirac, April 29, 1936 (CC). At that time, Schrödinger, who had left Germany in 1933, held a temporary fellowship at Oxford University.
38 Dirac to Tamm, May 14, 1936 (TDC).
39 Dirac to Bohr, June 9, 1936; Bohr to Dirac, July 2, 1936 (BSC). Dirac also visited Hungary and Denmark during the summer of 1936. He returned to Cambridge on October 1. See Dirac to Veblen, October 13, 1936 (LC).
40 Williams (1935), which was a refined quantum mechanical reworking of the scattering theory in Bohr (1915). For this theory and its role in the mid-1930s, see Galison (1987), 97–102. See also Wheeler's reminiscences: "Month after month in the fall and winter of 1934–35 Bohr hammered away at this question [the valid-

ity of quantum theory for high energies] with E. J. Williams and anybody else of us around who could contribute, establishing simple and battle-tested arguments that quantum electrodynamics cannot and does not fail at high energies" [Wheeler (1985), 225].

41 Dirac (1979A), 653.

42 Dirac (1939C).

43 Wentzel (1933). Dirac discussed Wentzel's work critically in his lectures at Princeton in the spring of 1935. "It is interesting," he said, "but one finds it difficult to give to it a physical meaning" (lecture notes, see note 8 above).

44 Dirac (1943 and 1946).

45 Dirac (1942A), 9.

46 Ibid., 8.

47 See Heisenberg (1972). In 1945, the American physicist John Blatt applied Dirac's theory in discussing the interaction of a proton with a meson field. He found effects that differed from earlier results. See Blatt (1945).

48 Pauli to Dirac, May 6, 1942 (PB3). I am indebted to K. v. Meyenn for sending me copies of the Pauli–Dirac correspondence of 1940–5 prior to the publication of PB3.

49 Pauli (1943 and 1946).

50 Pauli to Dirac, October 4, 1942 (PB3; Pauli's emphasis).

51 Pauli to Dirac, August 5, 1943 (PB3). And in Pauli to Oppenheimer, June 19, 1943, one finds: "With Dirac's 'hypothetical world' I was running into serious troubles (troubles of physics not of mathematics) . . ." [Reproduced in Smith and Weiner (1980), 259].

52 Pauli to Dirac, April 13, 1944.

53 Pauli (1964), 42.

54 Dirac (1939B), 416. The bra-ket notation was also worked out in Dirac (1943), but did not receive wide circulation until the third edition of *PQM* in 1947.

55 Dirac (1945A).

56 *PQM*, 3rd edn., p. 308.

57 Dirac (1948E), 12.

58 Dirac (1942A), 13.

59 Dirac (1948A). Dirac made the same point in his Princeton lecture the same year.

60 Dirac (1942A), 18.

61 See Pais (1986), 447–60, and Schweber (1986A).

62 Weisskopf (1983), 76.

63 Lamb (1983), 326.

64 Wigner (1947).

65 Kowarski interview 1970, (AIP).

66 Dirac, manuscript on "Elementary Particles and Their Interactions," September 1946. In later years, Dirac often repeated this belief.

67 Wigner to Feynman, February 26, 1946 (AIP).

68 For details, see Schweber (1986C), 468 and 483. Feynman's lecture took place on November 12, 1947. Harish-Chandra, an Indian physicist who at the time was a student of Dirac and later turned to a career in pure mathematics, was appointed Dirac's assistant during his stay at Princeton. In 1946, Harish-Chandra

reported that "Dirac is very impressed by Feynman and thinks he does some interesting things" [Schweber (1986C), 483].

69 See *Les Particules Élémentaires: Rapports et Discussions du Huitième Conseil de Physique*, Brussels: R. Stoops, 1950, p. 282.

70 Dirac (1953B).

71 Address by E. D. Adrian, President of the Royal Society, in *Proceedings of the Royal Society (London), A 216* (1953), v–vi.

72 Manuscript draft for talk given in Vancouver, September 6, 1949 (CC); a revised version was published as Dirac (1951A).

73 Dirac (1951A), 17.

74 Schweber (1986B), 299. The success of the young generation of quantum physicists tended to reinforce the general belief that physicists' creativity reaches its peak when they are in their twenties. Dirac is quoted to have expressed this feeling in the following verse: "Age is, of course, a fever chill / that every physicist must fear. / He's better dead than living still / when once he's past his thirtieth year." [See, e.g., Jungk (1958), 27, and Merton and Zuckerman (1973), 52.] However, since it appears unlikely that Dirac would express himself in verse, and since no source is given for the quotation, the story is probably apocryphal.

75 Dyson (1986), 103.

76 Dirac (1951A).

77 Dirac (1978A), 36.

78 Dirac (1965), 685.

79 Dirac (1981D), 129.

80 Dirac (1965), 685. On another occasion, Dirac stated that "the agreement with observation is presumably a coincidence, just like the original calculation of the hydrogen spectrum with Bohr orbits" [Dirac (1978C), 5].

81 Dirac (1965), 687, See also Dirac (1966B).

82 Dirac (1966A), 8.

83 Ibid., 113–43.

84 Dirac (1965), 690.

85 Dirac (1969C), 6.

86 Dirac (1983C), 745.

87 Dirac (1984C).

88 Dirac (1975E), 12.

89 Dirac (1971A, 1972A, and 1973F); Dirac (1978A), 56–69.

90 A linear relativistic equation with only positive-energy solutions was proposed in 1932 by the Italian physicist Ettore Majorana but remained virtually unknown for three decades. Dirac's equation is closely related to Majorana's. See Fradkin (1966).

91 Sudarshan, Mukunda, and Chiang (1982). The lack of interest in the theory is reflected in the small number of references to Dirac's paper; see Appendix I.

92 Dirac (1984B), 67.

Chapter 9

1 Oppenheimer (1934), 47. See Galison (1987), 107–10.

2. For references and a historical survey, see Rohrlich (1965), 8–25, Rohrlich (1973), and Pais (1972).

3 Fokker (1929).

4 See Sánchez-Ron (1983).

5 Dirac (1938B), 149. Cf. the query in Dirac (1928A), 610, as to "why nature should have chosen this particular model of the electron instead of being satisfied with the point charge."

6 Dirac (1938B), 149.

7 Ibid.

8 Ibid., 158.

9 Ibid., 159.

10 Ibid., 160; Dirac's emphasis.

11 Hoyle (1986), 179. By 1938, Dirac had stopped attending the meetings of the $\nabla^2 V$ Club. According to the minute book, he was not present at any of the thirteen meetings between February 28, 1936 and December 7, 1937, the latter date being the last one recorded in the minute book kept in the AHQP.

12 Pauli to Dirac, November 11, 1938 (PB2, 608).

13 Dirac to Bohr, December 5, 1938 (BSC).

14 Dirac (1939C).

15 Pauli to Heisenberg, April 27, 1939 (PB2, 639).

16 Pauli to Dirac, July 18, 1939 (PB2, 670). Dirac's letter of June 29, to which Pauli replied, has apparently been lost.

17 Lees (1939).

18 Lewis (1945); Dirac (1948C).

19 Eliezer (1946); Bhabha (1939A). Harish-Chandra was a student under Bhabha from 1943 to 1945 and worked with him on the theory of classical point-electrons. They published several papers on the subject.

20 Bhabha (1939B), 276.

21 Dirac, "Elementary Particles and Their Interactions" (manuscript, 1946). Eliezer expressed in his paper his "deepest gratitude to Prof. Dirac for his patient guidance and supervision" [Eliezer (1943), 179].

22 For views pro and con, see Earman (1976) and Grünbaum (1976).

23 Dirac (1951C).

24 Dirac (1949B), 393.

25 Dirac (1950A).

26 Dirac (1951B and 1958A). For a modern review of the subject, see Taylor (1987A). The revival of interest in Dirac's theory since the mid-1970s is illustrated by the citation profile given in Appendix I.

27 Dirac (1951C), 296.

28 Dirac (1951C), 293.

29 Dirac to Schrödinger January 9, 1952 (AHQP).

30 Dirac (1952B and 1954A).

31 Dirac (1954A), 438.

32 Dirac (1955B), 650.

33 Ibid., 659. See also Dirac (1963B).

34 Einstein (1922), 16 (first published as *Aether und Relativitätstheorie*, Berlin, 1919). See also Einstein (1918). Einstein's phrase "fundamental facts of mechanics" referred to the general theory of relativity, in which empty space is characterized by the components of the gravitational potential. For an account of Einstein's ether, and an attempt to revive it, see Kostro (1986).

35   Dirac (1954B), 146.
36   Dirac (1951D, 1953A, 1953C, and 1954B).
37   Dirac (1954B), 145.
38   Dirac (1951D), 907.
39   Dirac (1954B), 145.
40   Dirac (1953A), 890.
41   Ibid., 896.
42   "Ether, Abolished by Einstein, Restored," *New York Times,* February 9, 1951; "Briton Says Space Is Full of Ether," *New York Times,* February 4, 1952.
43   Bondi and Gold (1952); Dirac (1952A).
44   See Petroni and Vigier (1983) and the literature cited therein. According to Petroni and Vigier, the reintroduction of the ether "might well turn out to be one of Dirac's main contributions to the new era opened (in the authors' opinion) by Aspect's confirmation of the real existence of superluminal correlations in the physical world" (p. 255).
45   Dirac (1971D), 880.
46   Dirac (1960A). See also Chapter 8.
47   Dirac (1962A).
48   See Hasenfratz and Kuti (1978).

Chapter 10

1   Poincaré (1896); Thomson (1900), 396. "Magnetic fluids" had been discussed earlier, by Maxwell among others. However, in agreement with the standard view, Maxwell stressed that such entities were introduced only in a purely mathematical sense. See, e.g., Maxwell (1981), art. 380.
2   Villard (1905) studied so-called magneto-cathode rays, arising from cathode rays in a strong magnetic field, which he believed were a new kind of radiation, distinct from ordinary cathode rays.
3   Heaviside (1893–1912). See the discussion of pre-Dirac monopoles in Hendry (1983). "True magnetism" also appeared in Heinrich Hertz's version of electrodynamics, in which the divergence of the magnetic field was set equal to a magnetic charge.
4   See Hendry (1983) and references therein.
5   Dirac (1978E), 235.
6   Dirac (1931C), 60. For Eddington's "principle of identification" and further aspects of Dirac's philosophy of science, see Chapter 13.
7   Dirac (1913C), 62.
8   Eddington (1929).
9   For example, he wrote in 1978: "The problem of explaining this number [$e^2/\hbar c$] is still completely unsolved. . . . I think it is perhaps the most fundamental unsolved problem of physics at the present time, and I doubt very much whether any really big progress will be made in understanding the fundamentals of physics until it is solved" [Dirac (1978E), 236]. See also the similar statement appearing in Dirac's theory of the classicial electron, quoted in Chapter 9.
10   Heinsenberg to Dirac, March 27, 1935 (CC), trans. from German.
11   Dirac (1931C), 71.
12   Weyl (1929), Fock (1929).

13 Dirac (1931C), 66.
14 Ibid., 68.
15 London (1950), 152.
16 Dirac (1931C), 71.
17 "When I realized this [reciprocity], it was a great disappointment to me. Still I had to make the best of it, so I published my work as the theory of the monopole. That work has led to quite a lot of development.... But to me it remains just a disappointment. It is a disappointment because it gives no help in solving the fundamental problem of why $1/\alpha$ has just this one value" [Dirac (1983C), 748].
18 Dirac (1931C), 69.
19 Tamm (1931). "Apart from Dirac, who has become a close friend of mine, Blackett (whom I like very much) and Kapitza are the people I meet most of all" [Tamm to L. I. Mandelshtam, June 22, 1931, quoted in Kedrov (1984), 56]. Tamm stayed in Cambridge from May 9 to July 3, and then went on to Rostock to work with Jordan. See Dirac to Tamm March 9, 1931, and April 1, 1931 (TDC).
20 Quoted from Feinberg (1987), 303, who gives no date.
21 Dirac (1931C), 71.
22 "Blegdamsvejen Faust," as translated in Gamow (1966), 170–214; on p. 202. The Danish translation of Gamow's book, published in 1968, includes the original German version: "Es waren zwei Monopole, / Die hatten einander so lieb. / Sie konnten zusammen nicht kommen, / Denn Dirac war allzu tief."
23 "Single Magnetic Poles," *The Manchester Guardian,* November 8, 1931; "Lonely Magnetic Poles May Change Ideas of Universe," *Science News Letter* 21 (April, 16, 1932), 243. The lack of interest in Dirac's theory for almost thirty years is not well known among modern monopole physicists, some of whom believe that "Dirac's conjecture stimulated a flurry of theoretical papers on the expected properties of the hypothetical monopoles, and several experiments were undertaken to detect them" [Carrigan and Trower (1982), 91.]
24 Kapitza Club Minute Book, July 21, 1931: "Quantised Singularities in the Electromagnetic Field. Electrons and Magnons." The "magnon" of solid-state physics is a quantized spin wave and has nothing to do with monopoles; the name seems to have been suggested by L. Landau around 1940. See Walker and Slack (1970).
25 Altschuler, who was a research student under Tamm in 1933, recalls the event in Feinberg (1987), 24. If Altschuler's recollections are correct, the relevant letters have been lost. The only letter in the Tamm–Dirac correspondence that possibly relates to the episode is a letter from Tamm to Dirac of November 21, 1933. In this letter Tamm asked if there was any news concerning "Blackett's hairy caterpillars," presumably an allusion to a cloud chamber track.
26 Delbrück to T. S. Kuhn, March 13, 1962 (AHQP); Dirac, manuscript for monopole paper (BSC).
27 See Fischer, (1985), 57 and 187.
28 Pauli to Peierls, September 29, 1931 (PB2, 94). Notice that Pauli apparently thought of the monopole as a nuclear constituent in line with the neutrino, which in the letter is called a dipole neutron. Pauli's remark on the false statistics of the nitrogen nucleus refers to the fact that the nitrogen-14 nucleus was known to obey

Bose–Einstein statistics, whereas according to the then current model of the nucleus (fourteen protons and seven electrons) it ought to obey Fermi–Dirac statistics. The neutrino hypothesis removed this anomaly.

29 Nobel Memorandum 1933 (NA), trans. from Swedish.

30 See Kragh (1981B) for an analysis of monopole history.

31 Jordan (1938), 66. See also Jordan (1935).

32 Richardson (1931). See also Richardson, notes on "Dirac's Magnetic Poles," undated, 9 pp. (AHQP); and "Magnetic Atoms May Be Building-Blocks of Matter," *Science News Letter 20* (November 28, 1931), 345. Richardson's vague suggestion of monopoles in cosmology was prophetic: more than forty years later the magnetic pole became an essential part of cosmological theory.

33 Kestler to Dirac, August 15, 1935 (CC).

34 Langer (1932). The same idea was suggested by Tamm, who never published it. He assigned Altschuler to investigate how the idea would relate to the hyperfine structure of spectra. See Feinberg (1987), 24; and Tamm to Dirac, November 21, 1933 (TDC). A more elaborate version of Langer's suggestion was put forward by the Indian physicist Meghnad Saha in 1936 and again in 1948 [Saha (1936 and 1948)].

35 Dirac (1951A), 19.

36 Tuve (1933).

37 Ehrenhaft (1944). For a brief description of Ehrenhaft's magnetic experiments, see Kragh (1981B), 152–4. Ehrenhaft did not attempt to confirm Dirac's theory, of which he was at first unaware. In a letter of September 4, 1944, he asked Dirac: "Am I mistaken if I believe that in your work I have sometimes ago come across the idea of magnetic ions?" (Ehrenhaft collection, AIP).

38 Ehrenhaft to Dirac, February 19, 1937 (CC). Ehrenhaft did not accept Dirac's excuse, that his German was not good enough.

39 Ehrenhaft to Dirac, February 12, 1944 (Ehrenhaft collection, AIP). Ehrenhaft's letter was in reply to a letter from Dirac of January 26. At that time Ehrenhaft was trying, in vain, to make the *Proceedings of the Royal Society* accept his paper on "Decomposition of Water by the Magnet," which he sent to Dirac and other fellows of the Royal Society.

40 Dirac (1977A), 291.

41 Dirac (1948D), 817.

42 Cf. note 34 above.

43 Particles with both electric and magnetic charge were called dyons by Julian Schwinger, who developed a theory of such particles in the late sixties. See Schwinger (1969).

44 See Wentzel (1966). As pointed out in Hendry (1983), parts of the mathematics of Dirac's 1948 theory were foreshadowed by Whittaker (1921).

45 Dirac (1948D), 830.

46 Pauli to Bethe, March 8, 1949 (PB3).

47 Fermi, "Il Monopole di Dirac," in Fermi (1965), vol. 2, 780–8.

48 See Kragh (1981B) and Pickering (1981).

49 Dirac (1978A), 46.

50 Dirac (1977F), 432.

51 Dirac to Alvarez, February 3, 1978, quoted with permission of L. Alvarez.

Dirac's recollection that Blackett had evidence for the positron a year before Anderson's paper (that is, in late 1931) is not confirmed by other sources.
52   Dirac (1974C). See also Dirac (1977C).
53   Dirac (1974C), 747.
54   Dirac (1976A).
55   Ibid., 6.
56   Dirac (1978E), 241.
57   Dirac to Salam, November 11, 1981. Reproduced in fascimile in Craigie, Goddard, and Nahm (1983), iii, which also contains a reproduction of Dirac (1931C).
58   Cabrera (1982). See also Cabrera and Trower (1983).

Chapter 11

1   Dirac (1977B), 139.
2   Chandrasekhar (1973), 34. Interview with S. Chandrasekhar, May 1977, conducted by S. R. Weart (AIP).
3   Dirac to Bohr, May 6, 1932; Bohr to Dirac, November 14, 1932 (BSC).
4   Chandrasekhar interview, 1977 (AIP). See also Chandresekhar (1934) and Chandrasekhar (1987), 130–5.
5   Kapitza Club Minute Book, meeting of April 25, 1933 (CC). In a letter to Churchill College of September 1, 1979 (CC), Dirac recalled the lecture to have taken place "around 1930" and Lemaître to have emphasized there that he, contrary to Jeans, did not believe God influenced directly the course of atomic events. The following year "the club of remuneration creation of the universe" sent a greeting to Lemaître signed by Kapitza, Born, Cockroft, Walton, and Dirac; the postcard, dated April 17, 1934, is reproduced in Heller (1979), 208.
6   Dirac (1968A), 14.
7   Milne (1935), 104. See also Kragh (1982A), on which parts of the present chapter are based.
8   Summarized in Eddington (1936).
9   Minute Book of $\nabla^2 V$ Club, meeting of March 2, 1933 (CC).
10   Cosmo-physics is examed in Kragh (1982A).
11   Born to Einstein, October 10, 1944, as quoted in Born (1972), 160. Born referred to his "anti-Eddington and Milne essay," which was published the previous year; see Born (1943).
12   Dirac, Peierls, and Pryce (1942B), 193.
13   Eddington to Dingle, 1944, as quoted in Douglas (1956), 178. Schrödinger was the only physicist of eminence who supported Eddington and tried to develop his ideas along new lines. On Eddington's program and Schrödinger's fascination with it, see Rüger (1988).
14   Dirac (1937A), 323.
15   Dirac (1938A), 201.
16   Sambursky (1937),
17   Dirac's agrument for preferring $e^2 m^{-1} c^{-3}$ as the atomic unit of time was the following: It is roughly the geometric mean of all the simple atomic time units that can be constructed from e, m, M, c, and h; and it signifies the time required for light to traverse a classical electron of diameter $e^2 m^{-1} c^{-2}$. As George Temple,

professor at King's College, London, remarked, Dirac's choice of a time unit appeared ad hoc; other combinations were equally possible but would not yield a ratio comparable with the age of the universe ["Nature of Time. Professor Dirac's New Theory," *The Times* (London), February 23, 1937].

18 Dirac (1938A), 203.
19 Ibid., 204.
20 Dirac (1937A), 323.
21 Dirac (1938A), 204.
22 Dingle (1937A), 784.
23 Ibid.
24 Dirac (1937C).
25 Darwin (1937).
26 Dingle (1937B).
27 Gamow to "Phil," September 1, 1967, as reproduced in Gamow (1967B). "Phil" was presumably the geophysicist Philip H. Abelson, at the time editor of *Science*.
28 See the report in *The Observatory 62* (1939), 67–73. For press coverage, see *The Times* (London), February 23, 1937, and *New York Times,* February 19 and 20, 1937.
29 Chandrasekhar (1937). In the AIP interview of 1977, Chandrasekhar explained that the manuscript was a personal letter to Dirac and not intended for publication. Dirac suggested that it should be published and forwarded it to *Nature.*
30 Haldane (1939), 68 and 76.
31 Jordan (1937 and 1939).
32 Jordan (1952), 137.
33 For a representative view, see Yourgrau (1961). A historical survey of cosmonumerology is given in Barrow and Tipler (1986), 219–58.
34 Hoyle (1948), 372, and Hoyle (1980), 65.
35 Teller (1948).
36 Ibid, 802.
37 For these and other modern varying-G theories, see Wesson (1978).
38 Brans and Dicke (1961). Dicke had earlier discussed Dirac's theory in relation to Mach's principle in Dicke (1959).
39 Brans and Dicke (1961). 441.
40 See Dicke (1959).
41 Dirac (1961B), 441.
42 Dirac to Gamow, January 10,1961 (LC).
43 Alpher and Herman (1972), and Alpher to Barbara Gamow, September 4, 1968 (LC).
44 Gamow (1967A), 760.
45 Dirac to Gamow, April 15, 1967 (LC).
46 Gamow to Dirac, September 2, 1967 (LC). See also Gamow to Dirac, April 27, 1967, and Gamow to Dirac, August 28, 1967 (LC). Gamow suggested to Mott that the Cavendish should undertake experiments to test the consequences of his idea. "If the dependence of C on time is found, I will fly to Cambridge to celebrate with you and Dirac" [Gamow to Mott, September 4, 1967 (LC)].
47 Dirac to Gamow, October 7, 1967 (LC).

48 Gamow to Dirac, October 15, 1967 (LC).

49 Dirac to Gamow, November 20, 1967 (LC). See further objections from Gamow in Gamow to Dirac, November 24, 1967 (LC): "'Thus, sorry as I am, I must say that the prize to be paid for 'masking' the secular change of grav. const. is too high to be paid. And, even not being a comunist [sic], I always like to quote Vladimir Lenin who once said: 'Facts are stubborn things.'"

50 Dirac to Gamow, August 22, 1968 (LC).

51 Dirac (1972B), 58.

52 Weyl (1918 and 1921).

53 See Whittaker (1960), 188–9.

54 Dirac interview, 1965 (AHQP).

55 Dirac (1973E), 53; see also Dirac (1973A).

56 Dirac (1973A), 418.

57 Dirac (1938A), 206. In vague terms the Milne–Dirac hypothesis of the two metrics was foreshadowed by the evolutionist Thomas Huxley, who speculated in 1894 that the laws of physics might be a product of evolution from a universe "in which for example the laws of motion held good for some units and not for others, or for some units at one time and not another" [quoted from Barrow and Tipler (1986), 87].

58 Dirac (1973E), 46.

59 Ibid., 48. Dirac had privately returned to matter conservation in 1967; see his letter to Gamow of April 15, quoted in the text.

60 Dirac (1974B), 443. The concept of negative mass had earlier been considered in steady-state cosmology, for example, in Bondi (1957).

61 Dirac (1973C), 8.

62 Dirac (1973A and 1973C).

63 See, for example, Wesson (1978) and Adams (1982).

64 Dirac (1978B), 170.

65 Ibid.

66 Dirac (1979D), 9.

67 Ibid., 10.

68 Ibid. In fact, if the electron mass is used instead of the proton mass, the number becomes 8×10^{27}, which can hardly be considered to belong to the large number cluster of 10^{39}.

69 Halpern, Dirac's research assistant at Florida State University, recalled that "when the long-awaited evaluation of observations from the Viking lander on Mars last year indicated negative results, he [Dirac] told me that he felt like Einstein after Kaufmann's – at first – negative verifications of special relativity" [Halpern (1985), 258]. The background for this comment is that Walther Kaufmann in 1906 performed careful experiments on the mass variation of moving electrons. He concluded that the experimental results agreed with Abraham's nonrelativistic theory but categorically disproved the relativistic theory of Einstein.

70 Dirac (1958B).

71 Dirac (1962D, 1962E, 1964E, and 1968B). The Warsaw conference signaled a renewed interest in applying quantum theory to general relativity. Dirac, Møller, Wheeler, Fock, and Feynman, all mainly known as quantum physicists, participated together with, among others, Infeld, Synge, Bergmann, Penrose, and Bondi.

72   Dirac (1964E).

73   Dirac (1959B). The "graviton" did not appear in Dirac's article, but according to the *New York Times* of January 31, 1959, "Professor Dirac proposed that gravitational wave units be called gravitons. . . . [He] said he believed that his postulation at this time was in the same category as his postulation of positive electrons a quarter of a century ago." Dirac also spoke of gravitons half a year later when he gave a talk on gravitational waves at one of the Lindau meettings; see Dirac (1960C).

74   Dirac (1975B).

Chapter 12

1   Elsasser (1978), 51.

2   Mehra (1973), 819.

3   Ibid., 816, and also in Mehra (1972), 57. Another story referring to Dirac's shyness toward women is the following: Once, at a party in Copenhagen, Dirac suggested a theory according to which the face of a woman looks best at a certain optimum distance. He argued that at $d = \infty$ it is impossible to see the face, and at $d = 0$ it looks deformed because of the small aperture of the human eye; hence there must be a distance between these two values at which the face looks best (it is unknown whether Dirac's argument is valid only for women). When Gamow, interested in knowing the empirical basis for Dirac's theory, asked Dirac how close he had ever been to a woman's face, he answered, "Well, about that close," holding his palms about two feet apart. This story is told in Gamow (1966), 120.

4   Infeld (1980), 202. According to other sources, Dirac's vocabulary was considerably larger, consisting of five – and not just two – words: "Yes," "No," and "I don't know."

5   Dirac (1950C), 115.

6   Infeld (1980), 203.

7   Interview with D. Sciama, April 1978, conducted by S. Weart (AIP).

8   Peierls (1985), 122. A slightly different version is in Infeld (1980), 203. In Dalitz and Peierls (1986) the scene of the story is a journey across the Atlantic, and Dirac and the Frenchman share a cabin on the ship.

9   Feinberg (1987), 92.

10   Pais (1986), 25.

11   The story is referred to in Infeld (1980), 203, and in Gamow (1966), 122. According to Infeld, the event took place in the United States, whereas Gamow locates it at the University of Toronto. Fokker, on the other hand, recalled that it was during Dirac's visit to Leiden (in 1927 or 1928) and that Dirac had agreed with Ehrenfest to answer questions after the talk [Fokker interview, 1963 (AHQP)]. Of course, it is possible that Dirac acted in this way on more than one occasion.

12   Infeld (1978), 41.

13   Mott interview, 1963 (AHQP).

14   This story also exists in at least two different versions, in Eden and Polkinghorne (1972), 3, and in Mehra (1973), 816, told by Peierls. According to the Peierls version, the conversation took place during a walk near Cambridge.

15   Letter of March 8, 1931, in Mott (1986), 42.

16   Peierls (1987), 37.

17   Harish-Chandra (1987), 35.

18   Mehra (1973), 811.

19   *Sunday Dispatch,* November 19, 1933.

20   A probably incomplete list of the honors Dirac received is the following: Fellow of the Royal Society (1930), Hopkins Prize (1930), Corresponding Member of the USSR Academy of Science (1931), Nobel Prize (1933), Foreign Member of the American Philosophical Society (1938), Royal Medal (1939), Honorary Member of the Indian Academy of Science (1939), James Scott Prize (1939), Honorary Member of the Chinese Physical Society (1943), Honorary Member of the Royal Irish Academy (1944), Honorary Member of the Royal Society of Edinburgh (1946), Honorary Fellow of the National Istitute of Sciences, India (1947), Honorary Member of the American Physical Society (1948), Foreign Associate of the U.S. National Academy of Sciences (1949), Foreign Honorary Member of the American Academy of Arts and Sciences (1950), Member of Accademia delle Scienze di Torino (1951), Copley Medal (1952), Max Planck Medal (1952), Member of Academia das Ciencias de Lisboa (1953), Honorary Fellow of the Tata Institute (1958), Member of the Deutsche Akademie der Naturforscher Leopoldina (1958), Member of Accademia Nazionale dei Lincei (1958), Member of the Pontifical Academy of Science (1961), Member of the Royal Danish Academy of Sciences and Letters (1962), Helmholtz Medal (1964), Belfer Graduate School of Science Award (1965), Oppenheimer Prize (1969), Order of Merit (1973).

After Dirac's death, the International Centre for Theoretical Physics in Trieste instituted a Dirac medal to be given every year on his birthday, August 8. St. John's College in Cambridge endowed an annual lecture to be held in Dirac's memory. The first Dirac medals were awarded to Y. Zeldovich and E. Witten in 1985, and the first Dirac Memorial Lectures were given by R. Feynman and S. Weinberg in 1986.

21   Infeld (1980), 203.

22   Mott interview, 1963 (AHQP). Mott's observation is supported by two volumes of recollections of British physicists, in which Dirac appears only as a secondary figure. See Hendry (1984B) and Williamson (1987).

23   Eden and Polkinghorne (1972), 5.

24   Harish-Chandra (1987), 35.

25   Sciama interview, 1978 (AIP).

26   Casimir (1933), 72, and also interview with G. C. McVittie, March 1978, conducted by D. deVorkin (AIP).

27   See Shanmugadhasan (1987), who nonetheless "firmly believed that Dirac was the best kind of supervisor one could have" (p. 51).

28   Personal communication from Richard Eden (1988). Eden, who was supervised by Dirac from 1948 to 1950, has also recalled that on rare occasions Dirac could be genuinely interested and quite inquisitorial, namely, when the seminar dealt with one of his own subjects of research. See Eden and Polkinghorne (1972), 5.

29   Weisskopf (1973), 989.

30   As recalled by Serber in 1985, quoted in Crease and Mann (1986), 106.

31   Schrödinger to W. Wien, August 25, 1926, in Wien (1930), 74.

32 In 1943, Oppenheimer reported Wigner's appraisal of the young Feynman as "a second Dirac, only this time human" [Oppenheimer to Birge, November 4, 1943, as quoted in Smith and Weiner (1980), 269].

33 Dirac (1961A).

34 Dirac (1977B), 136.

35 Heinsenberg (1971B), 87.

36 Ibid.

37 Unpublished lecture on "Fundamental Problems in Physics"; cf. Dirac (1971D). The quotation is transcribed from a tape recording kindly provided by the *Ständiger Arbeitsausschuss für die Tagungen der Nobelpreisträger in Lindau.* Behram Kursunoglu, at the University of Miami, once asked Dirac if he believed in extraterrestrial life. Dirac answered, "If we do not find life in the universe other than on earth, then I must believe in the existence of god" [Kursunoglu and Wigner (1987), xv].

38 Dirac to Bohr, August 20, 1983 (BSC); Dirac to Tamm, April 2, 1935 (TDC).

39 Howarth (1978), 189. Crease and Mann (1986), 81, state that young Dirac had "violently colored political views," a description that to my knowledge lacks documentary support.

40 Mott (1986), 42. However, Peierls (1987), 35, recalled that he took Dirac to the theater during his stay in Göttingen in 1928.

41 Gamow (1966), 121. Dirac read, on advice, Tolstoy's *War and Peace.* It took him two years. See Salaman and Salaman (1986), 68.

42 Crowther (1970B), 107.

43 Mehra (1972), 52.

44 Alvarez (1987), 87.

45 Dirac to E. Salaman, January 10, 1953, quoted from Salaman and Salaman (1986), 67. Esther Salaman's husband, Myer Salaman, was a pathologist.

46 Dirac discussed chess problems with Heisenberg on their tour to Japan in 1929. After his return to Leipzig, Heisenberg wrote to Dirac: "You are wrong. . . . in the question of mating a King and a Knight with King and Castle; this is *not* possible according to the edition of 1926 of Dufresne's handbook of chess (the best book about theory of chess)" [Heisenberg to Dirac, December 7, 1929, as quoted in Brown and Rechenberg (1987), 141].

47 Tikhonov to Tamm, February 5, 1967, quoted in Feinberg (1987), 278.

Chapter 13

1 Dirac (1977B), 111. Of course, Dirac was not alone in his negative attitude toward philosophy. J. J. Thomson is reported to have regarded philosophy as "a subject in which you spend your time trying to find a shadow in an absolutely dark room," a view that was undoubtedly shared by many a Cambridge scientist.

2 Dirac interview, 1963 (AHQP).

3 Holton (1978).

4 Schrödinger (1956), 98.

5 For an attempt, in my view misleading, to interpret parts of Dirac's physics in terms of external factors, see Feuer (1977).

6 Bachelard (1976), 59 ff.

7 Margenau (1935). As will be elaborated in Chapter 14, Dirac considered "beauty" to be more important than "simplicity."

8 MacKinnon (1974) and Stöckler (1984).

9 Pauli (1919A), 749. Pauli's early use of the observability doctrine is discussed in Hendry (1983).

10 Pauli (1921), 206 (in the English translation of 1958).

11 Heisenberg (1925), as translated in Van der Waerden (1967), 262. See also Heisenberg to Pauli, June 24, 1925 (PB1, 227).

12 Dirac (1932B), 456. For Dirac's positivism, see also Chapter 4.

13 *PQM,* 4th edn. (1967 revision), p. 5.

14 Heisenberg (1925), as translated in Van der Waerden (1967), 261.

15 Margenau (1935), 87–88.

16 Dirac (1962B), 354.

17 Dirac (1973B), 759.

18 See Heelan (1975), who refers to Heisenberg's observability principle as "the principle of E-observability." Quantities, like the spin of the electron, which are E-observable are *measurable* but not necessarily observable in a positivistic (Machian) sense, and thus E-observability is a less restrictive criterion than positivist observability. Dirac adopted the observability doctrine in Heisenberg's version as the doctrine of E-observability.

19 According to Heisenberg's reconstruction of the conversation; see Heisenberg (1971A), 63.

20 Pauli to Schrödinger, January 27, 1955, translated from PB1, xxvii: "In evaluating a physical theory its logical and mathematical structure is (at least) as important as its relationship to experience (to me, personally, the first is even more important). When I think of where a theory is in need of improvement I never consider questions of measurability, but such deductions from the theory which are not mathematically correct (like infinities or divergencies)." Dirac could not have said it better.

21 Born (1936/37), 13. For a critical appraisal of Born's views on the philosphy of physics, see Vogel (1968).

22 Dirac (1942A), 18. See also Heisenberg in his uncertainty paper: "Physics should only describe formally the connection between experiences" [Heisenberg (1927), 197]. Heisenberg stressed the methodological analogy between quantum mechanics and the (general) theory of relativity. Just as the new concepts of space and time followed from the mathematics of the relativity theory, so the needed radical change in mechanical concepts "seems immediately to follow from the fundamental equations of quantum mechanics" (ibid., 173). Dirac completely agreed.

23 *The Mathematical Gazette 15* (1931), 505–6.

24 Dirac (1982C), 86.

25 By "complete" I do not refer to the technical meaning of "completeness" as it was discussed in the famous dispute between Bohr and Einstein, Podolsky, and Rosen in 1935. Dirac did not participate in this discussion, but it is likely that he would have agreed with Bohr in asserting that non-relativistic quantum mechanics was a complete, although not final, theory.

26 Dirac (1951A), 11.

27 Dirac (1984B), 65.

28 Rosenfeld recalled in 1967 that sometime around 1929 an eminent physicist told him: "In a couple of years we shall have cleared up electrodynamics; another couple of years for the nuclei, and physics will be finished. We shall then turn to biology" [Rosenfeld (1967), 114]. Peierl's recollections go in the same direction. Referring to a conference in Copenhagen about 1931, he experienced then "a general feeling among some people there, not everybody, that physics was almost finished" [interview with R. Peierls, June 1963, conducted by J. L. Heilbron (AHQP)].

29 Dirac (1929B), 714. Dirac referred to non-relativistic quantum mechanics, admitting that there were still "imperfections" with respect to the relativistic extension of the theory. In a partly autobiographical novel, written in the early thirties, C. P. Snow referred to Dirac's view of a completed physics. Snow, at that time a Cambridge scientist (in crystallography), wrote about Arthur Miles, his alter ego, attending a meeting in one of the Cambridge scientific clubs: "Suddenly, I heard one of the greatest mathematical physicists say, with complete simplicity: 'Of course, the fundamental laws of physics and chemistry are laid down for ever. The details have got to be filled up: we don't know anything of the nucleus; but the fundamental laws are there. In a sense, physics and chemistry are finished sciences.' . . .This man who spoke of 'finished sciences' was Newton's successor" [Snow (1934), 168].

30 Darrow (1933), 292.

31 Dirac (1939A), 126.

32 Ibid., 129.

33 Ibid.

34 Jeans (1930), 134.

35 Ibid., 141.

36 See Grythe (1982/1983).

37 Pauli to Dirac, December 21, 1943 (PB3). Dirac referred to Heisenberg's theory in an earlier letter to Pauli, which seems to have been lost.

38 See Dirac (1948A).

39 Dirac (1969C and 1970). Dirac did not believe in attempts to construct non-localizable theories such as Heisenberg's S-matrix theory; instead he studied the general properties of localizable quantum dynamical systems. See Dirac (1948B).

40 Dirac (1937B).

41 Dirac (1966A), preface.

42 Dirac (1969C), 4. Other physicists, closer to the field than Dirac, also objected that the bootstrap theory destroyed the unity of physics. See the quotations in Freundlich (1980), 275. Another reason for Dirac's dislike of the S-matrix program may have been that it was launched as an alternative to the Hamiltonian program, to which he was deeply committed.

43 Quoted from Lovejoy (1976), 272.

44 Rydberg argued from numerological considerations of the system of elements that there must exist a chemical element with atomic number zero. Referring to the apparent lack of empirical evidence for such an element, he argued: "One must try to find either the substance itself or the reason for its non-existence, for it [the non-existence] forms an exception to an otherwise generally valid

law" [Rydberg (1906), 17]. The Swedish scientist identified his new element with
the electron.
45 See Whittaker (1949), 58 ff., and Popper (1959), 69.
46 Bilaniuk and Sudarshan (1969), 44. Also, Barrow and Tipler (1986), 501,
write: "It is a general working principle in physics that what is not forbidden is
compulsory. . . ."
47 Eddington (1923), 222–3. In order to pass from mathematics to physics, the
quantities (like tensors) that describe the physical world must, according to the
principle of identification, be identified with purely mathematical quantities.
These consistitute "our world-building material" from which an ideal world,
functioning in the same way as the empirically known world, can be deduced.
Eddington's program was thus to establish identities between physical and mathematical quantities. "If we can do this completely," he wrote, "we shall have constructed out of the primitive relation-structure a world of entities which behave
in the same way and obey the same laws as the quantities recognized in physical
experiments. Physical theory can scarely go further than this" (p. 222).
48 Ford (1963), 122.
49 Peierls to Pauli, July 17, 1933 (PB2, 197).
50 Pauli and Weisskopf (1934), 713.
51 "Elementary Particles and Their Interactions," Princeton bicentennial
manuscript, 1946.
52 Milne (1935), 236.

Chapter 14

1 Kedrov (1979), 33. Yukawa's inscription (written in 1959) was "In essence,
Nature is simple," and Bohr (in 1961) chose the motto of his complementarity
principle, " Contraria non Contradictoria sed Complementa sunt."
2 Dirac (1939A), 122.
3 Ibid., 122.
4 Quoted from Goldberg (1976), 140.
5 Born (1956), 77.
6 Newton (1729), 398.
7 Dirac (1939A), 124.
8 Ibid., 124.
9 Ibid., 124.
10 Ibid., 125.
11 Ibid., 123.
12 Poincaré (1960), 20. Poincaré's considerations of the role of aesthetics in science were later elaborated by the mathematician J. Hadamard in Hadamard
(1954). No single work offers a general and satisfactory analysis of beauty in science, but various aspects are discussed in Curtin (1980), Wechsler (1978), Huntley
(1970), and Chandrasekhar (1987).
13 Poincaré (1960), 59. Following Poincaré, Hadamard advised the mathematicians to be guided by "that sense of scientific beauty, that special esthetic sensibility" [Hadamard (1954), 127].

14 Dirac (1979F), 39. Compare the view of Roger Penrose, an eminent British mathematical physicist: "I think that one's aesthetic judgments in mathematics are very similar to those in the arts. But in mathematics aesthetics is not only an end in itself, but also a means. . . . If you want to find a new way of solving a problem, you must feel your way around, in a sense, and look for pleasing and aesthetically attractive solutions. So in that way aesthetics can be a means towards solving a problem, rather than an end in itself. Of course, it is an end too: one really studies the subject of mathematics mainly for its beauty!" [Penrose (1979), 50]. This view seems to be close to the view held by Dirac.

15 Dirac, "Basic Beliefs and Fundamental Research," unpublished talk, University of Miami, 1972. Part of the talk is quoted in Dyson (1986).

16 Dirac (1951A), 12.

17 Dirac (1966B), 10.

18 Margenau (1933), 493.

19 Birkhoff to Kemble, March 3, 1933 (AHQP). The first part of this letter is reproduced in Chapter 5.

20 Kemble to Birkhoff, March 27, 1933 (AHQP). In his letter to Kemble, Birkhoff had pointed out that Dirac's treatment of the harmonic oscillator, as it appeared in the first edition of *PQM*, pp. 53–4, was partly mistaken.

21 Dirac (1977B), 112.

22 Dirac (1927A), 625. Heaviside, who introduced an earlier version of the δ-function in 1893, had an attitude toward mathematics that was even more antipurist than Dirac's. Trained as an engineer, Heaviside considered mathematics to be an experimental science.

23 Dirac (1965), 687. Also, in conversations with Mehra he said, "[I] have never been much interested in questions of mathematical logic or in any form of absolute measure or accuracy, an absolute standard of reasoning" [Mehra and Rechenberg (1982+), vol. 4, p. 12].

24 Dirac (1969B), 22. According to Dirac, Klein's lack of success with his five-dimensional theory was rooted in his being too ambitious and not following the piecemeal method; see Klein (1973), 164. Also, in an address in 1981, Dirac said: "A good many people are trying to find an ultimate theory which will explain all the difficulties, maybe a grand unified theory. I believe that is hopeless. It is quite beyond human ingenuity to think of such a theory" [Dirac (1983C), 743].

25 Heisenberg (1971A), 101. See also Heisenberg (1968), 46.

26 Dirac to Heisenberg, March 6, 1967, quoted from Brown and Rechenberg (1987), 148.

27 *PQM*, 1st edn., p. vi.

28 *PQM*, 1st edn., p. 1.

29 Dirac (1939A), 125.

30 Dirac (1937B), 48.

31 Dirac (1939A), 129.

32 Ibid.

33 Dirac (1982B).

34 Dirac (1963B), 53. The casual reference to God should not be taken as implying any religious commitment. Like Einstein in his famous statement about the dice-playing God, Dirac merely used the word as a substitute for "what caused the laws of nature." He usually referred to Nature instead of God.

35 "Basic Beliefs and Fundamental Research" (note 15, op. cit.).

36 Dirac (1970), 29.

37 Dirac in conversation with Mehra, 1968 or 1969, quoted from Mehra (1972), 59.

38 Quoted from Galison (1979), 100.

39 Minkowski (1915), 927, quoted in translation from Galison (1979), 97. The posthumosuly published paper was the manuscript of an address Minkowski delivered in 1907.

40 Dirac (1980B), 6.

41 Dirac (1979E), 17.

42 In 1943, Born considered Einstein's theory to be "a gigantic synthesis of a long chain of empirical results, not a spontaneous wave brain" and argued in general for the value of experiments and inductive methods even in the most abstract theories [Born (1943), 14]. However, Born was not always immune to the intellectual magic of Einstein's theory. In 1920, he saluted it for its "grandeur, the boldness, and the directness of the thought," which made the world picture "more beautiful and grander" [Born (1924), 5]. Chandrasekhar (1987) provides many other examples of the aesthetic evaluation of general relativity.

43 See Rosenthal-Schneider (1949).

44 Einstein (1934), 17–18.

45 See Pais (1982), 325 and 344.

46 Einstein to Weyl, September 5, 1921, quoted in Hendry (1983), 137. Einstein to Ehrenfest, August 18, 1925, quoted in Pais (1982), 244.

47 Einstein (1950), 17.

48 Conversation with Freeman Dyson, as quoted in Chandrasekhar (1979). Weyl's remark is consistent with the Dirac-Weyl doctrine only if "true" is taken to mean "empirically confirmed"; according to Dirac's principle of mathematical beauty, a real conflict between truth and beauty is inconceivable.

49 Weyl (1932), 11 and 21.

50 For a concise survey of the subject, see Merriell (1981).

51 Dirac (1954B), 143.

52 "Basic Beliefs and Fundamental Research" (note 15, op. cit.).

53 See Pyenson (1977).

54 See Heelan (1975), who examines Heisenberg's aesthetic commitments. Heelan concludes: "The fact of disagreement does not imply that criteria do not exist, only that there are different sensibilities, different 'esthetic' styles in science as in art. Criteria of this kind belong to the transcendental (or non-objective) conditions of possibility of theoretical scientific rationality" (p. 130). Heisenberg praised the role of beauty in science in Heisenberg (1971B).

55 Gamow (1967C), 192.

56 Gamow to Dirac, October 15, 1967 (LC).

57 As mentioned in Chapter 6, Dirac realized that for the solution of certain problems the Lagrangian formalism might be superior. See Dirac (1933A).

58 Dirac (1964A), 59.

59 Interview with E. Wigner, December 1963, conducted by T. S. Kuhn (AHQP).
60 Dirac (1978G), 16.
61 Dirac (1958B), See also Chapter 10.
62 Dirac (1963B), 46.
63 Dirac (1983C), 745.
64 "Basic Beliefs and Fundamental Research" (note 15, op. cit.).
65 Wigner (1928 and 1932).
66 Weyl (1929); Pauli (1933), 226.
67 In 1957, following the discovery of parity nonconservation, Weyl's equation was rehabilitated. Works by T. D. Lee, C. N. Yang, L. D. Landau, and A. Salam showed that the neutrino does in fact satisfy the two-component Weyl equation. Chandrasekhar considered the delayed success of the Weyl equation to be a case in support of the Dirac–Weyl doctrine; see Chandrasekhar (1979). However, this interpretation is unjustified since Pauli and other physicists did not reject Weyl's theory because of lack of experimental verification but because it contradicted an aesthetic principle, parity conservation. Writing of the events shortly after the Lee–Yang theory was established, Pais concluded with the lesson that "Once again principle has turned out to be prejudice . . ." [Pais (1986), 533]. In this connection it is worth pointing out that Dirac's generalized wave equation of 1936 did not satisfy parity invariance, a fact first pointed out by Yukawa and Sakata; see Dirac (1936B) and Yukawa and Sakata (1937).
68 For the history and literature of parity invariance, see Franklin (1979 and 1986).
69 See Wigner (1932).
70 Dirac (1937D), 81.
71 Franklin (1986), 24.
72 Pais (1986), 25.
73 Dirac (1949B), 393.

GENERAL BIBLIOGRAPHY

Adams, P. J. (1982), "Large number hypothesis, I," *International Journal of Theoretical Physics 21*, 607–32.

Alpher, R. A., and Herman, R. (1972), "Reflections on 'Big Bang' cosmology," pp. 1–14 in F. Reines, ed., *Cosmology, Fusion and Other Matters. George Gamow Memorial Volume,* New York: Colorado Associated Press.

Alvarez, L. W. (1987), *Alvarez. Adventures of a Physicist,* New York: Basic Books.

Anderson, C. D. (1932), "The apparent existence of easily deflectable positives," *Science 76,* 238–9.

(1933), "The positive electron," *Physical Review 43,* 491–4.

(1961), "Early work on the positron and muon," *Amercian Journal of Physics, 29,* 825–30.

Aramaki, S. (1987), "Formation of the renormalization theory in quantum electrodynamics," *Historia Scientiarum,* No. 32, 1–42.

Bachelard, G. (1976), *Nej'ets Filosofi,* Copenhagen: Vinten (Danish translation of *La Philosophie du Non,* first published 1940 in Paris).

Badash, L. (1985). *Kapitza, Rutherford, and the Kremlin,* New Haven: Yale University Press.

Baker, H. F. (1922). *Principles of Geometry,* vol. 1, Cambridge: Cambridge University Press.

Balibar, F. (1985), "Bohr entre Einstein et Dirac," *Revue d'Histoire des Sciences 38,* 293–307.

Barrow, J. D., and Tipler, F. J. (1986), *The Anthropic Cosmological Principle,* Oxford: Oxford University Press.

Beck, G., and Sitte, K. (1933), "Zur Theorie des β-Zerfalls," *Zeitschrit für Physik 86,* 105–19.

Becquerel, J. (1908), "Sur un phénomène attribuable á des électrons positifs," *Comptes Rendus 146,* 683–5.

Beller, M. (1985), "Pascual Jordan's influence on the discovery of Heisenberg's indeterminacy principle," *Archive for History of Exact Sciences, 33,* 337–49.

Belloni, L. (1978), "A note on Fermi's route to Fermi–Dirac statistics," *Scientia 113,* 421–30.

Bhabha, H. (1939A), "Classical theory of mesons," *Proceedings of the Royal Society of London A172*, 384–409.

(1939B), "The fundamental length introduced by the theory of the mesotron (meson)," *Nature 143*, 276–7.

Biedenharn, L. C. (1983), "The 'Sommerfeld Puzzle' revisited and resolved," *Foundations of Physics 13*, 13–34.

Bilaniuk, O., and Sudarshan, E. C. (1969), "Particles beyond the light barrier," *Physics Today*, May 1969, 43–51.

Birtwistle, G. (1928), *The New Quantum Mechanics*, Cambridge: Cambridge University Press.

Blackett, P. M. S., and Occhialini, G. P. S. (1933), "Some photographs of the tracks of penetrating radiation," *Proceedings of the Royal Society of London A139*, 699–726.

Blatt, J. (1945). "On the meson charge cloud around a proton," *Physical Review 67*, 205–16.

Bloch, F. (1931), [review of Dirac, *Die Prinzipien der Quantenmechanik*], *Physikalische Zeitschrift 32*, 456.

(1934), "Die physikalische Bedeutung mehrerer Zeiten in der Quantenelektrodynamik," *Physikalische Zeitschrift der Sowietunion 5*, 301–15.

Blochinzew, D. (1937), "Advance of theoretical physics in the Soviet Union in 20 years," *Physikalische Zeitschrift der Sowietunion 5*, 542–9.

Bohr, N. (1915), "On the decrease of velocity of swiftly moving electrified particles in passing through matter," *Philosophical Magazine 30*, 581–612.

(1928), "The quantum postulate and the recent development of atomic theory," *Nature 121*, 580–90.

(1932), "Atomic stability and conservation laws," *Covegno di Fisica Nucleare*, Rome: Reale Accademia d'Italia.

(1936), "Conservation laws in quantum theory," *Nature 138*, 25–6.

(1972+), *Niels Bohr, Collected Works*, gen. ed. E. Rüdinger, Amsterdam: North-Holland.

Bohr, N., Kramers, H. A., and Slater, J. C. (1924), "The quantum theory of radiation," *Philosophical Magazine 47*, 795–802.

Bondi, H. (1957), "Negative mass in general relativity," *Review of Modern Physics 29*, 423–8.

Bondi, H., and Gold, T. (1952), "Is there an aether?" *Nature 169*, 146.

Born, M. (1924), *Einstein's Theory of Relativity*, London: Methuen, 1924 (translation of German original from 1920).

(1928), "VI. Kongreß der Assoziation der russischen Physiker," *Die Naturwissenschaften 16*, 741–3.

(1936/1937), "Some philosophical aspects of modern physics," *Proceedings of the Royal Society (Edinburgh) 57*, 1–18.

(1943), *Experiment and Theory in Physics*, Cambridge: Cambridge University Press.

(1956), *Physics in My Generation*, London: Pergamon Press.

(1972), *Albert Einstein – Max Born, Briefwechsel 1916–1955*, Hamburg, Rowohlt.

(1978), *My Life. Recollections of a Nobel Laureate*, New York: Scribner.

Born, M., Heisenberg, W., and Jordan, P. (1925), "Zur Quantenmechanik, II," *Zeitschrift für Physik 35,* 557–615.

Born, M., and Infeld, L. (1934), "Foundations of the new field theory," *Proceedings of the Royal Society of London A144,* 425–51.

Born, M., and Jordan, P. (1925), "Zur Quantenmechanik," *Zeitschrift für Physik 34,* 858–88.

Bothe, W., and Maier-Leibnitz, H. (1936), "Eine neue experimentelle Prüfung der Photonenvorstellung," *Zeitschrift für Physik 102,* 143–55.

Brans, C., and Dicke, R. H. (1961), "Mach's principle and a relativistic theory of gravitation," *Physical Review 124,* 925–35.

Breit, G. (1926), "A correspondence principle in the Compton effect," *Physical Review 27,* 362–72.

Bromberg, J. (1975), "Remarks on the hole theory and Dirac's methodology," *Proceedings XIV International Congress of History of Science (1974),* vol. 2, 233–6.

(1976), "The concept of particle creation before and after quantum mechanics," *Historical Studies in the Physical Sciences 7,* 161–83.

Brown, L. M., and Hoddeson, L., eds. (1983), *The Birth of Particle Physics,* Cambridge: Cambridge University Press.

Brown, L. M., and Moyer, D. F. (1984), "Lady or tiger? The Meitner-Hupfeld effect and Heisenberg's neutron theory," *American Journal of Physics 52,* 130–6.

Brown, L. M., and Rechenberg, H. (1987), "Paul Dirac and Werner Heisenberg – a Partnership in Science," pp. 117–62 in Kursunoglu and Wigner (1987).

Bursian, V. (1927), "Notiz zu den Grundlagen der Dispersionstheorie von E. Schrödinger," *Zeitschrift für Physik 40, 708–13.*

Cabrera, B. (1982), "First results from a superconductive detector for moving magnetic monopoles," *Physical Review Letters 48,* 1378–81.

Cabrera, B., and Trower, W. P. (1983), "Magnetic monopoles: evidence since the Dirac conjecture," *Foundation of Physics 13,* 196–216.

Carrigan, R. A., and Trower, W. P. (1982), "Superheavy magnetic monopoles," *Scientific American,* April, 91–9.

Casimir, H. (1983), *Haphazard Reality. Half a Century of Science.* New York: Harper and Row.

(1985), "Paul Dirac 1902–1984," *Naturwissenschaftliche Rundschau 38,* 219–23.

Cassidy, D. C. (1981), "Cosmic ray showers, high energy physics, and quantum field theories: programmatic interactions in the 1930s," *Historical Studies in the Physical Sciences 12,* 1–40.

Chandrasekhar, S. (1934), "Stellar configurations with degenerate cores," *The Observatory 57,* 373–7.

(1937), "The cosmological constants," *Nature 139,* 757–8.

(1973), "A chapter in the astrophysicists's view of the universe," pp. 34–44 in Mehra, ed. (1973).

(1979), "Beauty and the quest for beauty in science," *Physics Today,* July, 25–30.

(1987), *Truth and Beauty,* Chicago: University of Chicago Press.

Cini, M. (1982), "Cultural traditions and environmental factors in the development of quantum electrodynamics 1925–1933," *Fundamenta Scientiae 3*, 229–53.

Condon, E. U., and Mack, J. E. (1930), "An interpretation of Pauli's exclusion principle," *Physical Review 35*, 579–82.

Craigie, N. S., Goddard, P., and Nahm, W., eds. (1983), *Monopoles in Quantum Field Theory*, Singapore: World Scientific.

Crawford, E., Heilbron, J. L., and Ullrich, R. (1987), *The Nobel Population 1901–1937*, Berkeley: Office for History of Science and Technology, University of California.

Crease, R. P., and Mann, C. C. (1986), *The Second Creation*, New York: Macmillan.

Crowther, J. G. (1970A), *British Scientists of the Twentieth Century*, London: Routledge & Kegan Paul.

(1970B), *Fifty Years with Science*, London: Barie & Jenkins.

Cunningham, E. (1915), *Relativity and the Electron Theory*, London: Longmans, Green & Co.

Curtin, D. W., ed. (1980), *The Aesthetic Dimension of Science*, New York: Philosophical Library.

Dalitz, R. H. (1987A), "Another side to Paul Dirac," 69–92 in Kursunoglu and Wigner, eds., (1987).

(1987B), "A biographical sketch of the life of professor P. A. M. Dirac, OM, FRS," pp. 3–29 in Taylor, ed. (1987B).

Dalitz, R. H., and Peierls, R. (1986), "Paul Adrien Maurice Dirac," *Biographical Memoirs of Fellows of the Royal Society, 32*, 139–85.

Darrigol, O. (1984A), "La genèse du concept de champ quantique," *Annales de Physiques 9*, 433–501.

(1984B), "A history of the question: can free electrons be polarized?" *Historical Studies in the Physical Sciences 15*, 39–80.

(1986), "The origin of quantized matter waves," *Historical Studies in the Physical Sciences 16*, 1–56.

(1988A), "The quantum electrodynamical analogy in early nuclear theory and the roots of Yukawa's theory," *Revue d'Histoire des Sciences 41*, 225–97.

(1988B), "Elements of a scientific biography of Tomonaga Sin-itiro," *Historia Scientiarum*, No. 35, 2–29.

Darrow, K. K. (1933), "Contemporary advances in physics, XXVI," *Bell System Technical Journal 12*, 288–330.

Darwin, C. G. (1927), "The electron as a vector wave," *Proceedings of the Royal Society of London A116*, 227–53.

(1928), "The wave equation of the electron," *Proceedings of the Royal Society of London A118*, 654–80.

(1935), "The quantum theory," *Nature 136*, 411–12.

(1937), "Physical science and philosophy," *Nature 139*, 1008.

De Broglie, L. (1924), "A tentative theory of light quanta," *Philosophical Magazine 47*, 446–58.

(1925), "Sur la fréquence propre de l'électron," *Comptes Rendus 180*, 498–500.

Delbrück, M. (1972), "Out of this world," pp. 280–8 in F. Reines, ed., *Cosmology, Fusion and other Matters. George Gamow Memorial Volume*, New York: Colorado Associated Press.

De Maria, M., and La Teana, F. (1982), "Schrödinger's and Dirac's unorthodoxy in quantum mechanics," *Fundamenta Scientiae 3*, 129–48.

De Maria, M., and Russo, A. (1985), "The discovery of the positron," *Rivista di Storia della Scienza 2* (2), 237–86.

Dicke, R. H. (1959), "Gravitation – an enigma," *American Scientist 47*, 25–40.

Dingle, H. (1937A), "Modern Aristotelianism," *Nature 139*, 784–6.

(1937B), "Deductive and inductive methods in science. A reply," *Nature 139*, 1011–12.

Dirac, M. (1987), "Thinking of my darling Paul," pp. 3–8 in Kursunoglu and Wigner, ed. (1987).

Dirac, P. A. M. (see Appendix II).

Dorfman, I. V. (1981), "Die Theorie Diracs und die Entdeckung des Positrons in der kosmischen Strahlung," *Zeitschrift für Geschichte der Naturwissenschaft, Technik und Medizin 18*, 50–7.

Douglas, A. V. (1956), *The Life of Arthur Stanley Eddington*, London: Thomas Nelson & Sons.

Dresden, M. (1987), *H. A. Kramers. Between Tradition and Revolution*, New York: Springer.

Dymond, E. G. (1934), "On the polarization of electrons by scattering, II," *Proceedings of the Royal Society of London A145*, 657–67.

Dyson, F. J., (1986), "Paul A. M. Dirac," *American Philosophical Society, Yearbook 1986*, 100–4.

Earman, J. (1976), "Causation: a matter of life and death," *Journal of Philosophy 73*, 5–25.

Earman, J., and Glymour, C. (1981), "Relativity and eclipses: the British eclipse expeditions of 1919 and their predecessors," *Historical Studies in the Physical Sciences 11*, 49–86.

Eddington, A. S. (1923), *The Mathematical Theory of Relativity*, Cambridge: Cambridge University Press.

(1928), *The Nature of the Physical World*, Cambridge: Cambridge University Press.

(1929), "The charge of an electron," *Proceedings of the Royal Society of London A122*, 358–69.

(1936), *Relativity Theory of Protons and Electrons*, Cambridge: Cambridge University Press.

Eden, R. J., and Polkinghorne, J. C. (1972), "Dirac in Cambridge," pp. 1–5 in Salam and Wigner, eds. (1972).

Ehrenhaft, F. (1944), "The magnetic current," *Nature 154*, 426–7.

Einstein, A. (1909), "Zum gegenwärtigen Stand des Strahlungsproblems," *Physikalische Zeitschrift 10*, 185–93.

(1916), "Strahlungs-Emission und -Absorption nach der Quantentheorie," *Verhandlungen der Deutschen Physikalischen Gesellschaft 18*, 318–23.

(1917), "Zur Quantentheorie der Strahlung," *Physikalische Zeitschrift 18*, 121–8.

(1918), "Dialog über Einwände gegen die Relativitätstheorie," *Die Naturwissenschaften 6*, 702.

(1922), *Sidelights on Relativity*, London: Methuen.

(1925), "Elektron und Allgemeines Relativitätstheorie," *Physica 5*, 330–4.

(1931), "Maxwell's influence on the development of the conception of physical reality," pp. 66–73 in *James Clerk Maxwell: a Commemoration Volume*, Cambridge: Cambridge University Press.

(1934), "On the method of theoretical physics," pp. 12–22 in Einstein, *Essays in Science*, New York: Philosophical Library.

(1950), "On the generalized theory of gravitation," *Scientific American 182*, April 13–17.

Eliezer, C. J. (1943), "The hydrogen atom and the classical theory of radiation," *Proceedings of the Cambridge Philosophical Society 39*, 173–80.

(1946), "The classical equations of motion of an electron," *Proceedings of the Cambridge Philosophical Society 42*, 278–85.

(1987), "Some reminiscences of professor P. A. M. Dirac," pp. 58–62 in Taylor, ed. (1987B).

Elsasser, W. M. (1978), *Memoirs of a Physicist in the Atomic Age*, Bristol: Adam Hilger.

Feinberg, E. L., ed. (1987), *Reminiscences about I. E. Tamm*, Moscow: Nauka Publishers.

Fellows, F. H. (1985), *J. H. Van Vleck: The early life and work of a mathematical physicist*, unpublished Ph.D. Thesis, University of Minnesota.

Fermi, E. (1926A), "Sulla quantizzione del gas perfetto monatomico," *Rendiconti Lincei 3*, 145–9.

(1926B), "Zur Quantelung des idealen einatomigen Gases," *Zeitschrift für Physik 36*, 902–12.

(1932), "Quantum theory of radiation," *Review of Modern Physics 4*, 87–132.

(1933), "Tentativo di una Teoria dell'Emissione dei Raggi 'Beta,'" *Ricerca Scientifica 2*, 491–5.

(1934), "Versuch einer Theorie der β-Strahlen, I," *Zeitschrift für Physik 77*, 161–71.

(1965), *Enrico Fermi. Collected Papers*, 2 vols., Rome: University of Chicago Press.

Feuer, L. S. (1977), "Teleological principles in science," *Inquiry 21*, 377–407.

Feynman, R. P. (1966), "The development of the space–time view of quantum electrodynamics," *Science 153*, 699–708.

Fierz, M., and Pauli, W. (1939), "On relativistic wave equations of particles of arbitrary spin in an electromagnetic field," *Proceedings of the Royal Society of London A173*, 211–32.

Fischer, P. (1985), *Licht und Leben. Ein Bericht über Max Delbrück, den Wegbereiter der Molekylarbiologie*, Konstanz: Universitätsverlag Konstanz.

Fock, V. (1929), "Geometrisierung der Diracschen Theorie des Elektrons," *Zeitschrift für Physik 57*, 261–77.

(1932), "Konfigurationsraum und Zweite Quantelung," *Zeitschrift für Physik 75*, 622–47.

Fock, V. and Podolsky, B. (1932A), "Zur Diracschen Quantenelektrodynamik," *Physikalische Zeitschrift der Sowietunion 1,* 798–800.
(1932B), "On the quantization of electro-magnetic waves and the interaction of charges on Dirac's theory," *Physikalische Zeitschrift der Sowietunion 1,* 801–17.
Fokker, A. D. (1929), "Eine invarianter Variationssatz für die Bewegung mehrerer elektrischer Massenteilchen," *Zeitschrift für Physik 58,* 386–93.
Ford, K. W. (1963), "Magnetic monopoles," *Scientific American,* December, 122–31.
Fowler, R. H. (1926), "On dense stars," *Monthly Notices of the Royal Astronomical Society of London 87,* 114–22.
Fradkin, D. M. (1966), "Comments on a paper by Majorana concerning elementary particles," *American Journal of Physics 34,* 314–18.
Franklin, A. (1979), "The discovery and nondiscovery of parity nonconservation," *Studies in the History and Philosophy of Science 10,* 21–57.
(1986), *The Neglect of Experiment,* Cambridge: Cambridge University Press.
Frenkel, Y. (1925), "Elektrodynamik punktförmiger Elektronen," *Zeitschrift für Physik 32,* 518–34.
(1928), "Zur Wellenmechanik des rotierenden Elektrons," *Zeitschrift für Physik 47,* 786–803.
Freundlich, Y. (1980), "Theory evaluation and the bootstrap hypothesis," *Studies in the History and Philosophy of Science 11,* 267–77.
Fürth, R. (1933), "Einige Bemerkungen zum Problem der Neutronen und positive Elektronen," *Zeitschrift für Physik 85,* 294–9.
Galison, P. L. (1979), "Minkowski's space–time: from visual thinking to the absolute world," *Historical Studies in the Physical Sciences 10,* 85–121.
(1983), "The discovery of the muon and the failed revolution against quantum electrodynamics," *Centaurus 26,* 262–316.
(1987), *How Experiments End,* Chicago: University of Chicago Press.
Gamow, G. (1934), "Ueber den heutigen Stand (20. Mai 1934) der Theorie des β-Zerfalls," *Physikalische Zeitschrift 35,* 533–42.
(1935A), "General stability-problems of atomic nuclei," pp. 60–71 in *International Conference on Physics, London 1934,* vol. 1, Cambridge: Cambridge University Press.
(1935B), "The negative proton," *Nature 135,* 858–61.
(1966), *Thirty Years That Shook Physics: the Story of Quantum Theory,* New York: Doubleday.
(1967A), "Electricity, gravity, and cosmology," *Physical Review Letters 19,* 759–61.
(1967B), "History of the universe," *Science 158,* 766–9.
(1967C), "Does gravity change with time?" *Proceedings of the National Academy of Sciences 57,* 187–93.
Gapon, E., and Iwanenko, D. (1934), "Alpha particles in light nuclei," *Comptes Rendus de l'Académie des Sciences de l'URSS 4,* 276–7.
Goldberg, S. (1976), "Max Planck's philosophy of nature and his elaboration of the special theory of relativity," *Historical Studies in the Physical Sciences 7,* 125–61.

Gordon, W. (1927), "Der Comptoneffekt nach der Schrödingerschen Theorie," *Zeitschrift für Physik 40,* 117–33.

Gould, P. L. et al. (1986), "Diffraction of atoms by light: the near-resonant Kapitza–Dirac effect," *Physical Review Letters 56,* 827–30.

Gowing, M. (1964), *Britain and Atomic Energy, 1939–1945,* London: Macmillan.

Grünbaum, A. (1976), "Is preacceleration of particles in Dirac's electrodynamics a case of backward causation?" *Philosophy of Science 43,* 165–201.

Grythe, I. (1982/83), "Some remarks on the early S-matrix," *Centaurus 26,* 198–203.

Hadamard, J. (1954), *The Psychology of Invention in the Mathematical Field,* New York: Dover Books.

Haldane, J. B. S. (1939). *The Marxist Philosophy and the Sciences,* New York: Random House.

Halpern, L. (1985), "Paul Adrien Maurice Dirac (1902–1984)," *Foundations of Physics 15,* 257–9.

Halpern, O., and Thirring, H. (1932), *The Elements of the New Quantum Mechanics,* London: Methuen.

Hanson, N. R. (1963), *The Concept of the Positron,* Cambridge: Cambridge University Press.

Harish-Chandra (1987), "My association with Professor Dirac," pp. 34–6, in Kursunoglu and Wigner, eds. (1987).

Hasenfratz, P., and Kuti, J. (1978), "The quark bag model," *Physics Reports 40,* 75–179.

Heaviside, O. (1893–1912), *Electromagnetic Theory,* 3 vols., London: "The Electrician" Printing and Publishing Co.

Heelan, P. A. (1975), "Heisenberg and radical scientific change," *Zeitschrift für Allgemeine Wissenschaftstheorie 6,* 113–36.

Heisenberg, E. (1984), *Inner Exile. Recollections of a Life with Werner Heisenberg,* Boston: Birkhäuser.

Heisenberg, W. (1925), "Ueber quantentheoretischer Umdeutung und mechanischer Beziehungen," *Zeitschrift für Physik 33,* 879–93.

(1926A), "Mehrkörperproblem und Resonanz in der Quantenmechanik," *Zeitschrift für Physik 38,* 411–26.

(1926B), "Schwankungserscheinungen und Quantenmechanik," *Zeitschrift für Physik 40,* 501–6.

(1927), "Ueber den anschaulichen Inhalt der quantentheoretischen Kinematik und Mechanik," *Zeitschrift für Physik 43,* 172–98.

(1934), "Bemerkungen zur Diraschen Theorie des Positrons," *Zeitschrift für Physik 90,* 209–31.

(1938), "Ueber die in der Theorie der Elementarteilchen auftretende universelle Länge," *Annalen der Physik 32,* 20–33.

(1968), "Theory, criticism and a philosophy," pp. 32–46 in *From a Life in Physics,* Vienna: IEAE.

(1971A), *Physics and Beyond,* New York: Harper & Row.

(1971B), "Die Bedeutung des Schönes in der exakten Naturwissenschaft," pp. 288–305 in *Schritte über Grenzen,* Munich: R. Piper & Co.

(1972), "Indefinite metric in state space," pp. 129–36 in Salam and Wigner, eds. (1972).

(1973), "Development of concepts in the history of quantum theory," pp. 264–75 in Mehra, ed. (1973).

Heisenberg, W. and Pauli, W. (1929), "Zur Quantendynamik der Wellenfelder," *Zeitschrift für Physik 56*, 1–61.

(1930), "Zur Quantendynamik der Wellenfelder, II," *Zeitschrift für Physik 59*, 168–90.

Heller, M. (1979), "Questions to infallible oracle," pp. 199–210 in M. Demianski, ed., *Physics of the Expanding Universe*, Berlin: Springer.

Hendry, J. (1981), "Bohr–Kramers–Slater: a virtual theory of virtual oscillators and its role in the history of quantum mechanics, *Centaurus 25*, 189–221.

(1983), "Monopoles before Dirac," *Studies in the History and Philosophy of Science 14*, 81–7.

(1984A), *The Creation of Quantum Mechanics and the Bohr-Pauli Dialogue*, Dordrecht: Reidel.

(1984B), *Cambridge Physics in the Thirties*, Bristol: Adam Hilger.

Hirosige, T. (1966), "Electrodynamics before the theory of relativity, 1890–1905," *Japanese Studies in History of Science 5*, 1–49.

Hoddeson, L., and Baym, G. (1980), "The development of the quantum mechanical electron theory of metals: 1900–28," *Proceedings of the Royal Society of London A371*, 8–23.

Hoddeson, L., Baym, G., and Eckert, M. (1987), "The development of the quantum-mechanical theory of metals, 1928–1933," *Review of Modern Physics 59*, 287–327.

Holton, G. (1978), *The Scientific Imagination*, Cambridge: Cambridge University Press.

Houston, W. V., and Hsieh, Y. M. (1934), "The fine structure of the Balmer lines," *Physical Review 45*, 263–72.

Howarth, T. (1978), *Cambridge between Two Wars*, London: Collins.

Hoyle, F. (1948), "A new model for the expanding universe," *Monthly Notices of the Royal Astronomical Society 108*, 372–82.

(1980), *Steady-State Cosmology Re-Visited*, Cardiff: University College Cardiff Press.

(1986), *The Small World of Fred Hoyle. An Autobiography*, London: Michael Joseph.

Huntley, H. E. (1970), *The Divine Proportion. A Study in Mathematical Beauty*, New York: Dover Books.

Infeld, L. (1978), *Why I Left Canada*, Montreal: McGill-Queen's Univeristy Press.

(1980), *Quest. An Autobiography*, 2nd edn., New York: Chelsea Publ. Co.

Iwanenko, D., and Ambarzumian, V. (1930), "Les électrons inobservables et les rayons β," *Comptes Rendus 190*, 582–4.

Iwanenko, D., and Landau, L. (1928), "Zur Theorie des magnetischen Elektrons," *Zeitschrift für Physik 48*, 340–8.

Jacobsen, J. C. (1936), "Correlation between scattering and recoil in the Compton effect," *Nature 138*, 25.

Jammer, M. (1966), *The Conceptual Development of Quantum Mechanics*, New York: McGraw-Hill.

Jeans, J. (1930), *The Mysterious Universe*, London: Cambridge University Press. (1931), "The annihilation of matter," *Nature 128*, 103–10.

Joffe, A. F. (1967), *Begegnungen mit Physikern*, Leipzig: Teubner.

Jordan, P. (1924), "Zur Theorie der Quantenstrahlung," *Zeitschrift für Physik 30*, 297–319.

(1927A), "Ueber eine neue Begründung der Quantenmechanik," *Zeitschrift für Physik 40*, 809–38.

(1927B), "Zur Quantenmechanik der Gasentartung," *Zeitschrift für Physik 44*, 473–80.

(1928), "Die Lichtquantentheorie," *Ergebnisse der Exakten Naturwissenschaften 7*, 158–208.

(1929), "Der gegenwärtige Stand der Quantenelektrodynamik," *Physikalische Zeitschrift 30*, 700–12.

(1932), "Zur Methode der zweiten Quantelung," *Zeitschrift für Physik 75*, 648–53.

(1935), "Zur Quantenelektrodynamik, III. Eichinvariante Quantelung und Diracsche Magnetpole," *Zeitschrift für Physik 97*, 535–7.

(1936), *Anschauliche Quantentheorie*, Berlin: Springer.

(1937), "Die physikalischen Weltkonstanten," *Die Naturwissenschaften 25*, 513–17.

(1938), "Ueber die Diracschen Magnetpole," *Annalen der Physik 32*, 66–70.

(1939), "Bermerkungen zur Kosmologie," *Annalen der Physik 36*, 64–70.

(1952), *Schwerkraft und Weltall*, Braunschweig: Vieweg.

Jordan, P., and Klein, O. (1927), "Zum Mehrkörperproblem der Quantentheorie," *Zeitschrift für Physik 45*, 751–65.

Jordan, P., and Pauli, W. (1928), "Zur Quantenelektrodynamik," *Zeitschrift für Physik 47*, 151–73.

Jordan, P., and Wigner, E. (1928), "Ueber das Paulische Aequivalenzverbot," *Zeitschrift für Physik 47*, 631–51.

Josephson, P. R. (1988), "Physics and Soviet-Western relations in the 1920s and 1930s," *Physics Today*, September, 54–61.

Jost, R. (1972), "Foundation of quantum field theory," pp. 61–78 in Salam and Wigner, eds. (1972).

Jungk, R. (1958), *Brighter than a Thousand Suns: A Personal History of the Atomic Scientists*, New York: Harcourt Brace.

Kedrov, B. M. ed. (1979), *Das Neutron*, Berlin (GDR): Akademie-Verlag.

Kedrov, F. B. (1984), *Kapitza: Life and Discoveries*, Moscow: MIR Publishers.

Kerber, G., et al. (1987), *Erwin Schrödinger*, Vienna: Fassbaender.

Klein, O. (1927), "Elektrodynamik und Wellenmechanik vom Standpunkt des Korrespondenzprinzips," *Zeitschrift für Physik 41*, 407–42.

(1929), "Die Reflexion von Elektronen an einem Potentialsprung nach der relativistischen Dynamik von Dirac," *Zeitschrift für Physik 53*, 57–65.

(1973), "Ur mitt Liv i Fysiken," *Svensk Naturvetenskap 1973*, 159–72.

Klein, O., and Nishina, Y. (1929), "Ueber die Streuung von Strahlung durch freie Elektronen nach der neuen relativistischen Quantendynamik von Dirac," *Zeitschrift für Physik 52*, 853–68.

Kostro, L. (1986), "Einstein's conception of the ether and its up-to-date applications in the relativistic wave mechanics," pp. 435–50 in W. M. Honig et al., eds. *Quantum Uncertainties*, New York: Plenum Press.

Kragh, H. (1981A), "The genesis of Dirac's relativistic theory of electrons," *Archive for History of Exact Sciences 24*, 31–67.

   (1981B), "The concept of the monopole: a historical and analytical case-study," *Studies in the History and Philosopy of Science 12*, 141–72.

   (1982A), "Cosmo-physics in the thirties: towards a historiography of Dirac cosmology," *Historical Studies in the Physical Sciences 13*, 69–108.

   (1982B), "Erwin Schrödinger and the wave equation: the crucial phase," *Centaurus 26, 154–97*.

   (1984), "Equation with many fathers: the Klein–Gordon equation in 1926," *American Journal of Physics 52*, 1024–33.

   (1985), "The fine structure of hydrogen and the gross structure of the physics community, 1916–26," *Historical Studies in the Physical Sciences 15*, 67–126.

Krajcik, R. A., and Nieto, M. M. (1977), "Historical developmet of the Bhabha first-order relativistic wave equations for arbitrary spin," *American Journal of Physics 45*, 818–22.

Kramers, H. A. (1925), "Eenige opmerkingen over de quantenmechanica van Heisenberg," *Physica 5*, 369–76.

   (1934), "On the classical theory of the spinning electron," *Physica 1*, 825–8.

   (1957), *Quantum Mechanics*, Amsterdam: North-Holland.

Kramers, H. A., and Heisenberg, W. (1925), "Ueber die Streuung von Strahlen durch Atome," *Zeitschrift für Physik 31*, 681–708.

Kursunoglu, B. N., and Wigner, E. P., eds. (1987), *Paul Adrien Maurice Dirac. Reminiscences about a Great Physicist*, Cambridge: Cambridge University Press.

Lamb, W. (1983), "The fine structure of hydrogen," pp. 311–28 in Brown and Hoddeson, eds. (1983).

Lanczos, K. (1926), "Ueber eine feldmässige Darstellung der neuen Quantenmechanik," *Zeitschrift für Physik 35*, 812–36.

Landau, L., and Peierls, R. (1931), "Erweiterung des Unbestimmtheitsprinzips für die relativistische Quantentheorie," *Zeitschrift für Physik 69*, 56–66.

Langer, R. M. (1932), "The fundamental particles," *Science, 76*, 294–5.

Langevin, A. (1966), "Paul Langevin et les congrès de physique Solvay," *La Pensée*, no. 129, 3–32, and no. 130, 89–104.

Larmor, J. (1894), "A dynamical theory of the electric and luminiferous medium," *Philosophical Transactions of the Royal Society 185*, 719–822.

Lees, A. (1935) "The electric moment of an electron," *Proceedings of the Cambridge Philosophical Society 31*, 94–7.

   (1939), "The electron in classical general relativity theory," *Philosophical Magazine 28*, 385–95.

Lenard, P. (1909), "Ueber Lichtemission und deren Erregung," *Sitzungsberichte Heidelberger Akademie*, dritte Abhandlung.

Lewis, T. (1945), "Some criticism of the theory of point electrons," *Philosophical Magazine 36*, 533–41.

Lipschitz, R. (1884), *Untersuchungen über Summen von Quadraten*, Bonn: F. Cohen.

Lodge, O. (1906), *Electrons*, London: Bell and Sons.

(1922), "Speculation concerning the positive electron," *Nature 110*, 696–7.

(1925), *Ether and Reality*, New York: George Doran Co.

London, F. (1950), *Superfluids. Macroscopic Theory of Superconductivity*, New York: Wiley.

Lovejoy, A. O. (1976), *The Great Chain of Being*, Cambridge, Mass.: Harvard University Press (1st edn. 1936).

Lützen, J. (1982), *The Prehistory of the Theory of Distributions*, Berlin: Springer.

MacKinnon, E. M. (1974), "Ontic commitments of quantum mechanics," pp. 225–308 in R. S. Cohen and M. W. Wartofsky, eds., *Logic and Epistemological Studies in Contemporary Physics*, Dordrecht: Reidel.

(1982), *Scientific Explanation and Atomic Physics*, Chicago: University of Chicago Press.

Mandel, H. (1930), "Einige vergleichende Bemerkungen zur Quantentheorie des Elektrons," *Zeitschrift für Physik 60*, 782–94.

Margenau, H. (1933), *Mathematical Gazette 17*, 493.

(1935), "Methodology of modern physics," pp. 52–89 in Margenau, *Physics and Philosophy: Selected Essays*, Dordrecht: Reidel, 1978 (first published in *Philosophy of Science 2*, 1935, 48–72, 164–87).

Maxwell, J. C. (1891), *Treatise on Electricity and Magnetism*, 3rd edn., Oxford: Clarendon Press.

Mehra, J. (1972), "The golden age of theoretical physics: P. A. M. Dirac's scientific works from 1924–1933," pp. 17–59 in Salam and Wigner, eds. (1972).

ed. (1973), *The Physicist's Conception of Nature*, Dordrecht: Reidel.

Mehra, J., and Rechenberg, H. (1982+), *The Historical Development of Quantum Theory*, 5 vols., Berlin: Springer.

Merriell, D. (1981), "On the idea of beauty," pp. 184–223 in *The Great Ideas Today*, Chicago: Encyclopedia Britannica.

Merton, R. K., and Zuckerman, H. (1973), "Age, aging, and age structure in science," pp. 497–559 in R. K. Merton, *Sociology of Science*, Chicago: University of Chicago Press.

Milne, E. A. (1935), *Relativity, Gravitation and World-Structure*, Oxford: Oxford University Press.

Minkowski, H. (1915), "Das Relativitätsprinzip," *Annalen der Physik 47*, 927–38.

Mituo, T. (1974), "Methodological approaches in the development of the meson theory of Yukawa in Japan," pp. 24–38 in S. Nakayama et al., eds., *Science and Society in Modern Japan*, Cambridge, Mass.: MIT Press.

Mohorovlclc, S. (1934), "Möglichkeit neuer Elemente und ihre Bedeutung für die Astrophysik," *Astronomische Nachrichten 253*, 93–108.

Møller, C. (1963), "Nogle erindringer fra livet på Bohrs institut i sidste halvdel af tyverne," pp. 54–64 in *Niels Bohr, et Mindeskrift*, Copenhagen: Gjellerup. Special issue of *Fysisk Tidsskrift 60*.

Morrison, M. (1986), "More on the relationship between technically good and conceptually important experiments: a case study," *British Journal for Philosophy of Science 37*, 101–22.

Mott, N. (1986), *A Life in Science,* London: Taylor & Francis.
  (1987), "Reminiscences of Paul Dirac," pp. 231–4 in Kursunoglu and Wigner, eds. (1987).
Moyer, D. F. (1981A), "Origins of Dirac's electron, 1925–1928," *American Journal of Physics 49,* 944–9.
  (1981B), "Evaluation of Dirac's electron," *American Journal of Physics 49,* 1055–62.
  (1981C), "Vindication of Dirac's electron," *American Journal of Physics 49,* 1120–5.
Nernst, W. (1907), *Theoretische Chemie,* 5th edn., Stuttgart: Ferdinand Enke.
Newton, I. (1729). *Principia Mathematica,* London. Trans. by A. Motte, Ed. by F. Cajori, Berkeley: University of California Press, 1946.
Nikolsky, K. (1932), "The interaction of charges in Dirac's theory," *Physikalische Zeitschrift der Sowietunion 2,* 447–52.
Oppenheimer, J. R. (1930A), "Note on the theory of the interaction of field and matter," *Physical Review 35,* 461–77.
  (1930B), "On the theory of electrons and protons," *Physical Review 35,* 562.
  (1930C), "Two notes on the probability of radiative transitions," *Physical Review 35,* 939–47.
  (1931), [review of Dirac, *Principles of Quantum Mechanics*], *Physical Review 37,* 97.
  (1934), "Are the formulae for the absorption of high energy radiations valid?" *Physical Review 47,* 44–52.
Oppenheimer, J. R., and Carlson, J. F. (1932), "The impacts of fast electrons and magnetic neutrons," *Physical Review 41,* 763–92.
Oppenheimer, J. R., and Furry, W. H. (1934), "On the theory of the electron, negative and positive," *Physical Review, 45,* 245–62.
Pais, A. (1972), "The early history of the theory of the electron: 1897–1947," pp. 79–93 in Salam and Wigner, eds. (1972).
  (1982), *'Subtle is the Lord . . .': The Science and the Life of Albert Einstein,* Oxford: Oxford University Press.
  (1986), *Inward Bound. Of Matter and Forces in the Physical World,* Oxford: Clarendon Press.
  (1987), "Playing with equations, the Dirac way," pp. 93–116 in Kursunoglu and Wigner, eds. (1987).
Pauli, W. (1919A), "Mercurperihelbewegung und Strahlenablenkung in Weyls Gravitationstheorie," *Verhandlungen der Deutschen Physikalischen Gesellschaft 21,* 742–50.
  (1919B), "Zur Theorie der Gravitation und der Elektrizität von Herman Weyl," *Physikalische Zeitschrift 20,* 457–67.
  (1921), *Theory of Relativity,* Oxford: Pergamon Press, 1958 (trans. of "Relativitätstheorie," *Encyklopädie der mathematischen Wissenschaften,* vol. V19, Leipzig: Teubner, 1921).
  (1923), "Ueber das thermische Gleichgewicht zwischen Strahlung und freien Elektronen," *Zeitschrift für Physik 18,* 272–86.
  (1927), "Zur Quantenmechanik des magnetischen Elektrons," *Zeitschrift für Physik 43,* 601–23.

(1931), [review of Dirac, *The Principles of Quantum Mechanics*] *Die Naturwissenschaften 19*, 188–9.

(1933), "Die allgemeinen Prinzipien der Wellenmechanik," pp. 771–938 in *Handbuch der Physik* 24/1, Berlin: Springer.

(1943), "On Dirac's new method of field quantization," *Review of Modern Physics 15*, 175–207.

(1946), "Dirac's Feldquantizierung und Emission von Photonen kleiner Frequenzen," *Helvetica Physica Acta 19*, 234–7.

(1964), "Exclusion principle and quantum mechanics," Nobel Lecture delivered December 13, 1946, in Stockholm; pp. 27–43 in *Nobel Lectures, Physics 1942–1962*, Amsterdam: Elsevier.

(1979), *Wissenschaftlicher Briefwechsel mit Bohr, Einstein, Heisenberg u.a.*, eds. K. von Meyenn, A. Hermann, and V. F. Weisskopf, vol. 1 (1919–29), Berlin: Springer.

(1985), *Wissenschaftlicher Briefwechsel mit Bohr, Einstein, Heisenberg u.a.*, ed. K. von Meyenn, vol. 2 (1930–9), Berlin: Springer.

Pauli, W., and Solomon, J. (1932), "La théorie unitaire d'Einstein et Mayer et les équations de Dirac," *Journal de Physique et Radium* (7), *3*, 452–63, 582–9.

Pauli, W. and Weisskopf, V. F. (1934), "Ueber die Quantisierung der skalaren relativistischen Wellengleichung," *Helvetica Physica Acta 7*, 709–31.

Pearson, K. (1891), "Ether squirts," *American Journal of Mathematics 13*, 309–70.

Peierls, R. (1936), "Interpretation of Shankland's experiment," *Nature 137*, 904.

(1985), *Bird of Passage. Recollections of a Physicist*, Princeton: Princeton University Press.

(1987), "Address to Dirac Memorial Meeting, Cambridge," pp. 35–7 in Taylor, ed. (1987).

Penrose, R. (1979), Interview, pp. 34–50 in P. Buckley and F. D. Peat, eds., *A Question of Physics: Conversations in Physics and Biology*, London: Routledge & Kegan Paul.

Petiau, G. (1936), "Contribution à la théorie des équations d'ondes corpusculaire," *Memoires de l'Académie Science Royale de Belgique 16*, 1–116.

Petroni, C., and Vigier, J. P. (1983), "Dirac's aether in relativistic quantum mechanics," *Foundations of Physics 13*, 253–86.

Pickering, A. (1981), "Constraints on controversy: the case of the magnetic monopole," *Social Studies in Science 11*, 63–93.

Placintéanu, J. J. (1933), "Considérations théoriques sur la constitution des neutrons, électrons positifs et photons. Existence des protons négatifs," *Comptes Rendus 197*, 549–52.

Poincaré, H. (1896), "Remarques sur une expérience de M. Birkeland," *Comptes Rendus 123*, 530–3.

(1960), *Science and Method*, New York: Dover Books (first published in Paris in 1908).

Pokrowski, G. I. (1929), "Ueber die Wahrscheinlichkeitsgesetz bei dem Zerfall radioaktiver Stoffe sehr kleiner Konzentrations," *Zeitschrift für Physik 58*, 706–9.

Popper, K. R. (1959), *The Logic of Scientific Discovery*, New York: Basic Books.

Proca, A. (1936), "Sur la théorie ondulatoire des électrons positifs et negatifs," *Comptes Rendus 7*, 347–53.

Przibram, K., ed., (1963), *Briefe zur Wellenmechanik*, Vienna: Springer-Verlag.

Pyenson, L. (1977), "Herman Minkowski and Einstein's special theory of relativity," *Archive for History of Exact Sciences 17*, 71–95.

Richardson, O. W. (1931), "Isolated quantised magnetic poles," *Nature 128*, 582.

Rohrlich, F. (1965), *Classical Charged Particles*, Reading, Mass.: Addison-Wesley.

(1973), "The electron: development of the first elementary particle theory," pp. 331–69 in Mehra, ed. (1973).

Rosenfeld, L. (1932), "Ueber eine mögliche Fassung der Diracschen Programms zur Quantenelektrodynamik und deren formalen Zusammenhang mit der Heisenberg-Paulischen Theorie," *Zeitschrift für Physik 76*, 729–34.

(1967), "Niels Bohr in the thirties," pp. 114–36 in S. Rozental, ed., *Niels Bohr. His Life and Work as Seen by His Friends and Colleagues*, New York: Wiley.

Rosenthal-Schneider, I. (1949), "Presuppositions and anticipations in Einstein's physics," pp. 131–46 in P. Schilpp, ed., *Albert Einstein: Philosopher–Scientist*, New York: Tudor Publ. Co.

Rüger, A. (1988), "Atomism from cosmology: Erwin Schrödinger's work on wave mechanics and space–time structure," *Historical Studies in the Physical Sciences 18*, 377–401.

Rydberg, J. R. (1906), *Elektron. Der erste Grundstoff*, Lund: Håkan Ohlsson.

Saha, M. N. (1936), "On the origin of mass in neutrons and protons," *Indian Journal of Physics 10*, 141–53.

(1948), "Note on Dirac's theory of magnetic poles," *Physical Review 75*, 1968.

Salam, A., and Wigner, E. P., eds. (1972), *Aspects of Quantum Theory*, Cambridge: Cambridge University Press.

Salaman, E., and Salaman, M. (1986), "Remembering Paul Dirac," *Endeavour*, May, 66–70.

Sambursky, S. (1937), "Static universe and nebular red shift," *Physical Review 52*, 335–8.

Sánchez-Ron, J. M. (1983), "The problems of interaction: on the history of the action-at-a-distance concept in physics," *Fundamenta Scientiae 4*, 55–76.

Schrödinger, E. (1926A), "Quantisierung als Eigenwertproblem (erste Mitteilung)," *Annalen der Physik 79*, 361–76.

(1926B), "Quantisierung als Eigenwertproblem (vierte Mitteilung)," *Annalen der Physik 81*, 109–40.

(1931A), "Zur Quantendynamik des Elektrons," *Sitzungsberichte Preussischen Akademie der Wissenschaften 3*, 63–72.

(1931B), "Spezielle Relativitätstheorie und Quantenmechanik," *Sitzungsberichte Preussischen Akademie der Wissenschaften 3*, 238–47.

(1940), "Maxwell's and Dirac's equations in the expanding universe," *Proceedings of the Royal Irish Academy, Section A 46A*, 25–47.

(1956), *Science, Theory and Man*, London: Allen & Unwin.

Schuster, A. (1898), "Potential matter – a holiday dream," *Nature 58*, 367.

Schweber, S. S. (1986A), "Shelter Island, Pocono, and Oldstone: the emergence of American quantum electrodynamics after World War II," *Osiris 2*, 265–302.

(1986B), "The empiricist temper regnant: theoretical physics in the United States 1920–1950," *Historical Studies in the Physical Sciences 17*, 55–98.

(1986C), "Feynman and the visualization of space–time processes," *Review of Modern Physics 58*, 449–508.

Schwinger, J., ed. (1958), *Selected Papers on Quantum Electrodynamics*, New York: Dover Books.

(1969), "A magnetic model of matter," *Science 165*, 757–61.

Shankland, R. (1936), "An apparent failure in the photon theory of scattering," *Physical Review 49*, 8–13.

Shanmugadhasan, S. (1987), "Dirac as a supervisor and other remembrances," pp. 48–57 in Taylor, ed. (1987).

Singh, J. (1970), *Great Ideas and Theories of Modern Cosmology*, New York: Dover.

Slater, J. C. (1927), "Radiation and absorption on Schrödinger's theory," *Proceedings of the National Academy of Sciences 13*, 7–12.

(1929), "The theory of complex spectra," *Physical Review 34*, 1293–322.

(1967), "Quantum physics in America between the wars," *International Journal of Quantum Chemistry 1*, 1–23.

(1975), *Solid-State and Molecular Theory: A Scientific Biography*, New York: John Wiley & Sons.

Small H., ed. (1981), *Physics Citation Index 1920–1929*, Philadelphia: Institute for Scientific Information.

(1986), "Recapturing physics in the 1920s through citation analysis," *Czechoslovachian Journal of Physics B36*, 142–7.

Smith, A. C., and Weiner, C., eds. (1980), *Robert Oppenheimer, Letters and Recollections*, Cambridge, Mass.: Harvard University Press.

Snow, C. P. (1934), *The Search*, New York: Charles Scribner's Sons.

(1981), *The Physicists*, London: Macmillan.

Stöckler, M. (1984), *Philosophische Probleme der Relativistischen Quantenmechanik*, Berlin: Duncker & Humblot.

Strachan, C., ed. (1969), *The Theory of Beta-Decay*, Oxford: Pergamon.

Stuewer, R. H. (1983), "The nuclear electron hypothesis," pp. 19–68 in W. R. Shea, ed., *Otto Hahn and the Rise of Nuclear Physics*, Dordrecht: Reidel.

Sudarshan, E. C. G., Mukunda, N., and Chiang, C. C. (1982), "Dirac's new relativistic wave equation in interaction with an electromagnetic field," *Proceedings of the Royal Society of London A379*, 103–7.

Sutherland, W. (1898), "Cathode, Lenard, and Röntgen rays," *Philosophical Magazine 47*, 269–84.

Tamm, I. (1930), "Ueber die Wechselwirkung der Freien Elektronen nach der Diracschen Theorie des Elektrons und nach der Quantenelektrodynamik," *Zeitschrift für Physik 62*, 545–68.

(1931), "Die verallgemeinerten Kugelfunktionen und die Wellenfunktionen im Feld eines Magnetpoles," *Zeitschrift für Physik 71*, 141–50.

Taylor, J. G. (1987A), "Constrained dynamics," pp. 114–23 in Taylor, ed. (1987B).

ed. (1987B), *Tributes to Paul Dirac,* Bristol: Adam Hilger.

Teller, E. (1948), "On the change of physical constants," *Physical Review 73,* 801–2.

Thomson, J. J. (1900), *Elements of the Mathematical Theory of Electricity and Magnetism,* Cambridge: Cambridge University Press.

(1907), *The Corpuscular Theory of Matter,* London: Constable and Co.

Tolman, R. C. (1930), "Discussion of various treatments which have been given to the non-static line-element for the universe," *Proceedings of the National Academy of Sciences 16,* 582–94.

Tomonaga, S.-I. (1973), "Development of quantum electrodynamics," pp. 404–12 in Mehra, ed. (1973); reprint of Nobel Lecture, 1966.

Tuve, M. A. (1933), "Search by deflection-experiments for the Dirac isolated magnetic pole," *Physical Review 43,* 770–1.

Van der Waerden, B. L., ed. (1967), *Sources of Quantum Mechanics,* New York: Dover Books.

Van Vleck, J. H. (1932), *The Theory of Electric and Magnetic Susceptibilities,* Oxford: Oxford University Press.

(1934A), "A new method of calculating the mean value of $1/r^2$ for Keplerian systems in quantum mechanics," *Proceedings of the Royal Society of London A143,* 679–81.

(1934B), "The Dirac vector model in complex spectra," *Physical Review 45,* 405–19.

(1968), "My Swiss visits of 1906, 1926, and 1930," *Helvetica Physica Acta 41,* 1234–7.

(1972), "Travels with Dirac in the Rockies," pp. 7–16 in Salam and Wigner, eds. (1972).

Villard, P. (1905), "Les rayons magnétocathodiques," *Revue Générale des Sciences 16,* 405–7.

Vogel, H. (1968), *Physik und Philosophie bei Max Born,* Berlin (GDR): Akademie-Verlag.

Walker, C. T., and Slack, G. A. (1970), "Who named the −ON's?" *American Journal of Physics 38,* 1380–9.

Waller, I. (1930), "Bemerkungen über die Rolle der Eigenenergie des Elektrons in der Quantentheorie der Strahlung," *Zeitschrift für Physik 62,* 673–6.

Weart, S. R., And Szilard, G. W. (1978), *Leo Szilard: His Version of the Facts,* Cambridge, Mass.: Harvard University Press.

Wechsler, J., ed. (1978), *On Aesthetics in Science,* Cambridge, Mass.: MIT Press.

Weinberg, S. (1977), "The search for unity: notes for a history of quantum field theory," *Dædalus 106,* Fall 1977, 17–35.

Weiner, C., ed. (1972), *Exploring the History of Nuclear Physics,* New York: American Institute of Physics.

Weisskopf, V. F. (1934), "Ueber die Selbstenergie des Elektrons," *Zeitschrift für Physik 89,* 27–39.

(1972), *Physics in the Twentieth Century: Selected Essays,* Cambridge, Mass.: MIT Press.

(1973), "My life as a physicist," pp. 983–97 in A. Zichichi, ed., *Properties of the Fundamental Interactions,* Bologna: Editrice Compositori.

(1983), "Growing up with field theory: the development of quantum electrodynamics," pp. 56–81 in Brown and Hoddeson, eds. (1983).

Weisskopf, V. F., and Wigner, E. (1930), "Berechnung der natürlichen Linienbreite auf Grund der Diracschen Lichttheorie," *Zeitschrift für Physik 63,* 54–73.

Wentzel, G. (1926), "Die mehrfach periodischen Systeme in der Quantenmechanik," *Zeitschrift für Physik 37,* 80–94.

(1933), "Ueber die Eigenkräfte der Elementarteilchen," *Zeitschrift für Physik 86,* 479–94, 635–45.

(1960), "Quantum theory of fields (until 1947)," pp. 48–77 in M. Fierz and W. F. Weisskopf, eds., *Theoretical Physics in the Twentieth Century,* New York: Interscience.

(1966), "Comments on Dirac's theory of magnetic monopoles," *Progress in Theoretical Physics, Suppl.,* nos. 37–8, 163–74.

Werskey, G. (1979), *The Visible College: A Collective Biography of British Scientists and Socialists of the 1930s,* London: Allen Lane.

Wesson, P. S. (1978), *Cosmology and Geophysics,* Bristol: Adam Hilger.

Weyl, H. (1918), "Gravitation and Elektrizität," *Sitzungsberichte der Preussischen Akademie der Wissenschaften,* 1918, 465–78.

(1921), *Raum, Zeit, Materie,* 4th edn., Berlin: Springer.

(1929A), "Gravitation and the electron," *Proceedings of the National Academy of Sciences 15,* 323–34.

(1929B), "Elektron und Gravitation," *Zeitschrift für Physik 56,* 330–52.

(1931), *The Theory of Groups and Quantum Mechanics,* New York: Dover Books (German edn. 1928).

(1932), *The Open World. Three Lectures on the Metaphysical Implications of Science,* New Haven: Yale University Press.

Wheeler, J. A. (1985), "Physics in Copenhagen in 1934 and 1935," pp. 221–6 in A. P. French and P. J. Kennedy, eds., *Niels Bohr. A Centenary Volume,* Cambridge, Mass.: Harvard University Press.

Whitley, S. (1984), "Review of the gas centrifuge until 1962," *Review of Modern Physics 56,* 41–98.

Whittaker, E. T. (1917), *A Treatise on the Analytical Dynamics of Particles and Rigid Bodies,* 2nd edn., Cambridge: Cambridge University Press.

(1921), "On tubes of electromagnetic force," *Proceedings of the Royal Society of Edinburgh 42,* 1–23.

(1949), *From Euclid to Eddington,* Cambridge: Cambridge University Press.

(1960), *A History of the Theories of Aether and Electricity,* vol. 2, New York: Harper.

Wick, G. C. (1934), "Induced radioactivity," *Rendiconti Lincei 19,* 319–24.

Wien, W. (1930), *Aus dem Leben und Wirken eines Physikers,* Leipzig: Teubner.

Wigner, E. P. (1928), "Ueber die Erhaltungssätze in der Quantenmechanik," *Nachrichten von der Kgl. Gesellschaft der Wissenschaften zu Göttingen, Mathematisch-Physikalische Klasse,* 1928, 375–81.

(1932), "Ueber die Operation der Zeitumkehr in der Quantenmechanik," *Nachrichten von der Kgl. Gesellschaft der Wissenschaften zu Göttingen, Mathematisch-Physikalische Klasse*, 1932, 546–59.

(1939), "On unitary representations of the inhomogeneous Lorentz group," *Annals of Mathematics 40*, 149–204.

(1947), *Physical Science and Human Values*, Princeton: Princeton University Press.

(1969), "The unreasonable effectiveness of mathematics in the natural sciences," pp. 123–40 in T. L. Saaty and F. J. Weyl, eds., *The Spirit and the Uses of Mathematical Sciences*, New York: McGraw-Hill.

(1973), "Relativistic equations in quantum mechanics," pp. 320–31 in Mehra, ed. (1973).

(1984), "Interview de Eugene P. Wigner sur sa vie scientifique," *Archive Internationale d'Histoire des Sciences 34*, 177–217.

Williams, E. J. (1935), "Correlation of certain collision problems with radiation theory," *Kongelige Danske Videnskabernes Selskab, Matematisk-Fysiske Meddelelser, 13*, no. 1, 1–50.

(1936), "Conservation of energy and momentum in atomic processes," *Nature 137*, 614–15.

Williamson, R., ed. (1987), *The Making of Physicists*, Bristol: Adam Hilger.

Wood, R. W. (1908), "On the existence of positive electrons in the sodium atom," *Philosophical Magazine 15*, 274–9.

Yourgrau, W. (1961), "Some problems concerning fundamental constants in physics," pp. 319–41 in H. Feigl and G. Maxwell, eds., *Current Issues in the Philosophy of Science*, New York: Holt, Rinehart & Winston.

Yukawa, H., and Sakata, S. (1937), "Note on Dirac's generalized wave equations," *Proceedings of the Mathematical-Physical Society of Japan 19*, 91–5.

INDEX OF NAMES

INDEX OF SUBJECTS

(*See also* Dirac, Paul A. M. in Name Index.)